JOINT EVOLUTION OF BLACK HOLES AND GALAXIES

Series in High Energy Physics, Cosmology and Gravitation

Gravitation: From the Hubble Length to the Planck Length
Edited by I Ciufolini, E Coccia, V Gorini, R Peron and N Vittorio

Neutrino Physics
K Zuber

The Galactic Black Hole: Lectures on General Relativity and Astrophysics
H Falcke and F W Hehl

**The Mathematical Theory of Cosmic Strings: Cosmic Strings
in the Wire Approximation**
M R Anderson

Geometry and Physics of Branes
Edited by U Bruzzo, V Gorini and U Moschella

Modern Cosmology
Edited by S Bonometto, V Gorini and U Moschella

Gravitation and Gauge Symmetries
M Blagojevic

Gravitational Waves
Edited by I Ciufolini, V Gorini, U Moschella and P Fré

Classical and Quantum Black Holes
Edited by P Fré, V Gorini, G Magli and U Moschella

Pulsars as Astrophysical Laboratories for Nuclear and Particle Physics
F Weber

**The World in Eleven Dimensions: Supergravity, Supermembranes
and M-Theory**
Edited by M J Duff

Particle Astrophysics
Revised paperback edition
H V Klapdor-Kleingrothaus and K Zuber

Non-Accelerator Particle Physics
Revised edition
H V Klapdor-Kleingrothaus and A Staudt

**Ideas and Methods of Supersymmetry and Supergravity
or A Walk Through Superspace**
Revised edition
I L Buchbinder and S M Kuzenko

Series in High Energy Physics, Cosmology and Gravitation

JOINT EVOLUTION OF
BLACK HOLES AND GALAXIES

Edited by

M Colpi
Department of Physics
University of Milano Bicocca, Italy

V Gorini
Department of Physics and Mathematics
University of Insubria at Como, Italy

F Haardt
Department of Physics and Mathematics
University of Insubria at Como, Italy

U Moschella
Department of Physics and Mathematics
University of Insubria at Como, Italy

CRC Press
Taylor & Francis Group
Boca Raton London New York

CRC Press is an imprint of the
Taylor & Francis Group, an **informa** business
A TAYLOR & FRANCIS BOOK

CRC Press
Taylor & Francis Group
6000 Broken Sound Parkway NW, Suite 300
Boca Raton, FL 33487-2742

First issued in paperback 2019

© 2006 by Taylor & Francis Group, LLC
CRC Press is an imprint of Taylor & Francis Group, an Informa business

No claim to original U.S. Government works

ISBN-13: 978-0-7503-0999-8 (hbk)
ISBN-13: 978-0-367-39116-4 (pbk)

Library of Congress Card Number 2005054924

Library of Congress Cataloging-in-Publication Data

Joint evolution of black holes and galaxies / edited by Monica Colpi ... [et al].
      p. cm. -- (Series in high energy physics, cosmology, and gravitation)
  Includes bibliographical references and index.
  ISBN 0-7503-0999-7
  1. Black holes (Astronomy) 2. Galaxies. 3. Cosmic physics. I. Colpi, Monica. II. Series.

QB843.B55J65 2006
523.8'875--dc22   2005054924

**Visit the Taylor & Francis Web site at
http://www.taylorandfrancis.com**

**and the CRC Press Web site at
http://www.crcpress.com**

Contents

Preface

Black holes are among the most mysterious objects that the human mind has been capable of imagining. They are pure mathematical constructions, tools for exploiting the fundamental laws of physics, but also astronomical sources, part of our cosmic landscape. For this reason, black holes play a key role in contemporary research in astrophysics as well as in basic theoretical physics.

During the Spring 2003 SIGRAV, the Italian Society of Relativity and Gravitation promoted the organization of a doctoral school on *The Joint Evolution of Black Holes and Galaxies,* which took place at the Alessandro Volta Center for Scientific Culture, in the beautiful setting of Villa Olmo in Como, Italy. The black holes of this school are the "astrophysical" black holes that inhabit nearby galaxies and distant quasars and that are studied here to unveil the deep symbiotic relationship they have with their environment. This volume brings together eight contributions from leading authorities, providing a valuable review for professional astronomers and graduate students interested in the subject. A brief introduction endeavors to uncover the *fil rouge* linking black holes to galaxies, in a cosmological context.

The school was made possible thanks to the support of SIGRAV, the University of Milano Bicocca, the University of Insubria, the Department of Physics and Mathematics of the Insubria University, and the National Institute of Nuclear Physics (INFN). We are grateful to all members of the scientific organizing committee and to the secretarial conference staff of the Center Alessandro Volta, in particular to Chiara Stefanetti and Francesca Gamba for their assistance, precious help and kindness. A heartful thanks to Dr. Piergiorgio Casella for his invaluable help in editing this book.

Monica Colpi
Vittorio Gorini
Francesco Haardt
Ugo Moschella

Editors

Monica Colpi is associate professor in stellar astronomy and relativistic astrophysics at the Department of Physics G. Occhialini of the University of Milano Bicocca, Italy. She graduated in physics in 1982 and earned her Ph.D. degree in 1987 at the University of Milano. She has been a visiting scientist at Cornell University and postdoctoral fellow at the International School for Advanced Study in Trieste. Since 1991 she has been a permanent staff member of the University of Milano. Monica Colpi is author of numerous scientific papers published in international journals, in the field of compact objects and high energy astrophysics. She has contributed relevant studies on the theory of accretion onto black holes and neutron stars; on the secular/dynamical/internal evolution of neutron stars in the Milky Way, from magnetars to pulsars and x-ray sources; on boson stars; on the dynamics of massive black holes in merging galaxies and dense stellar systems. Monica Colpi is a member of the board of the Italian Society of General Relativity and Gravitational Physics and co-chairperson of the Society's Graduate School in Contemporary Relativity and Gravitational Physics, which takes place biannually at the Alessandro Volta Center for Scientific Culture in Villa Olmo, Como, Italy.

Francesco Haardt is associate professor of cosmology and astrophysics at the Department of Physics and Mathematics of the University of Insubria, Como. He graduated with a Ph.D. in astrophysics in 1994 at the International School for Advanced Studies, in Trieste, working on corona models for accreting black holes. He moved to the Space Telescope Science Institute in Baltimore, Maryland, and then to the Chalmers Institute of Technology in Sweden. Now he is a permanent staff member at the University of Insubria, Como. Professor Haardt has written numerous scientific papers published in international journals, mainly in the fields of high energy astrophysics and physical cosmology. He has contributed relevant studies on models of x-ray emission from Seyfert galaxies, quasars and galactic black holes; theoretical studies of the intergalactic medium at high redshift; and recently, the evolution of massive black hole binaries in stellar and gaseous environments.

Vittorio Gorini is professor of theoretical physics at the Department of Physics and Mathematics of the University of Insubria, Como. He is the author or coauthor of numerous papers published in international journals, and his research interests have ranged over a variety of subjects in theoretical and mathematical physics. Professor Gorini has been a Humboldt fellow at the Institute of Theoretical Physics at the University of Marburg, Germany and a research fellow at the Center for Particle Theory at the University of Texas, Austin. He has also been an associate professor of theoretical physics at the University of Milan, Italy, and has been for five years professor of mathematical methods of physics at the University of Bari, Italy. Professor Gorini is a board member of the Italian Society of General Relativity and Gravitational Physics and co-chairman of the Society's Graduate School in Contemporary Relativity and Gravitational Physics which takes place biannually at the Alessandro Volta Center for Scientific Culture in Villa Olmo, Como, Italy. He is coeditor of several books arising from the school's proceedings.

Ugo Moschella graduated in physics at Bologna University in 1985 and obtained his Ph.D. at the International School for Advanced Studies in Trieste in 1991. Subsequently, he carried out postdoctoral research at the Catholic University of Louvain-la-Neuve, Belgium; the University of Paris 7 - Denis Diderot; the Theoretical Physics Division of the Atomic Energy Agency - Saclay, France; and the Institute of High Scientific Studies - Bures-sur-Yvette, France. Presently he is associate professor of theoretical physics at the Department of Physics and Mathematics of the University of Insubria, Como. Since 1997 he has been the scientific secretary of the International Doctoral School in Relativity and Gravitation, which is held yearly at Villa Olmo in Como. Since 2000 he has been a member of the scientific board of the Italian Society of Gravitation. Ugo Moschella's main research interests are in quantum field theory. His contributions are concerned with the infrared problem in gauge theory, quantum field theory on curved spacetimes with special attention to de Sitter and anti-de Sitter field theories, and the role of diffeomorphism groups in quantum physics.

Contributors

Tom Abel
Stanford Linear Accelerator
Kavli Institute for Astroparticle
Physics and Cosmology
Menlo Park, California

Alfonso Cavaliere
University of Rome Tor Vergata
Department of Physics
Roma, Italy

Andrea Ferrara
International School for Advanced
Studies
Trieste, Italy

Laura Ferrarese
Rutgers State University
Department of Physics and
Astronomy
Piscataway, New Jersey

Alberto Franceschini
University of Padova
Department of Astronomy
Padova, Italy

Francesco Haardt
University of Insubria
Department of Physics and
Mathematics
Como, Italy

Andrea Lapi
University of Rome Tor Vergata
Department of Physics
Roma, Italy

Piero Madau
University of California
Observatories
Astronomy Department
Santa Cruz, California

Simon Portegies Zwart
University of Amsterdam
Astronomical Institute
Anton Pannekoek
Amsterdam, The Netherlands

Emanuele Ripamonti
University of Groningen
Kapteyn Astronomical Institute
Groningen, The Netherlands

Ruben Salvaterra
International School for Advanced
Studies
Trieste, Italy

Rachel S. Somerville
Space Telescope Science Institute
Baltimore, Maryland

Introduction

Massive black holes weighing from a million up to several billion suns have long been suspected to be the central powerhouse of energetic phenomena taking place in distant quasars. Since the early discovery of active galactic nuclei (AGNs), the accretion paradigm has been at the heart of our interpretation of massive black holes, as "real" sources, in the universe. Energy in the form of radiation, high velocity plasma outflows, and ultra-relativistic jets, can be extracted efficiently from the strong gravitational field of a black hole, when gas is accreting profusely, from the sub-parsec scale of a galactic nucleus down to the scale of the black hole horizon. Black holes are tiny objects: their horizon is less than or comparable to an astronomical unit and for this reason it is still a challenge to explore the volume where the space–time is severely curved by the extreme gravity of the black hole. Evidence of space–time warping around spinning black holes has been found recently, following the discovery of double horn-shaped iron lines in the x-ray spectra of a number of Seyfert nuclei, but a clear sign of an horizon that would bring unmistakable evidence of the existence of massive black holes in active nuclei is still lacking.

Beside the search of this unique signature that next generation x-ray telescopes will likely uncover, a major goal of contemporary astronomy is the search of "quiescent", inactive black hole "candidates" in the cores of nearby galaxies through the measure of "dark" mass potentially present, using stars and gas clouds as dynamical probes. For a long time, black hole masses in AGNs and quasars, have been inferred only indirectly, using as chief argument the concept of Eddington limited accretion. A direct measurement, until recently, was lacking.

Since the late 1960s it was also recognized that quasars and luminous AGNs were undergoing strong cosmic evolution: nuclear activity was common in the past and declined with cosmic time. No quasar-like objects live in our local universe, but only a modest activity, the fading tail, is observed in few nearby galaxies. From simple considerations on the life-cycle of quasars, there has always been the suspicion that, at high redshifts, accretion ignited briefly in many if not all galaxies, leading to the commonly held premise that most of the galaxies we see in our local universe should host in their nucleus a massive black hole, relic of an earlier active phase.

For many years, quasars have been a subject explored primarily by AGN researchers. But a recent breakthrough, due largely to the impact of the Hubble Space Telescope, has provided spectacular proof of the long-standing black hole paradigm, leading to the first measurement of the mass of inactive "black hole candidates" in the cores of a number of close-by galaxies with a significant bulge component. Astronomers, including Laura Ferrarese, discovered in addition, and more importantly, the existence of a tight "correlation" between the mass of the black hole and the dispersion velocity of the stars in the bulge. This is surprising, given the tiny size of the black hole sphere of gravitational influence compared to the size and mass of the bulge. How can this happen? Black hole masses seem to be not only tightly coupled to the large-scale properties of their host galaxies as they stand, but even more profoundly, to the "life-cycle" of the galaxies themselves. The suspicion is that during the formation of galaxies, a universal mechanism is at play, able to deposit, in the nuclear regions, large amounts of gas to fuel and grow the black hole to such an extent that its feedback, i.e., its large-scale injection of energy, acts on the galaxy, vaporizing the cold gas, and sterilizing it against star formation. In such a cosmic grand design, the black holes do not fail to exert a major influence on the dynamical and thermal properties of the galaxy. Major mergers among pre-galactic structures could be at the heart of this symbiotic relationship. They have been advocated to explain both the ignition of a powerful AGN, and the formation of a dynamically violently perturbed galaxy remnant hosting a massive spheroid with its black hole. The concordance of the "bright" quasar cosmic evolution with the overall star-formation history is also strongly suggestive of a link between the assembly of stars in galaxies and black hole fueling and ignition.

Galaxy formation is an exquisite cosmological problem: the cooling and collapse of baryon in dark matter halos, clustering hierarchically, is the prime element for understanding galaxy formation and evolution. The time of first appearance of black holes in obscured dusty environments is still unknown: whether they formed at redshift $z \sim 20$, from seed black holes of intermediate mass (between $100 M_\odot - 10^5 M_\odot$), relic of the first stars, or formed later, in massive virialized structures that developed sufficiently deep potential wells to favor and promote the collapse of giant gas clouds into large black holes (as predicated in models where baryons cluster in an anti-hierarchically fashion). What is known is that bright quasars hosting a billion solar mass black holes should be in place, plausibly in very massive galaxy spheroids, at redshift $z \sim 6$ when the universe was still young. The new paradigm of the joint evolution of black holes and galaxies that would emerge from considerations relying on our current knowledge of the universe and understanding of the physics of AGNs, now finds its foundation on observational ground. Black holes, previously believed to play a passive role, are now "in action" and may affect profoundly their cosmic environment.

The co-evolution of black holes and galaxies embraces a variety of astrophysical phenomena that are now becoming of major scientific interest: they cover issues on black hole formation, clustering, fueling and evolution, issues on galaxy formation and evolution, with feedback playing a key role. These processes can be studied collectively, and the cumulative deposition of energy through accretion and stellar feedback that lights up our cosmic backgrounds (from the far-infrared to hard x-rays) is a clear example. Many of these issues are contained and illustrated in this book.

Laura Ferrarese opens this volume with a review of the status of the research on massive black holes in galaxies, as seen from an observational standpoint. Hunting black holes in our local universe and in AGNs is a challenging experience that requires the elaboration of sophisticated techniques that are described in detail. Ferrarese introduces the reader to the scaling relation between the black hole mass and the stellar dispersion velocity of the host galaxy, a relation that has proven of great value in the study of the demography of massive black holes. Tracing the mass function of present-day massive black holes can bring light into the accretion history, when compared with what is known about the quasar era. Ferrarese indicates the next steps that are necessary to improve our ability in spatially resolving the sphere of influence of the black holes. This will allow us to diagnose the morphology of gas and the dynamics of stars at only few thousand Schwarzschild radii away from the black hole, and, in addition, to explore the yet poorly sampled range of black hole masses below a million solar mass, in the domain of dwarf elliptical galaxy hosts.

Alberto Franceschini brings our attention to AGNs and galaxies, describing their rich and complex phenomenology. It is a review of our current observational understanding of high redshift and low redshift AGNs and on the ultra-luminous and hyper-luminous infrared galaxies, undergoing major mergers and surges of intense star formation. The link between these two classes seems profound. Combining key observational facts and theoretical considerations Franceschini describes type II AGNs in dust enshrouded infrared galaxies, and, in a broader context, highlights links and differences between star dominated galaxies and galaxies where the energy source from the accreting black hole is more important. The recent discovery, with Chandra and XMM, of a distinct evolution, with cosmic time, of luminous AGNs, contrasted with the more gentle still ongoing evolution of low luminosity AGNs is reviewed. The lecture ends with a survey of the cosmic backgrounds, and the coeval evolution of starbursts and AGNs.

The book continues with a theoretical dissertation by Rachel Somerville that introduces the reader to the realm of physical cosmology. Somerville illustrates the modern paradigm of hierarchical structure formation that arises from a theory in which most of the matter in the universe is in an as yet unidentified form of cold dark matter (CDM). She reviews dark matter halos, i.e., their internal characteristics as well as their collective properties, introducing the semi-analytical techniques used to model the overall assembly of galaxies and galaxy halos. Key questions on galaxy formation and evolution within the CDM paradigm are addressed. The main physical processes that are responsible in shaping galaxies are reviewed, including gastrophysics, star formation, and supernova feedback.

The essay of Cavaliere and Lapi is focused on galaxy clusters and groups, as seen in x-rays. The authors describe the hot plasma pervading these bound structures and our current knowledge of the origin of the correlation between the x-ray luminosity and temperature of the intra-cluster medium. The authors show that heating mechanisms that involve mergers, shocks, and energy feedback from supernovae are insufficient to explain the entropy level observed in the intra-cluster medium, and suggest the need of a substantial contribution by the central AGN.

The volume continues with Tom Abel and Emanuele Ripamonti who introduce us to the very high redshift universe. After touching the foundation of our cosmological concordance ΛCDM model, and the theory of the non linear growth of primordial density perturbations, the authors explore the physics of formation of the first luminous objects. The dynamics of condensation of baryons in the first virialized halos and the complex chemistry of metal-free primordial gas are illustrated using cosmological hydrodynamical simulations and analytical tools. The formation of population III stars is followed in detail, from the early development of an opaque embrion up to the phase where accretion becomes important, a phase that may ultimately end with the formation of a very massive star, providing a pathway for the creation of seed intermediate-mass black holes, in dark halos.

Piero Madau and Francesco Haardt bring us into the cosmic history of the diffuse intergalactic medium (IGM). It is an exploration of the high redshift universe, from the time of decoupling of matter with radiation to the still uncertain epoch of reionization. After an introductory chapter on the physics of recombination and ionization of the IGM, the authors explore the role played by the first stars, the first miniquasars, in reionizing the universe, and describe the clumpy IGM.

The essay of Andrea Ferrara and Ruben Salvaterra contains an introduction to the physics of shock waves and of thermal instabilities, and a first inventory of all relevant feedback mechanisms operating since the cosmic dawn. They discuss the over-cooling problem in collapsing primordial galaxies, and dwarf galaxies as laboratories for the study of the chemical enrichment, and stellar and radiative feedback.

With the lecture of Simon Portegies Zwart, we return to black holes, and their ecology in star clusters. After reviewing key relevant timescales related to two-body relaxation, dynamical friction, and stellar evolution, the author discusses the main role played by mass segregation and runaway collisions in forming very massive stars, even in the metal polluted environment of a young forming star cluster. These super-stars end their lives as intermediate mass black holes, that may be unveiled as ultra-luminous x-ray sources in star-forming regions. A similar pathway may have been operating in globular clusters where there is a current search for intermediate-mass black holes. Portegies Zwart last speculates on the potential role played by intermediate mass black holes in the construction of the massive ones in galaxy nuclei.

The future will certainly hold new surprises, and black holes will stand out as key sources to explore our cosmic dawn, as well as our local universe.

Monica Colpi

Chapter 1

Observational Evidence for Supermassive Black Holes

Laura Ferrarese
Department of Physics and Astronomy
Rutgers State University
Piscataway, New Jersey

We discuss the current status of supermassive black hole research, as seen from a purely observational standpoint. Since the early 1990s, rapid technological advances—most notably the launch of the Hubble Space Telescope, the commissioning of the Very Long Baseline Array, and improvements in near-infrared speckle imaging techniques—have not only given us incontrovertible proof of the existence of SuperMassive Black Holes (SMBHs), but have unveiled fundamental connections between the mass of the central singularity and the global properties of the host galaxy. It is thanks to these observations that we are now, for the first time, in a position to understand the origin, evolution, and cosmic relevance of these fascinating objects.

1.1 Introduction

Observational evidence for black holes—in particular, the supermassive variety—has proven remarkably elusive, in spite of the fact that the theoretical foundation was in place immediately following the 1915 publication of Albert Einstein's theory of general relativity. As World War I raged across Europe and Russia, Karl Schwarzschild's 1916 solution of Einstein's field equations led to the unsettling but inescapable conclusion that for a star of given mass, there exists a finite, critical radius at which light reaches an infinite gravitational redshift, and therefore infinite time dilation. That real stars can indeed attain such a critical radius was later demonstrated in a series of seminal papers (Chandrasekhar 1931, 1935; Landau 1932; Oppenheimer & Volkoff 1938; Oppenheimer & Snyder 1939).

The first, long-awaited detection of a stellar mass black hole (in the rapidly variable x-ray source Cygnus X-1, Brucato & Kristian 1972; Bolton 1972; Lyutyi, Syunyaev, & Cherepashchuk 1973; Mauder 1973; Rhoades & Ruffini 1974) was thus the glorious confirmation of 60 years of theoretical groundwork. By contrast, the widespread acceptance of supermassive black holes in galactic nuclei—not several solar masses in size, but rather millions of solar masses—was propelled forward by an increasingly overwhelming body of observational evidence, but to this day, the benefits of a solid theoretical background are lacking.

Although many of the present-day SMBH detections are in quiescent or weakly active galaxies (for reasons that will become apparent shortly), the SMBH paradigm evolved exclusively in the context of Active Galactic Nuclei (AGN). The first clue came with Carl Seyfert's early 1940s identification of galaxies with "unusual nuclei." Fifty years later, one of Seyfert's original galaxies, NGC4258, was the first for which the existence of a SMBH was conclusively demonstrated (Miyoshi et al. 1995).

The significance of Seyfert's work was not fully appreciated until the 1950s, when studies of the newly discovered radio sources made it abundantly clear that new and extraordinary physical processes were at play in the nuclei of galaxies. The most unequivocal evidence came from M87 (Figure 1.1) and Cygnus A (Figure 1.2). In 1954, Baade & Minkowski associated the bright radio source Vir A with the seemingly unremarkable giant elliptical galaxy at the dynamical center of the Virgo cluster. They also identified the double radio source Cygnus A (Jennison & Das Gupta 1953) with a galaxy at the center of a rich cluster at z=0.057 (D $\sim$ 250 Mpc for $H_0 = 70$ km s^{-1}Mpc^{-1}). In both cases, the presence of radio lobes on either side of the visible galaxy and, in the case of M87, of a bright, optical narrow jet, strongly suggested that the radio emission might be due to relativistic particles ejected in opposite direction from the nucleus. Furthermore, optical spectra of the nuclei revealed strong extended emission lines with unusual line strengths. Forty years later, the presence of a supermassive black hole in M87 was firmly established using the Hubble Space Telescope (Ford et al. 1994; Harms et al. 1994).

As observations continued to be collected, it became apparent that the menagerie of AGN is as diverse as could be imagined. Quasars, radio galaxies, Seyfert nuclei, Blazars, Low Ionization Nuclear Emission Regions (LINERs), and BL Lacertae objects, to name a few, are set apart from each other by both the detailed character of their nuclear activity and the traits of their host galaxies. Underneath this apparent diversity, however, lie three revealing common properties. First, AGN are extremely compact. Flux variability—a staple of all AGN—confines the size of the AGN to within the distance light can travel in a typical variability timescale. In

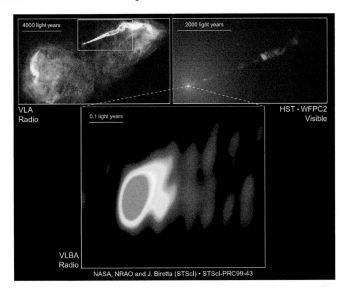

Figure 1.1. VLA (top left), HST (top right), and VLBI (bottom) images of M87. The images show a bright radio and optical source at the center of the galaxy, and a helical jet emanating from the nucleus. The radio and optical emission is synchrotron radiation. Courtesy of NASA, National Radio Astronomy Observatory/National Science Foundation, John Biretta (STScI/JHU), and Associated Universities, Inc.

many cases, x-ray variability is observed on timescales of less than a day and flares on timescales of minutes (e.g., MCG 6-30-15, McHardy 1988). Second, the spectral energy distribution is decisively non-stellar; roughly speaking, the AGN power per unit logarithmic frequency interval is constant over seven decades in frequency, while stars emit nearly all of their power in a frequency range a mere factor of three wide. Third, AGN bolometric luminosities are remarkably large—at least comparable to, and often several orders of magnitude larger than, the luminosity of the entire host galaxy. This implies that, whatever its true nature, the central engine must be massive—at least 10^6 M$_\odot$ for it not to become unbound by its own outpouring of energy (see Equation 1.1).

As early as 1963—the year Maarten Schmidt realized that the 13th magnitude "star" associated with the radio source 3C273 (Figure 1.3) was at an astoundingly (at the time) large redshift of $z = 0.158$—much pondering was devoted to the idea that the energy source of AGN is gravitational in nature (Robinson, Schild, & Schucking 1965; Zel'dovich & Novikov 1964; Salpeter 1964; Lynden-Bell 1969). The paradigm—invoking the existence of a gravitational singularity that converts into energy much of the mat-

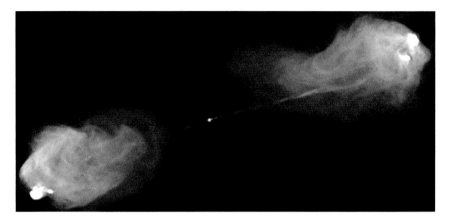

Figure 1.2. VLA 1.4 6cm image of the bright radio source Cygnus A. The jets transport energy from the nucleus to the radio lobes at distance of ~220 kpc, before being stopped by the intergalactic medium surrounding the galaxy. Image courtesy of NRAO/AUI.

ter falling into it—has to this day survived as the only viable explanation for AGN behavior (Rees 1984). This conclusion is bolstered by the fact that some AGN display the markedly relativistic signature of a strong field regime, most notably superluminal motions of the radio jets and the broadening of the low excitation x-ray iron $K\alpha$ emission line (Reynolds & Nowak 2003).

Although the arguments discussed above apply strictly to AGN, modern SMBH searches have targeted almost exclusively quiescent or weakly active nearby galaxies. There are two good reasons for this. First, quiescent galaxies are *expected* to host SMBHs. The cumulative mass density in SMBHs needed to explain the energetics of high redshift, powerful quasars falls short, by at least two orders of magnitude, of that required to power local AGN (Padovani, Burg, & Edelson 1990; Ferrarese 2002a). The unaccounted SMBHs must therefore reside in local, quiescent galaxies.[1] Second, although compelling, AGN signatures provide only a smoking gun for the existence of SMBHs. Only the telltale Keplerian signature imprinted by the SMBH on the motions of gas and stars in its immediate vicinity can produce definitive proof. Such signatures can be resolved only in the most nearby galaxies, very few of which are AGNs.

After introducing some formalism (§1.2) and some general considera-

[1] This conclusion opens new questions. For example, there is an abundant supply of gas and dust in galactic centers, so it is not a simple task to prevent a SMBH from shining as brightly as a quasar (Fabian & Canizares 1988). Indeed, a definitive answer to the question of why some SMBHs seem to be eerily quiet has yet to be found (Rees et al. 1982; Narayan & Yi 1994; Blandford & Begelman 1999; Di Matteo et al. 1999).

Figure 1.3. A pair of Hubble Space Telescope images of the quasar 3C 273, taken with the Wide Field and Planetary Camera 2 (left), and the Advanced Camera for Surveys (right). In the latter image, the bright nucleus was placed behind an occulting finger, to reveal the light from the host galaxy. The WFPC2 image clearly shows the optical synchrotron jet extending 50 Kpc from the nucleus. For comparison, the jet in M87 is only 2 kpc long. Credit: NASA, the ACS Science Team and ESA.

tions (§1.3), we review the techniques at the disposal of SMBH hunters, in particular, dynamical modeling of stellar (§1.4) and gas (§1.5) kinematics, and reverberation mapping (§1.6). In §1.7, we discuss scaling relations between SMBHs and their host galaxies. Such relations have proven invaluable in the study of SMBH demographics, formation, and evolution. The demography of SMBHs is reviewed in §1.8; lastly, in §1.9, we conclude by reviewing the current status of SMBH research and outline a few of the outstanding issues.

1.2 Some Useful Formalism

For convenience, we present in this section some terminology and equations which that will recur in the remainder of this review.

An important measure of the accretion rate onto a BH of mass $M_{\bullet}$

is provided by the Eddington luminosity, i.e. the luminosity at which radiation pressure on free electrons balances the force of gravity. Because the force due to radiation pressure has exactly the same inverse square dependence on distance as gravity, but does not depend on mass, L_E is independent of distance but depends on $M_\bullet$:

$$L_E = \frac{4\pi G M m_p c}{\sigma_T} \sim 1.3 \times 10^{46} \left(\frac{M_\bullet}{10^8\, M_\odot} \right) \text{erg s}^{-1} \qquad (1.1)$$

where m_p is the proton rest mass and σ_T is the Thomson cross-section. Above the Eddington luminosity, the source is unable to maintain steady spherical accretion (although the presence of magnetic fields can considerably complicate the picture; Begelman 2001).

Related to the Eddington luminosity is the Salpeter time

$$t_S = \frac{\sigma_T c}{4\pi G m_p} \sim 4 \times 10^8\, \epsilon \text{ yr} \qquad (1.2)$$

t_S can be interpreted in two, equivalent ways. It would take a black hole radiating at the Eddington luminosity a time t_S to dissipate its entire rest mass. Also, the luminosity (and mass) of a black hole accreting at the Eddington rate with constant $\dot{M}/M$ will increase exponentially, with e-folding time t_S. ϵ is the efficiency of conversion of mass into energy, and depends on the spin of the black hole, varying between 6% if the black hole is not spinning, and 42% if the black hole is maximally spinning.

The "boundary" of a (non-rotating) black hole of mass $M_\bullet$ is a spherical surface called the event horizon, the radius of which is given by the Schwarzschild (or gravitational) radius:

$$r_{Sch} = \frac{2G M_\bullet}{c^2} \sim 3 \times 10^{13} \left(\frac{M_\bullet}{10^8 M_\odot} \right) \text{cm} \sim 2 \left(\frac{M_\bullet}{10^8\, M_\odot} \right) \text{A.U.} \qquad (1.3)$$

At the Schwarzschild radius the gravitational time dilation goes to infinity and lengths are contracted to zero.

The radius r_{st} of the last stable orbit, inside which material plunges into the black hole, depends on the black hole angular momentum, being smaller for spinning Kerr black holes. For a non rotating Schwarzschild black hole:

$$r_{st} = \frac{6G M_\bullet}{c^2} = 3 r_{Sch} \qquad (1.4)$$

The photon sphere, of radius $1.5 r_{Sch}$, is defined as the surface at which gravity bends the path of photons to such an extent that light orbits the hole circularly.

For a Kerr (rotating) black hole there are two relevant surfaces, the event horizon, and the static surface, which completely encloses it. At the static surface, space–time is flowing at the speed of light, meaning that a particle would need to move at the speed of light in a direction opposite to the rotation of the hole in order to be stationary. In the region of space within the static surface and the event horizon, called the ergosphere, the rotating black hole drags space around with it (frame dragging) in such a way that all objects must corotate with the black hole. For a maximally rotating black hole, the radius of the last stable orbit is

$$r_{st} = \frac{1.2GM_{\bullet}}{c^2} \tag{1.5}$$

Because of the dependence of r_{st} on the black hole spin, the latter can be inferred provided a measure of the former, and an estimate of the black hole mass, are available, for instance from rapid flux variability (see also §1.6). In the case of supermassive black holes inhabiting galactic nuclei, the "sphere of influence" is defined as the region of space within which the gravitational potential of the SMBH dominates over that of the surrounding stars. Its radius is given by:

$$r_h \sim G \ M_{\bullet}/\sigma^2 \sim 11.2 \ (M_{\bullet}/10^8 \ \mathrm{M}_{\odot})/(\sigma/200 \ \mathrm{km \ s^{-1}})^2 \ \mathrm{pc} \tag{1.6}$$

where σ is the velocity dispersion of the surrounding stellar population. Beyond a few thousand Schwarzschild radii from the central SMBH, but within the sphere of influence, the motion of stars and gas is predominantly Keplerian (relativistic effects are minimal), with a component due to the combined gravitational potential of stars, dust, gas, dark matter, and anything else contributing mass to within that region. Beyond the sphere of influence, the gravitational dominance of the SMBH quickly vanishes.

1.3 General Considerations

A SMBH that forms or grows in a galactic nucleus will produce a cusp in the stellar density (Peebles 1972; Young 1980; Quinlan, Hernquist, & Sigurdsson 1995; van der Marel 1999). Unfortunately, as demonstrated very effectively by Kormendy & Richstone (1995), the growth, or even the presence of a SMBH is not a necessary condition for "light cusps" to form. Moreover, even when originally present, central density cusps can be destroyed during galaxy mergers, as a consequence of the hardening of the SMBH binary that forms at the center of the merger product (Milosavljevic & Merritt 2001).

The dynamical signature imprinted by a SMBH on the motion of surrounding matter is, however, unique. Within the sphere of influence, a Keplerian rotation or velocity dispersion of stars or gas is unambiguous proof of the existence of a central mass concentration. The ultimate test as to its nature (a singularity or a dense star cluster?) can only reside in the detection of relativistic velocities within a few Schwarzschild radii. Only observations of the Fe Kα emission line in Type 1 AGNs might give us a chance of peering within the relativistic regime of a SMBH (Table 1.1), although this is still considered to be a controversial issue.

In the absence of the ultimate, relativistic signature, the case for the detected masses to be confined within a singularity becomes stronger as the corresponding mass-to-light ratio and mass density increase. Maoz (1998) provided rough calculations of the lifetime of a dark cluster against evaporation and collapse. For any choice of the cluster's mass and density (or, equivalently, radius and density) such lifetime depends on the cluster's composition. Maoz considered the case of clusters of brown dwarfs (with masses down to 3×10^{-3} M$_\odot$), white dwarfs, neutron stars, and stellar black holes. His results are reproduced in Figure 1.4, where the maximum lifetime attainable by any such cluster is plotted as a function of the cluster's density and radius (an upper limit to which is given by the scale probed by the data), the latter normalized to the Schwarzschild radius. The case for a SMBH is tight, according to this simple argument, when the observations imply densities and masses for which the dark cluster lifetime is short compared to the age of the galaxy. When all SMBH detections claimed to date are considered (Table 1.5.1.2), this condition is verified only in the case of the Milky Way and NGC 4258. For all other galaxies, although we will tacitly assume for the rest of this chapter that the detected masses are indeed SMBHs, dark clusters can provide viable explanations.

Table 1.1 summarizes typical central mass densities inferred for each of the methods which are (or might become) available to measure SMBH masses in galactic nuclei. The Fe Kα emission line arises from material within a few Schwarzschild radii of the central SMBH and seems to be an almost ubiquitous feature in the x-ray spectra of Seyfert 1 galaxies (Nandra et al. 1997; Reynolds 1997; but see also Done, Madejski, & Zycki 2000; Gondoin et al. 2001a, b). It provides a powerful diagnostic of the properties of space–time in strong gravitational fields. Future generations of x-ray satellites (most notably the European mission XEUS and NASA's Constellation X) will reveal whether the line responds to flares in the x-ray continuum (a point which is debated, Nandra et al. 1997, 1999; Wang, Wang, & Zhou 2001; Takahashi, Inoue, & Dotani 2002). If so, it will open the use of reverberation mapping (§1.6) to measure the mass of the central SMBH. We will discuss Fe Kα observations no further in this chapter, as an excellent recent review can be found in Reynolds & Nowak (2003).

In its current application, reverberation mapping (§1.6) targets the Broad Line Region (BLR) of Type 1 AGNs (Blandford & McKee 1982; Peterson 1993, 2002). It is currently the least secure, but potentially most powerful of the methods which we will discuss, probing material within only a few hundred Schwarzschild radii from the singularity, a factor 10^3

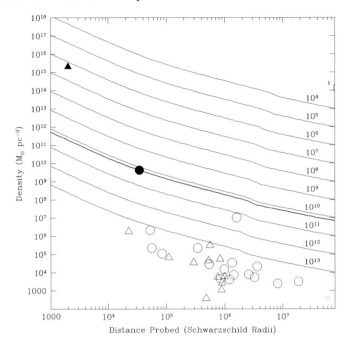

Figure 1.4. The innermost radius (normalized to the Schwarzschild radius) probed by current experiments which have led to the detection of SMBHs in nearby galactic nuclei, plotted against the inferred mass density within that region. The detection in the Milky Way (§1.4.1) is shown as the solid triangle, water maser detections (§1.5.1) are shown as solid circles, detections based on stellar (§1.4.2) and gas (§1.5.2) dynamics are shown as open circles and triangles respectively. The solid curves show the maximum lifetime of a dark cluster against collision and evaporation, using the prescription in Maoz (1998), from 10^4 to 10^{13} years. The thick solid line represents a lifetime of 15 billion years.

closer than can be reached by stellar and gas dynamical studies using HST. As discussed in detail in §1.6, although there is compelling, albeit indirect, observational evidence that the BLR motion is Keplerian, the kinematics and morphology of the BLR have not yet been mapped directly. In the Keplerian hypothesis, the SMBH mass is derived from the observed "average" size and velocity of the BLR. If the method does indeed measure masses (and the evidence that this is indeed the case is growing stronger by the day), the inferred central densities leave no doubt that such masses are indeed confined in a singularity.

To date, the only secure detections of SMBHs (as opposed to dense clusters of stars or exotic particles) come from stellar proper motion in the galactic center and the H_2O megamaser study of the nearby Seyfert

Table 1.1. Probing the Centers of Galaxies

Method & Telescope	Scale (R_S)	No. of SMBH Detections	$M_\bullet$ Range ($M_\odot$)	Typical Densities ($M_\odot$ pc^{-3})
Fe Kα line (XEUS, ConX)	3-10	0	N/A	N/A
Reverberation Mapping (Ground based optical)	600	36	$10^6 - 4 \times 10^8$	$\gtrsim 10^{10}$
Stellar Proper Motion (Keck, NTT, VLT)	1000	1	4×10^6	4×10^{16}
H$_2$O Megamasers (VLBI)	10^4	1	4×10^7	4×10^9
Gas Dynamics (optical) (Mostly HST)	10^6	11	$7 \times 10^7 - 4 \times 10^9$	$\sim 10^5$
Stellar Dynamics (Mostly HST)	10^6	17	$10^7 - 3 \times 10^9$	$\sim 10^5$

The columns give all methods that can (or, in the case of the Fe Kα line emission, might) be used to estimate SMBH masses, and the telescopes needed for the observations; the typical distance from the singularity of the material probed by each method; the number of SMBH detections claimed based on that method; the range in the detected SMBH masses; and the corresponding implied central mass density.

2 galaxy NGC 4258 (§1.4.1 and §1.5.1 respectively). The applicability of either method is however limited (to one galaxy, the Milky Way, in the case of proper motion studies). The most prolific methods are based on optical stellar and gas dynamical studies, generally carried out using the Hubble Space Telescope (HST, §1.4.2 and §1.5.2). On the downside, these methods can rarely reach closer than several million Schwarzschild radii from the singularity; the implied central densities are always far lower than needed to conclude that the mass is indeed collapsed into a SMBH.

Table 1.3 lists all galaxies for which a SMBH detection has been claimed based on stellar proper motion, H$_2$O megamasers, or optical stellar and gas dynamical studies (reverberation mapping detections will be listed in §1.6). Cases for which the analysis did not lead to a successful determination of the SMBH mass (according to the original investigators) are grouped at the end of the table. For each galaxy, we list the Hubble type, distance (mostly from Tonry et al. 2001), SMBH mass and reference, central bulge velocity dispersion, total and extinction corrected blue magnitude (from the RC3, de Vaucouleurs et al. 1991, corrected for galactic extinction using the reddening maps of Schlegel et al. 1998), and fraction of the total light judged to be in the hot stellar component (from Fukugita et al. 1998). In the last column, the ratio r_h/r_{res} between the radius of the

SMBH sphere of influence (Equation 1.6) and the spatial resolution of the data is given as a rough indicator of the quality of the SMBH mass estimate. All studies that have addressed the issue (Ferrarese & Merritt 2000; Merritt & Ferrarese 2001b&c; Graham et al. 2001; Ferrarese 2002a; Marconi & Hunt 2003; Valluri et al. 2004) have concluded that resolving the sphere of influence is an important (although not sufficient) factor: not resolving r_h can lead to systematic errors on $M_\bullet$ or even spurious detections.

This rather intuitive fact explains the dominance of HST in this field. As an example, consider our close neighbor, the Andromeda galaxy. At a fiducial stellar velocity dispersion of $\sigma \sim 160$ km s^{-1}, and assuming a SMBH with mass $\sim 3 \times 10^7$ M$_\odot$ (unfortunately, the well-known presence of a double nucleus does not allow for an accurate determination of the SMBH mass, Bacon et al. 2001), the radius of the sphere of influence is 5.2 pc, or 1″.4 at a distance of 770 kpc. This is resolvable from the ground. However, if Andromeda were just a factor two or three farther, ground-based observations would be unable to address the question of whether its nucleus hosts a SMBH. In the absence of maser clouds or an active nucleus, HST data would offer the only viable option. At the distance of the Virgo cluster, 15 Mpc, the sphere of influence of a $\sim 3 \times 10^7$ M$_\odot$ SMBH would shrink to a projected radius of 0″.07, not only well beyond the reach of any ground based telescope, beyond even HST capabilities. Overall, as will be discussed in §1.9, the number of galaxies for which the SMBH sphere of influence can be resolved with ground-based optical observations can be counted on the fingers of one hand. HST has enabled that number to be increased by well over an order of magnitude.

1.4 Resolved Stellar Dynamics

1.4.1 Stellar Proper Motion Studies: The Galactic Center

The case for a massive object at the Galactic Center has been building since the detection in the 1970s of strong radio emission originating from the innermost 1 pc (Balick & Brown 1974; Ekers et al. 1975). Not only is the source, dubbed Sgr A*, extremely compact (VLBI observations at 86 GHz set a tight upper limit of 1 A.U. to its size; Doeleman et al. 2001), but the absence of any appreciable proper motion implies that it must also be very massive. In the most recent study on the subject, Reid et al. (2003) argue that Sgr A* must be in excess of 4×10^5 M$_\odot$, thus excluding that it might consist of a compact cluster of stellar objects. Because of its proximity (8.0 ± 0.4 kpc, Eisenhauer et al. 2003), the Galactic Center can be studied at a level of detail unimaginable in any other galaxy. Proper motions of the star cluster surrounding Sgr A* can be detected using near-infrared (near-IR) speckle imaging techniques. Ongoing monitoring studies conducted for the past ten years at, first, the ESO New Technology Telescope (NTT)

Table 1.2. Complete List of SMBH Mass Detection Based on Resolved Dynamical Studies

Object	Hubble Type	Distance (Mpc)	$M_\bullet$ $(10^8 M_\odot)$	$M_\bullet$ Ref. & Method	σ (km s^{-1})	$M_{B,T}^0$ (mag)	$L_{B,bulge}/L_{B,total}$	r_h/r_{res}
MW	SbI-II	0.008	$0.040^{+0.003}_{-0.003}$	1,PM	100±20	-20.08±0.50	0.34	1700
N4258	SAB(s)bc	7.2	$0.390^{+0.034}_{-0.034}$	2,MM	138±18	-20.76±0.15	0.16	880
N4486	E0pec	16.1	$35.7^{+10.2}_{-10.2}$	3,GD	345±45	-21.54±0.16	1.0	34.6
N3115	S0	9.7	$9.2^{+3.0}_{-3.0}$	4,SD	278±36	-20.19±0.20	0.64	22.8
I1459	E3	29.2	$26.0^{+11.0}_{-11.0}$	5,SD	312±41	-21.50±0.32	1.0	17.0
N4374	E1	18.7	$17^{+12}_{-6.7}$	6,GD	286±37	-21.31±0.13	1.0	10.3
N4697	E6	11.7	$1.7^{+0.2}_{-0.3}$	7,SD	163±21	-20.34±0.18	1.0	10.2
N4649	E2	16.8	$20.0^{+4.0}_{-6.0}$	7,SD	331±43	-21.43±0.16	1.0	10.1
N221	cE2	0.8	$0.025^{+0.005}_{-0.005}$	8,SD	76±10	-15.76±0.18	1.0	10.1
N5128	S0pec	4.2	$2.0^{+3.0}_{-1.4}$	9,GD	145±25	-20.78±0.15	0.64	8.41
M81	SA(s)ab	3.9	$0.70^{+0.2}_{-0.1}$	10,GD	174±17	-20.42±0.26	0.33	5.50
N4261	E2	31.6	$5.4^{+1.2}_{-1.2}$	11,GD	290±38	-21.14±0.20	1.0	3.77
N4564	E6	15.0	$0.56^{+0.03}_{-0.08}$	7,SD	153±20	-19.00±0.18	1.0	2.96
CygA	E	240	$25.0^{+7.0}_{-7.0}$	12,GD	270±87	-20.03±0.27	1.0	2.65
N2787	SB(r)0	7.5	$0.90^{+6.89}_{-0.69}$	13,GD	210±23	-18.12±0.39	0.64	2.53
N3379	E1	10.6	$1.35^{+0.73}_{-0.73}$	14,SD	201±26	-19.94±0.20	1.0	2.34
N5845	E*	25.9	$2.4^{+0.4}_{-1.4}$	7,SD	275±36	-18.80±0.25	1.0	2.28
N3245	SB(s)b	20.9	$2.1^{+0.5}_{-0.5}$	15,GD	211±19	-20.01±0.25	0.33	2.10
N4473	E5	15.7	$1.1^{+0.5}_{-0.8}$	7,SD	188±25	-19.94±0.14	1.0	1.84
N3608	E2	22.9	$1.9^{+1.0}_{-0.6}$	7,SD	206±27	-20.11±0.17	1.0	1.82
N4342	S0	16.7	$3.3^{+1.9}_{-1.1}$	16,GD	261±34	-17.74±0.20	0.64	1.79
N7052	E	66.1	$3.7^{+2.6}_{-1.5}$	17,GD	261±34	-21.33±0.38	1.0	1.53
N4291	E3	26.2	$3.1^{+2.3}_{-0.8}$	7,SD	269±35	-19.82±0.35	1.0	1.52
N6251	E	104	$5.9^{+2.0}_{-2.0}$	18,GD	297±39	-21.94±0.28	1.0	1.19
N3384	SB(s)0-	11.6	$0.16^{+0.01}_{-0.02}$	7,SD	151±20	-19.59±0.15	0.64	1.12
N7457	SA(rs)0-	13.2	$0.035^{+0.011}_{-0.014}$	7,SD	73±10	-18.74±0.24	0.64	0.92

Object	Hubble Type	Distance Mpc	$M_\bullet$ $10^8 M_\odot$	$M_\bullet$ Ref. & Method					Notes
N1023	S0	11.4	$0.44^{+0.06}_{-0.06}$	7,SD	201 ± 14	-20.20 ± 0.17	0.64	0.89	
N821	E6	24.1	$0.37^{+0.24}_{-0.08}$	7,SD	196 ± 26	-20.50 ± 0.21	1.0	0.74	
N3377	E5	11.2	$1.00^{+0.9}_{-0.1}$	7,SD	131 ± 17	-19.16 ± 0.13	1.0	0.74	
N2778	E	22.9	$0.14^{+0.08}_{-0.09}$	7,SD	171 ± 22	-18.54 ± 0.33	1.0	0.39	
Galaxies for which the dynamical models might be in error.									
Circinus	SA(s)b	4.2	$0.017^{+0.003}_{-0.003}$	19,MM					disk inclination angle not constrained
N4945	SB(s)cd	3.7	$0.014^{+0.007}_{-0.005}$	20,MM					no 2-D velocity field
N1068	Sb	23.6	$0.17^{+0.13}_{-0.07}$	21,MM					maser disk is self gravitating
N4459	SA(r)0+	16.1	$0.70^{+0.13}_{-0.13}$	13,GD					disk inclination angle not constrained
N4596	SB(r)0+	16.8	$0.8^{+0.4}_{-0.4}$	13,GD					disk inclination angle not constrained
N4594	SA(s)a	9.8	$10.0^{+10.0}_{-7.0}$	22,SD					no 3-integral models
N224	Sb	0.77	$0.35^{+0.25}_{-0.25}$	23,SD					double nucleus
N4041	Sbc	16.4	< 0.2	24,GD					disk might be dynamically decoupled

Notes to Table 1.2: The columns give the galaxy's Hubble type; distance (from Tonry et al. 2001 whenever available; derived from the heliocentric systemic velocity and $H_0 = 75$ km s^{-1} Mpc^{-1} in all other cases); black hole mass, reference (coded below) and method of detection (PM = stellar proper motion, GD = gas dynamics, SD = stellar dynamics, MM = H_2O megamasers); central bulge velocity dispersion; total, extinction corrected blue magnitude (from the RC3, de Vaucouleurs et al. 1991, corrected for galactic extinction using the reddening maps of Schlegel et al. 1998); fraction of the total light judged to be in the hot stellar component (from Fukugita et al. 1998); the ratio of the diameter of the SMBH sphere of influence to the spatial resolution of the data. References: 1. Ghez et al. 2003 – 2. Miyoshi et al. 1995 – 3. Macchetto et al. 1997 – 4. Emsellem et al. 1999 – 5. Cappellari et al. 2002 – 6. Bower et al. 1998 – 7. Gebhardt et al. 2003 – 8. Verolme et al. 2002 – 9. Marconi et al. 2001 – 10. Devereux et al. 2003 – 11. Ferrarese et al. 1996 – 12. Tadhunter et al. 2003 – 13. Sarzi et al. 2001 – 14. Gebhardt et al. 2000a – 15. Barth et al. 2001 – 16. Cretton & van den Bosch 1999 – 17. van der Marel & van den Bosch 1998 – 18. Ferrarese & Ford 1999 – 19. Greenhill et al. 2003 – 20. Greenhill, Moran & Herrnstein 1997 – 21. Greenhill et al. 1996 – 22. Kormendy et al. 1988 – 23. Bacon et al. 2001 – 24. Marconi et al. 2003.

and the Keck Telescope and, more recently, the ESO Very Large Telescope (VLT), have reached a staggering $0''.003$ (0.1 Mpc) astrometric accuracy in the stellar positions (Eckart et al. 1993; Ghez et al. 1998, 2000, 2003; Schödel et al. 2003). Proper motion has been measured for over 40 stars within $1''.2$ of Sgr A*; deviations from linear motion have been detected for eight stars, and four stars in particular have passed the pericenter of their orbits since monitoring began (Ghez et al. 2003; Schödel et al. 2003). In the three cases for which accurate orbits can be traced, the stars orbit Sgr A* with periods between 15 and 71 years, reaching as close as 87 A.U. from the central source. Using a simultaneous multiorbital solution, Ghez et al.(2003) derive a best fit central mass of $(4.0 \pm 0.3) \times 10^6$ M$_\odot$. The implied central mass density of 4×10^{16} M$_\odot$ Mpc^{-3} provides virtually incontrovertible evidence that the mass is indeed in the form of a singularity. Detailed information about the nature of Sgr A* can be found in the excellent review by Melia & Falcke (2001).

1.4.2   Integrated Stellar Dynamics

Historically, modeling the kinematics of stars in galactic nuclei has been the method of choice to constrain the central potential, and for good reason: stars are always present, and their motion is always gravitational. But, as with every method, there are downsides. Stellar absorption lines are faint, and the central surface brightness, especially in bright ellipticals, is low (Crane et al. 1993; Ferrarese et al. 1994; Lauer et al. 1995; Rest et al. 2001). Acquiring stellar kinematical data therefore often entails walking a fine line between the need for high spatial resolution and the need for high spectral signal-to-noise (S/N). The latter benefits from the large collecting area of ground-based telescopes, while the former demands the use of HST in all but a handful of cases. Theoretical challenges arise from the fact that the stellar orbital structure is unknown and difficult to recover from the observables. Although dynamical models have reached a high degree of sophistication (Verolme et al. 2002; Gebhardt et al. 2003; van de Ven et al. 2003), the biases and systematics that might affect them have not been fully investigated and could be severe (Valluri et al. 2004).

The fact that galaxies are collisionless stellar systems (e.g., Binney & Tremaine 1987) greatly simplifies the theoretical treatment. Each star can be seen as moving in the total gravitational potential $\Phi(\vec{x}, t)$ of the system; thus, the ensemble is completely described by a continuity equation for the Distribution Function (DF) $f(\vec{x}, \vec{v}, t)$ (i.e., the number of stars that occupy a given infinitesimal volume in phase-space):

$$\frac{\delta f}{\delta t} + \vec{v} \cdot \vec{\nabla} f - \vec{\nabla}\Phi(\vec{x}, t) \cdot \frac{\delta f}{\delta \vec{v}} = 0 \,. \tag{1.7}$$

Equation 1.7 is known as the Collisionless Boltzmann Equation (CBE).

The gravitational potential $\Phi(\vec{x}, t)$ is linked to the total mass density ρ (to which both stars and the putative SMBH contribute) by the Poisson equation:

$$\nabla^2 \Phi(\vec{x}, t) = 4\pi G \rho(\vec{x}, t) . \tag{1.8}$$

The DF can be reconstructed from knowledge of the stellar mass density and the six components of the streaming velocity and velocity dispersion, all of which are related to integrals of the DF over velocity (modulo the stellar mass-to-light ratio, which can be taken as a multiplicative factor). From the DF, the total gravitational potential follows from the CBE, and the total mass density (and hence the SMBH mass, since the stellar mass density is known from the observed surface brightness profile) from the Poisson equation.

Unfortunately, not all components of the velocity tensor can be extracted from observational data: it is only at the cost of making significant simplifications that the CBE and Poisson equation can be solved analytically. Under the assumption of a system in steady state, spherically symmetric, and isotropic, Sargent et al. (1978) "detected" a central $\sim 5 \times 10^9$ M$_\odot$ dark mass within the inner 110 pc of Virgo's cD galaxy, M87. Soon after, Binney & Mamon (1982), and later Richstone & Tremaine (1985), demonstrated that equally acceptable fits to Sargent et al.'s data could be obtained in the assumption of a constant mass-to-light ratio and no central SMBH, provided that the velocity dispersion tensor was allowed to be anisotropic. M87 is thus the textbook example of the "mass-to-light ratio/velocity dispersion anisotropy degeneracy" that affects stellar dynamical studies. Indeed, the presence of a SMBH in M87 could not be firmly established even using more recent, state of the art data (van der Marel 1994), including a full analysis of the Line of Sight Velocity Distribution (LOSVD). The LOSVD, defined as the integral of the DF along the line of sight and the two tangential components of the velocity (normalized to the projected surface brightness profile), is observationally reflected in the shape of the absorption line profile. It is sensitive to the level of anisotropy of the system (van der Marel & Franx 1993; Gerhard 1993), potentially allowing one to break the degeneracy.

In practice, a comprehensive study of the LOSVD requires higher S/N than available (or obtainable) for the vast majority of galaxies. It therefore remains generally true that stellar dynamical models are best applied to rapidly rotating systems (most faint ellipticals are in this class, e.g., Kormendy & Richstone 1992), or systems hosting small nuclear stellar disks (Scorza & Bender 1995; van den Bosch & de Zeeuw 1996; van den Bosch, Jaffe, & van der Marel 1998; Cretton & van den Bosch 1999). As the streaming velocity components become more dominant, terms depending on the velocity dispersion have a lesser influence in the CBE.

Since the 1978 work of Sargent et al., dynamical models have become more general and increasingly complex. Simple isotropic systems are described by a DF that depends on one integral of motion, namely the total energy of the system. More general axisymmetric anisotropic systems that admit two integrals of motion have been widely used (e.g., Magorrian et al. 1998) but are also inadequate. Two-integral models predict major axis velocity dispersions that are larger than observed, and require identical velocity dispersions in the radial and vertical direction—a condition not verified, for instance, in the solar neighborhood. Furthermore, numerical simulations show that most orbits are not completely described by two integrals of motion: a third integral must exist, although an analytical description is not known. Three-integral models, which are now routinely applied to stellar kinematics (Verolme et al. 2002; Gebhardt et al. 2003), cannot be solved analytically and are handled numerically as follows.

- The observed surface brightness profile is deprojected to obtain the luminosity density. The deprojection is not unique, and the galaxy inclination angle i must be assumed.
- A mass density profile is constructed by assuming a (generally constant with radius) stellar mass-to-light ratio γ_{star} and a central point mass $M_\bullet$. The gravitational potential Φ is then derived from the Poisson equation.
- Given a grid of cells in position space, initial conditions for a set of orbits are chosen. For each orbit, the equations of motion are integrated over many orbital periods. A tally of how much time each orbit spends in each cell provides a measure of the mass contributed by each orbit to each cell.
- Non-negative weights are determined for each orbit such that the summed mass and velocity structure in each cell, when integrated along the line of sight, reproduces the observed surface brightness and kinematical constraints.

In principle, three-integral models provide a completely general and unconstrained description of a stellar system. In practice, though, some assumptions must still be made. Only axisymmetric systems have been modeled to date, although steps toward a formalism for triaxial systems have recently been taken (van de Ven et al. 2003). Even in axisymmetric systems, the extra degree of freedom introduced by treating the galaxy's inclination angle as a free parameter cannot generally be constrained given the observables. With one exception (Verolme et al. 2002), the inclination angle is therefore assumed a priori (e.g., Gebhardt et al. 2003). A recent study by Valluri et al. (2004) claims that potentially severe systematics in $M_\bullet$ are introduced, unless the number of orbits used in the modeling is at least an order of magnitude larger than the number of observational constraints. More disturbing still is the claim that even in the case of the best observational datasets, three-integral models admit too many degrees of freedom to constrain $M_\bullet$, which remains undetermined to within a factor of several. Given the potentially severe implications of these findings, it is to be hoped that more groups will pursue similar investigations of the systematics associated with three-integral models and come to a clear resolution of the issue. There are currently 17 detections of SMBHs in galactic nuclei based on three-integral models applied to stellar kinematics, as listed in Table 1.3.

1.5 Gas as a Tracer of the Gravitational Potential

In their 1978 paper, Sargent et al. note that the [OII]λ3727 emission doublet in the spectrum of M87 consists of a broadened, unresolved component, plus a narrower, asymmetric feature. Although they "appreciate the dangers of associating the broad lines with gas clouds swirling around a massive object, especially in view of the possibility of ejection or infall," Sargent et al. point out that the width of the narrow line, 600 km s^{-1}, corresponds to the stellar velocity dispersion measured at the same radius. Interpreting the broadening of the unresolved component as due to Keplerian rotation, they also noticed that the implied mass was of the same order as that derived from the stellar dynamical modeling.

In the Galactic Center, gas kinematics hinted at the presence of a central mass concentration well before the spectacularly conclusive proper motion studies described in §1.4.1. The sharp increase in the gas velocity from $\sim$ 100 km s^{-1} at 1.7 pc to 700 km s^{-1} at 0.1 pc implies a virial mass of a few 10^6 M$_\odot$ within this radius (Wollman et al. 1977; Lacy et al. 1980; Lacy, Townes, & Hollenbach 1982; Crawford et al. 1985; Serabyn & Lacy 1985; Mezger & Wink 1986). Although often interpreted as evidence for a central black hole, concerns over the possibility of gas outflows or inflows ultimately limited the credibility of such claims (e.g., Genzel & Townes 1987).

Indeed, gas, unlike stars, can easily be accelerated by non-gravitational forces, and dynamical studies based on gas kinematics have often been quickly—and sometimes unjustly—[2]dismissed by the establishment. Now thanks to two critical observations, this is no longer the case. In the early 1990s, HST images revealed a small ($\sim$ 230 pc), regular, cold dust and gas disk in the E2 galaxy NGC4261 (Jaffe et al. 1993; see Figure 1.5). The disk morphology strongly suggests a regular velocity field controlled by the galaxy gravitational potential (§1.5.2). Soon after, Very Long Baseline Array (VLBA) observations unveiled an even smaller and colder molecular disk in the Seyfert 2 galaxy NGC4258. Along with the Milky Way,

---

[2] For instance, Kormendy & Richstone (1995) remark that although the H$_2$O rotation curve of NGC4258 "looks Keplerian ... as usual it is not certain that gas velocities measure mass."

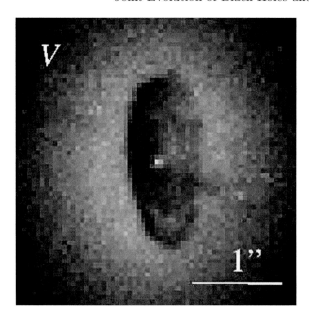

Figure 1.5. An HST Wide Field and Planetary Camera 2 (WFPC2) image of the nuclear dust disk in NGC4261, taken with the F555W filter ($\sim$ Johnson's V). The disk, which surrounds the bright, non-thermal, unresolved nucleus, is 230 pc across and roughly perpendicular to the galaxy's radio jets. Hα+[NII] narrow band images reveal an ionized gas component associated with the inner parts of the disk. Adapted from Ferrarese, Ford, & Jaffe (1996).

NGC4258 has given us the only undeniable proof of the existence of a SMBH in a galactic nucleus and the most reliable and elegant determination of its mass (§1.5.1.1).

1.5.1   Water Maser Disks

A galaxy at the distance and with the stellar velocity dispersion of NGC4258 would need to host a SMBH in excess of a few 10^8 M$_\odot$ for HST to resolve its sphere of influence. NGC4258, however, is blessed by the presence of circumnuclear clouds emitting strong water masers. At a frequency of 22 GHz, the maser emission can be studied in the radio at spatial resolutions a factor of $\sim$ 200 higher than can be achieved using HST, instantly pushing the ability to detect a SMBH to correspondingly smaller masses.

The discovery of nuclear megamasers trailed that of galactic masers (Cheung et al. 1969), which are associated with star forming regions and late-type stellar envelopes, by about ten years (dos Santos & Lepine 1979; Gardner & Whiteoak 1982; Claussen, Heiligman, & Lo 1984; Henkel et al. 1984). From the beginning, it was clear that nuclear and galactic masers resulted from completely different physical phenomena. The 22 GHz emission line luminosity in the prototypical Seyfert 2 galaxy NGC1068, for instance, exceeds the luminosity of a typical galactic water maser by a factor of $\sim 3.5 \times 10^5$. Early Very Large Array (VLA) and Very Long Baseline Interferometry (VLBI) observations revealed that nuclear megamaser sources are extremely compact, probably sub-parsec scale (Claussen & Lo 1986). Before long, it was concluded that the energy needed to excite the masers must ultimately stem from the active nucleus. The emerging picture is one in which the masers arise when x-rays, originating from the innermost parts of the accretion disk surrounding the central SMBH, illuminate and heat a torus of dense circumnuclear gas and dust (Neufeld, Maloney, & Conger 1994). At present, nuclear water masers are known in 30 Seyfert 2 galaxies and LINERs (Table 1.5.1.2).

1.5.1.1 NGC4258

H_2O megamasers at the systemic velocity of NGC4258 were discovered by Claussen et al. (1984). Ten years later, using the Nobeyana 45 m telescope, Nakai, Inoue, & Miyoshi (1993) detected additional megamaser emission at $\sim \pm 1000$ km s^{-1} relative to NGC4258's systemic velocity. Nakai et al. (1993) proposed three possible explanations for the observations: a circumnuclear molecular torus in Keplerian orbit around a $\sim 10^8$ M$_\odot$ central object, bi-directional outflow, and stimulated Raman scattering of the emission features near systemic velocity.

Nakai et al. (1993)'s results precipitated a great deal of theoretical and observational activity. Almost simultaneously, three independent teams (Watson & Wallin 1994; Haschick, Baan, & Peng 1994; Greenhill et al. 1995b) arrived at the conclusion that the maser emission must originate in a rapidly rotating Keplerian disk viewed nearly edge-on. The incriminating piece of evidence, besides the existence of the high velocity features, was the discovery that the systemic emission was subject to both temporal and spatial frequency changes, $d(\Delta\bar{v})/dt = 6$ km s^{-1} yr^{-1} and $d(\Delta\bar{v})/d\alpha = 280$ km s^{-1} mas^{-1}, respectively (Haschick et al. 1994; Greenhill et al. 1995a, b). In the Keplerian, edge-on disk scenario, the high velocity lines arise from masing along the lines of sight through the two tangent points in the disk, while the systemic features originate from clouds along the arc of the disk, which is seen projected against the nucleus. The frequency of the systemic features should therefore change with time as the clouds are carried around the nucleus, and should also depend on the exact position of the clouds, as observed.[3]

---

[3] The change should be imperceptible for the high velocity features, for which no drift has, in fact, been observed.

To reproduce the observed values of $d(\Delta \bar{v})/dt$ and $d(\Delta \bar{v})/d(\alpha)$, Watson & Wallin (1994) and Haschick et al. (1994) estimated that the rotational velocity of the disk must be $v_0 \sim 700 - 900$ km s^{-1} at 0.1 pc from the center.[4] The agreement of their derived v_0 with the observed values of the high velocity satellite lines (740 to 980 km s^{-1} and -760 to -920 km s^{-1}) strongly suggested, although it did not prove, that the disk is in simple Keplerian motion around a central mass of 10^7 M$_\odot$.

The confirmation came with the high resolution (0.6×0.3 mas at a position angle of $\sim 7°$) VLBA observations by Miyoshi et al. (1995) of the maser emission, which spatially resolved the morphological and kinematical structure of the clouds and allowed a more detailed study of the physical properties of the molecular disk (Moran et al. 1996; Neufeld & Maloney 1995; Neufeld et al. 1994; Herrnstein et al. 1996). The maser clouds were found to be distributed in an annulus with inner and outer bounds of 0.13 and 0.25 pc, respectively, from the center of the galaxy. The annulus is no more than 0.0003 pc thick, inclined at an angle of $83° \pm 4°$ relative to the line of sight, and slightly warped.[5] The rotational velocity of the high-velocity clouds increases inwards from 770 ± 2 km s^{-1} to 1080 ± 2 km s^{-1}, following a Keplerian rotation curve to very high accuracy. Miyoshi et al. (1995) calculate that the absence of measurable perturbations in the $r^{-0.5}$ Keplerian behavior of the lines requires $M_{disk} < 4 \times 10^6$ M$_\odot$, and a central binding mass of $(3.90 \pm 0.34) \times 10^7$ M$_\odot$ (for a distance of 7.2 Mpc). There is no ambiguity in these results; their reliability by far surpasses those of any stellar dynamical studies (§1.4.2).

Taking the 0.13 pc inner radius of the disk as an upper limit to the size of the central object, Miyoshi et al. (1995) find that the mass density is $\rho > 4 \times 10^9$ M$_\odot$ pc^{-3}, larger than the density in the densest known globular clusters, where $\rho < 10^5$ M$_\odot$ pc^{-3}. It is also several orders of magnitude larger than measured in any other galactic nucleus for which a SMBH has been claimed, with the exception of the Milky Way (see Table 1.1). Stronger constraints still can be obtained by assuming that the central source has size less than the angular extent of the systemic lines on the sky (~ 0.6 mas $= 0.02$ pc). This requires $\rho > 1.4 \times 10^{12}$ M$_\odot$ pc^{-3}, leaving virtually no doubt that the detected mass belongs to a central black hole, with the masers orbiting at 5500 Schwarzschild radii.

---

[4] Watson & Wallin (1994) assumed a distance $D = 6.6$ Mpc, while Haschick et al. (1994) used $D = 7$ Mpc. Subsequent high resolution VLBA observations, combined with the measurements of the angular size of the disk, the central mass, and the observed temporal and spatial variation of the systemic lines, allowed Herrnstein et al. (1999) to derive a geometric distance to NGC4258 of $D = 7.2 \pm 0.3$ Mpc.

[5] Interestingly, Neufeld & Maloney (1995) argue that warping is a necessary condition for the maser emission to ensue.

1.5.1.2   Beyond NGC4258

The spectacular success of the NGC4258 campaign motivated several systematic surveys explicitly targeted at detecting high-velocity masers. Although a few candidates have been identified, like all good things, H_2O megamasers do not appear to be common, and of those that are found, the ones suitable for dynamical studies are rarer still. The first large survey was conducted by Braatz, Wilson, & Henke (1994, 1996). Targeting 354 galaxies (mostly within 10 Mpc), including Seyfert galaxies, LINERs, and radio galaxies, they found a 7% detection rate (13 sources) among Seyfert 2 galaxies and LINERS, but no detections in Seyfert 1 nuclei. This is not surprising, since the masers are expected to be beamed along the major plane of the molecular torus and therefore away from the observer's line of sight in type 1 objects. Greenhill, Moran, & Herrnstein's (1997) survey of 26 AGN with the 70 m antenna of the NASA Deep Space Network produced only one detection, in NGC3735. One additional detection stemmed from a survey of 131 AGN with the Parkes Observatory (Greenhill et al. 2002).

The most recent, and most promising, results come from the survey of Greenhill et al. (2003). Benefiting from higher sensitivity and wider wavelength coverage than previous surveys, this study led to the discovery of seven new sources among 160 nearby ($cz < 8100$ km s^{-1}) AGN surveyed with the NASA Deep Space Antenna. More exciting still, two of the sources exhibit high-velocity masers and are promising targets for VLBI follow-up.

All nuclear maser detections are summarized in Table 1.5.1.2. Of the 13 galaxies with high velocity maser emission, the masers were spatially resolved, using VLBI observations, in NGC1068, NGC4945, NGC3079, Circinus, and (marginally) NGC5793. Unfortunately, in none of these galaxies did the dynamical analysis prove to be as clean as in the case of NGC4258. Only in Circinus did some of the maser clouds seem to follow a regular Keplerian curve, implying a central mass of $(1.7 \pm 0.3) \times 10^8$ M$_\odot$. For the other galaxies, the masses listed in Table 1.5.1.2 were derived *assuming* Keplerian motion. In the case of Mrk1419, a mass estimate was derived from the temporal change in the frequency of the systemic features.

1.5.2   HST Observations of Gas and Dust Disks

In the early 1990s, while still in its aberrated state, HST was snapping the first pictures of early-type galaxies. Although a far cry from VLBI standards, HST could improve on the resolution of ground-based telescopes available at the time by a factor of 10. One of the first early-type galaxies to be observed was NGC4261, a well-known Fanaroff–Riley type I radio galaxy (e.g., Jones et al. 2000) at a distance of ~ 30 Mpc. The images (Jaffe et al. 1993; see Figure 1.5) showed a well-defined, small (~ 230 pc across) disk of gas and ionized gas, roughly perpendicular to the radio jets.

Hopes were immediately raised that the regular morphology of the disk might be the signature of material in cold, circular rotation in the galaxy gravitational potential.

Narrow band Hα+[NII] images of M87 also showed a nuclear disk-like structure, although not as regular as the one in NGC4261 (Ford et al. 1994). Spectra taken with the $0''.26$ Faint Object Spectrograph (FOS) at five different locations showed that the gas velocity reached ± 500 km s^{-1} at 12 pc on either side of the nucleus (Harms et al. 1994). Interpreted as Keplerian motion, this requires a binding mass of $(2.4 \pm 0.7) \times 10^9$ M$_\odot$. The limited amount of kinematical information available to Harms et al. did not allow for very sophisticated modeling: the Keplerian motion was assumed, and the inclination angle of the disk was fixed to the value estimated from the disk morphology. The correctness of these assumptions was, however, verified a few years later by a much more extensive kinematical survey of the disk using HST's Faint Object Camera (Macchetto et al. 1997). The disk was sampled at over 30 (roughly) independent positions within $0''.5$ of the nucleus at a spatial resolution of 3 pc. Detailed dynamical modeling required a $(3.2 \pm 0.9) \times 10^9$ M$_\odot$ SMBH.

A kinematical study of NGC4261 followed in 1996 (Ferrarese, Ford, & Jaffe), claiming a $(4.9 \pm 1.0) \times 10^8$ M$_\odot$ SMBH. At the same time, it started to become apparent that nuclear dust/gas disks are not uncommon in early-type galaxies. The most recent and statistically complete study (Tran et al. 2001) finds "NGC4261-type" disks in 18% of early-type galaxies, or almost 40% of early-type galaxies with Infrared Astronomical Satellite (IRAS) detected 100 μm emission. Tran et al. (2001) find that all galaxies hosting dust disks show signs of nuclear activity and that in almost all cases, the disks are aligned with the major axes of the galaxies in which they reside. The origin of the dust (the two possibilities being internal from stellar mass lost or external from merging or interactions with other galaxies) remains a mystery (Ferrarese & Ford 1999; Tran et al. 2001).

By 1996, it had become apparent that the dust disks presented a powerful means of constraining the central potential in early-type galaxies, and many such studies began to be published (see Table 1.3 for a complete list). Stellar and gas dynamical studies are complementary: gas disks are often found in bright, pressure supported elliptical galaxies for which stellar dynamical studies are particularly problematic. Unfortunately, the large amount of dust obscuration associated with the disks makes the observations of stellar features more challenging than usual, and so far it has been possible to compare the results of the two methods in only one case, IC1459 (Cappellari et al. 2002), to which we will return later in this section. The possibility of using gas kinematics to constrain the central potential in spiral galaxies is also being explored (Marconi et al. 2003; Sarzi et al. 2001)

Constraining the central mass using gas dynamical data proceeds through the following steps:

- A total mass density is constructed, accounting for the contribution of the stellar population (derived by deprojecting the stellar surface brightness profile and assuming a mass-to-light ratio, which is treated as a free parameter), the mass in dust/gas (measured from multicolor observations of the disk; Ferrarese & Ford 1999[6]), and the central SMBH.

- The circular velocity at each radius is simply determined by the mass enclosed within that radius and is then projected along the line of sight. The inclination angle of the disk is generally (but not always; see Harms et al. 1994; Sarzi et al. 2001) treated as a free parameter. When the inclination cannot be constrained from the kinematical data, it is fixed to the value estimated from the images under the (probably incorrect) assumption that the disk is flat with a constant line of nodes. Additional free parameters are the projected position of the kinematical center of the disk (relative to the center of the observed kinematical map), the position angle of the line of nodes (again relative to the observed map), and the systemic velocity of the disk.

- The models are artificially "degraded" to simulate the observing conditions. Accounting for the finite size of the slit width when this is larger than the instrumental resolution is particularly important (Maciejewski & Binney 2001). This step, which is very nicely described in Barth et al. (2001), depends on the emission line surface brightness within the disk, as well as on the intrinsic and instrumental velocity dispersion of the line. The reason for this is that the contribution to the velocity observed at any given position is given by the integrated contribution at each neighboring position, with a weight that depends on the broadening and strength of the line at each point.

- The free parameters in the model, namely the inclination angle of the disk, the location of its center, the position angle of the line of nodes, the systemic velocity of the disk, the mass-to-light ratio of the stellar population, and the mass of the central SMBH, are varied until the best fit to the data is obtained.

In the case of gas dynamical studies, a two-dimensional velocity field is essential if one wishes to verify the critical underlying assumption that the gas motion is gravitational. It must be pointed out that as long as the motion is gravitational, the gas does not need to be in circular rotation in a geometrically thin disk for the analysis to be performed. Since the very first studies, it was apparent that the emission lines are broadened beyond what is expected from simple thermal broadening and instrumental effects (e.g., Ferrarese et al. 1996; Ferrarese & Ford 1999; van der Marel & van den Bosch 1998; Verdoes Kleijn et al. 2000; Barth et al. 2001). The

---

[6] The mass of the disk itself generally turns out to be too small (of the order 10^{5-6} $M_\odot$) to be of any dynamical relevance (e.g., Ferrarese & Ford 1999; Barth et al. 2001).

origin of the broadening is not well understood and has been interpreted either as micro-turbulence within the disk (Ferrarese & Ford 1999; van der Marel & van den Bosch 1998; Marconi et al. 2003), or as evidence of non-circular motion, as could be expected if the gas is fragmented into collisionless clouds that move ballistically, providing hydrostatic support against gravity (Verdoes Kleijn et al. 2000; Barth et al. 2001; Cappellari et al. 2002). In their analysis of NGC3245, Barth et al. (2001) conclude that the mass of the central SMBH should be increased by 12% to account for non-circular motions. In either case, turbulence and asymmetric drifts can and have been successfully incorporated in the modeling (e.g., Verdoes Kleijn et al. 2000; Barth et al. 2001), although doing so requires high S/N data with a wide spatial coverage.

While the entire method is invalidated if the gas motion is dominated by non-gravitational forces, the signatures of non-gravitational motions can, and have been, recognized in the data. In NGC4041, a quiescent Sbc spiral, Marconi et al. (2003) remark that the systematic blueshift of the disk relative to systemic velocity might be evidence that the disk is kinematically decoupled. They conclude that only an upper limit, of 2×10^7 M$_\odot$, can be put on the central mass. Cappellari et al. (2002) detected evidence of non-gravitational motions in IC1459, for which the ionized gas shows no indication of rotation in the inner $1''$. IC1459 is the only galaxy for which a SMBH mass estimate exists based both on gas and stellar kinematics (Cappellari et al. 2002). Three-integral models applied to the stellar kinematics produce $M_\bullet = (2.6 \pm 1.1) \times 10^9$ M$_\odot$, while the gas kinematics produces estimates between a few $\times 10^8$ and 10^9 M$_\odot$, depending on the assumptions made regarding the nature of the gas velocity dispersion. Unfortunately, the authors express strong reservations as to the reliability of either mass estimate. As mentioned above, there is evidence that the gas motion might not be completely gravitational, which would invalidate the method entirely. On the other hand, the HST stellar spectra "do not show any obvious evidence for the presence of a BH," and the "BH mass determination via the stellar kinematics should be treated with caution" (Cappellari et al. 2002).

1.6 Tackling the Unresolvable: Reverberation Mapping

Beyond $cz \sim 10000$ km s^{-1}, SMBH mass estimates based on resolved stellar (§1.4.2) or gas (§1.5.2) kinematics become unfeasible, even in the heftiest of cases. While a way to measure black hole masses in more distant, quiescent galaxies has yet to be devised, for type I AGN and quasars, an alternative already exists.

According to the unification scheme of AGNs (Antonucci 1993; Urry

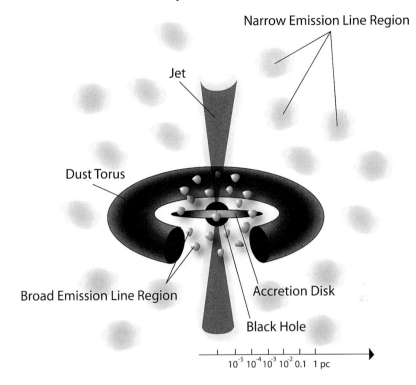

Figure 1.6. A schematic view of the central region of an AGN.

& Padovani 1995; Figure 1.6), the central black hole and accretion disk are completely surrounded by an obscuring, geometrically thick molecular torus. The existence of the torus is strongly supported by polarization maps (e.g., Watanabe et al. 2003), infrared observations (e.g., Reunanen et al. 2003), as well as by the H_2O maser observations discussed in §1.5.1. The geometry and extent of the torus are dictated by the requirement that, to account for the well-known polarization properties of Seyfert 1 and Seyfert 2 galaxies, the Broad Line Region (BLR) must be entirely enclosed within it, while the Narrow Line Region (NLR) — which extends from a few to several thousand parsecs (Pogge 1989) — must be largely outside. According to the standard model, the BLR consists of many (10^{7-8}, Arav et al. 1997, 1998; Dietrich et al. 1999), small, dense ($N_e \sim 10^{9-11}$ cm^{-3}), cold ($T_e \sim 2 \times 10^4$ K) photoionized clouds (Ferland et al. 1992; but see also, e.g., Smith & Raine 1985, 1988; Pelletier & Pudritz 1992; Murray et al. 1995; Murray & Chiang 1997; Collin-Souffrin et al. 1988). Even if direct observational constraints on the BLR geometry are lacking, the emerging picture is that of a BLR which extends from a few light days to several tens of light weeks from the central engine. Although a factor

10 to 100 times larger than the inner accretion disk, and three orders of magnitude larger than the Schwarzschild radius of the SMBH that powers the AGN activity, the BLR is, and will remain, spatially unresolved even using space-based instrumentation. In the early 1980s it was realized that, while preventing direct imaging, the closeness of the BLR to the ionizing continuum source might open a new, unconventional way to map its properties. Short-term continuum variability is a staple of AGNs (and indeed is undisputed evidence of the compactness of the central engine); if the delayed response of the line emitting region to changes in the continuum source is regulated by light-travel time from the latter to the former, then the time lag between variations in the ionizing continuum and in the flux of each velocity component of the line emission directly probes the spatial structure and emissivity properties of the BLR. The technique is widely known as "reverberation" or "echo" mapping (Blandford & McKee 1982; Peterson 1993; Peterson 2002). It requires long-term, careful monitoring of both continuum and broad emission lines, and it can of course only be applied to those AGNs in which the BLR is directly observable, namely Type 1 Seyferts and quasars.

Once the structure of the BLR is mapped, the SMBH mass is only a short step away. If the motion of the clouds is gravitational, the binding mass can be rather trivially derived from the virial theorem: $M = fr\sigma^2/G$, where r is the radius of the BLR, and σ the mean velocity. The factor f depends on the geometry and kinematics of the BLR. For instance, if the BLR is in Keplerian motion, f is 3/2 for randomly oriented, isotropic, parabolic orbits, but infinity for circular, coplanar orbits in the plane of the sky.

The technical and theoretical challenges associated with reverberation mapping are detailed in Peterson (2002) and Krolik (2001). Methodical UV/optical monitoring of AGN started in the early 1980s (see Peterson 2002). As a result of these programs, it was conclusively demonstrated that, although variations in AGN continuum fluxes are aperiodic, they correlate tightly with variations in the broad emission line flux. The less than optimal temporal sampling and duration of the experiments have so far prevented "2D reverberation mapping," i.e., monitoring of the line response as a function of both time and line of sight velocity. Doing so requires multi-wavelength monitoring campaigns of duration and temporal sampling that far exceed not only those of any program attempted so far (Krolik 2001; Peterson 2002; Ulrich & Horne 1996), but also of any program that could realistically be attempted from the ground, calling for a dedicated space-craft mission (Peterson & Horne 2003). However, time delays between flux variations in the continuum (generally optical) and emission line (generally Hβ, and integrated over the emission line width) have been measured with varying degrees of accuracy for three dozen AGN. This process is generally referred to as 1D reverberation mapping; although it does not lead to a complete morphological and kinematical picture of the central regions of AGN, it does produce a measure of the emissivity weighted size of the BLR. With this information, but without a direct confirmation of the geometry of the BLR, and the assumption of Keplerian motion, the binding mass can still be derived as $M = fr\sigma^2/G$, but with potentially severe reservations (see Krolik 2001).

Even with these daunting caveats, the importance of reverberation mapping as a mass estimator cannot be overstated. First, the method probes regions only 1000 times beyond the Schwarzschild radius, reaching at least a thousand times closer to the central engine than methods based on resolved kinematics. The implied mass densities (Table 1.1) are in excess of 10^{10} $M_\odot$ Mpc^{-3}, leaving little doubt that, if the measurements are correct, the mass must indeed belong to the singularity. Second, even the nearest type I AGN targeted by reverberation mapping cannot be easily probed using standard techniques, precisely because the bright AGN continuum overwhelms the dynamical signature of the gas and dust needed for resolved dynamical studies. Although type I AGN comprise only 1% of the total galaxy population, exploring the SMBH mass function in as many and as diverse galactic environments as possible is essential to understand the evolutionary histories of SMBHs. Third, unlike SMBH masses derived from resolved dynamics, those inferred from reverberation mapping do not depend on the cosmology adopted: since the BLR size is measured directly in physical units from the time delay, the distance to the host galaxy, or the value of the Hubble constant, does not enter the analysis. This has the stunning consequence that the redshift to which reverberation mapping measurements can be pushed is limited only by the requirement that the AGN must be bright enough to be easily detected. Fourth, reverberation mapping is intrinsically unbiased with respect to the mass of the SMBH, since spatially resolving the sphere of influence (whose size depends on the SMBH mass) is irrelevant. Very small SMBHs can be probed as long as the monitoring can be carried out on short enough timescales, and vice versa for very massive SMBHs. Virial masses and the associated random uncertainties inferred from the time lags are listed with the appropriate references in Table 1.4.

Table 1.3. Luminous 22 GHz H_2O Masers in the Nuclei of Active Galaxies

Galaxy	Type	Distance (Mpc)	Reference	High v Masers?	Rot. Curve?	SMBH Mass ($M_\odot$)
M51	S2/L	9.6	Hagiwara et al. 2001a	yes	no	
NGC 1052	L	20	Braatz et al. 1996	no	no	
NGC 1068	S2	16	Greenhill et al. 1996	yes	yes	$\sim 1.5 \times 10^7$
NGC 1386	S2	12	Braatz et al. 1996	no	no	
NGC 2639	S2	44	Braatz et al. 1996	no	no	
NGC 2824	S?	36	Greenhill et al. 2003a	no	no	
NGC 2979	S2	36	Greenhill et al. 2003a	no	no	
NGC 3079	S2	16	Trotter et al. 1998	yes	yes	$\sim 10^6$
NGC 3735	S2	36	Greenhill et al. 1997	no	no	
NGC 4258	S2	7.2	Greenhill et al. 1995	yes	yes	$(3.9 \pm 0.34) \times 10^7$
NGC 4945	S2	3.7	Greenhill et al.1997	yes	yes	$\sim 10^6$
NGC 5347	S2	32	Braatz et al. 1996	no	no	
NGC 5506	S2	24	Braatz et al.1996	no	no	
NGC 5643	S2	16	Greenhill et al.2003	no	no	
NGC 5793	S2	50	Hagiwara et al.2001b	yes	yes	$\sim 10^7$
NGC 6240	L	98	Hagiwara et al. 2002	yes	no	
NGC 6300	S2	15	Greenhill et al. 2003a	no	no	
NGC 6929	S2	78	Greenhill et al. 2003a	yes	no	
IC1481	L	83	Braatz et al. 1996	no	no	
IC2560	S2	38	Ishihara et al. 2001	yes	no	2.8×10^6
Mrk1	S2	65	Braatz et al. 1996	no	no	
Mrk348	S2	63	Falcke et al. 2000	yes	no	
Mrk1210	S2	54	Braatz et al. 1996	no	no	

Mrk1419	S2	66	yes	Henkel et al.2002	no	$\sim 10^7$
Circinus	S2	4	yes	Greenhill et al. 2003a	yes	$(1.7 \pm 0.3) \times 10^6$
ESO269-G012	S2	65	yes	Greenhill et al. 2003a	no	
IRASF18333-6528	S2	57	no	Braatz et al. 1996	no	
IRASF22265-1826	S2	100	no	Braatz et al. 1996	no	
IRASF19370-0131	S2	81	no	Greenhill et al. 2003a	no	
IRASF01063-8034	S2	53	no	Greenhill et al. 2003a	no	

The columns: the galaxy name; AGN type (L = LINER, S2 = Seyfert 2); distance (from the heliocentric velocity, assuming $H_0 = 75$ km s^{-1} Mpc^{-1}); whether high velocity emission has been detected; whether the emission has been spatially resolved; and the estimated SMBH mass.

Table 1.4. Reverberation Mapping Radii and Masses[a]

Object	z	R_{BLR} (lt-days)	$\lambda L_\lambda(5100\text{Å})$[b] 10^{44}ergs s^{-1}	v_{FWHM}(rms) km s^{-1}	M(rms) $10^7 M_\odot$
3C 120	0.033	42^{+27}_{-20}	0.73 ± 0.13	2210 ± 120	$3.0^{+1.9}_{-1.4}$
3C 390.3	0.056	$22.9^{+6.3}_{-8.0}$	0.64 ± 0.11	10500 ± 800	37^{+12}_{-14}
Akn 120	0.032	$37.4^{+5.1}_{-6.3}$	1.39 ± 0.26	5850 ± 480	$18.7^{+4.0}_{-4.4}$
F 9	0.047	$16.3^{+3.3}_{-7.6}$	1.37 ± 0.15	5900 ± 650	$8.3^{+2.5}_{-4.3}$
IC 4329A	0.016	$1.4^{+3.3}_{-2.9}$	0.164 ± 0.021	5960 ± 2070	$0.7^{+1.8}_{-1.6}$
Mrk 79	0.022	$17.7^{+4.8}_{-8.4}$	0.423 ± 0.056	6280 ± 850	$10.2^{+3.9}_{-5.6}$
Mrk 110	0.035	$18.8^{+6.3}_{-6.6}$	0.38 ± 0.13	1670 ± 120	$0.77^{+0.28}_{-0.29}$
Mrk 335	0.026	$16.4^{+5.1}_{-3.2}$	0.622 ± 0.057	1260 ± 120	$0.38^{+0.14}_{-0.10}$
Mrk 509	0.034	$76.7^{+6.3}_{-6.0}$	1.47 ± 0.25	2860 ± 120	$9.2^{+1.1}_{-1.1}$
Mrk 590	0.026	$20.0^{+4.4}_{-2.9}$	0.510 ± 0.096	2170 ± 120	$1.38^{+0.34}_{-0.25}$
Mrk 817	0.031	$15.0^{+4.2}_{-3.4}$	0.526 ± 0.077	4010 ± 180	$3.54^{+1.03}_{-0.86}$
NGC 3227	0.0038	$12.0^{+14.9}_{-10.0}$	0.0202 ± 0.0011	4360 ± 1320	3.6 ± 1.4
NGC 3516	0.0088	$7.4^{+5.4}_{-2.6}$		3140 ± 150	1.68 ± 0.33
NGC 3783	0.0097	$10.4^{+4.1}_{-2.3}$	0.177 ± 0.015	2910 ± 190	0.87 ± 0.11
NGC 4051	0.0023	$5.9^{+3.1}_{-2.0}$	0.00525 ± 0.00030	1110 ± 190	$0.11^{+0.08}_{-0.05}$
NGC 4151	0.0033	$3.0^{+1.8}_{-1.4}$	0.0720 ± 0.0042	5230 ± 920	$1.20^{+0.83}_{-0.70}$
NGC 4593	0.0090	$3.1^{+7.6}_{-5.1}$		4420 ± 950	0.66 ± 0.52
NGC 5548	0.017	$21.2^{+2.4}_{-0.7}$	0.270 ± 0.053	5500 ± 400	$9.4^{+1.7}_{-1.4}$
NGC 7469	0.016	$4.9^{+0.6}_{-1.1}$	0.553 ± 0.016	3220 ± 1580	$0.75^{+0.74}_{-0.75}$
PG 0026	0.142	113^{+18}_{-21}	7.0 ± 1.0	1358 ± 91	$2.66^{+0.49}_{-0.55}$
PG 0052	0.155	134^{+31}_{-23}	6.5 ± 1.1	4550 ± 270	$30.2^{+8.8}_{-7.4}$
PG 0804	0.100	156^{+15}_{-13}	6.6 ± 1.2	2430 ± 42	$16.3^{+1.6}_{-1.5}$
PG 0844	0.064	$24.2^{+10.0}_{-9.1}$	1.72 ± 0.17	2830 ± 120	$2.7^{+1.1}_{-1.0}$
PG 0953	0.239	151^{+22}_{-27}	11.9 ± 1.6	2723 ± 62	$16.4^{+2.5}_{-3.0}$
PG 1211	0.085	101^{+23}_{-29}	4.93 ± 0.80	1479 ± 66	$2.36^{+0.56}_{-0.70}$
PG 1226	0.158	387^{+58}_{-50}	64.4 ± 7.7	2742 ± 58	$23.5^{+3.7}_{-3.3}$
PG 1229	0.064	50^{+24}_{-23}	0.94 ± 0.10	3490 ± 120	$8.6^{+4.1}_{-4.0}$
PG 1307	0.155	124^{+45}_{-80}	5.27 ± 0.52	5260 ± 270	33^{+12}_{-22}
PG 1351	0.087	227^{+149}_{-72}	4.38 ± 0.43	950 ± 130[c]	$3.0^{+2.1}_{-1.3}$
PG 1411	0.089	102^{+38}_{-37}	3.25 ± 0.28	2740 ± 110	$8.8^{+3.3}_{-3.2}$
PG 1426	0.086	95^{+31}_{-39}	4.09 ± 0.63	5520 ± 340	37^{+13}_{-16}
PG 1613	0.129	39^{+20}_{-14}	6.96 ± 0.87	2500 ± 140	$2.37^{+1.23}_{-0.88}$
PG 1617	0.114	85^{+19}_{-25}	2.37 ± 0.41	3880 ± 650	$15.4^{+4.7}_{-5.5}$
PG 1700	0.292	88^{+190}_{-182}	27.1 ± 1.9	1970 ± 150	5.0^{+11}_{-10}
PG 1704	0.371	319^{+184}_{-285}	35.6 ± 5.2	400 ± 120	$0.75^{+0.63}_{-0.81}$
PG 2130	0.061	200^{+67}_{-18}	2.16 ± 0.20	3010 ± 180	$20.2^{+7.1}_{-2.4}$

[a] Data for the PG quasars are from Kaspi et al. (2000), while data for all other objects (with the exceptions noted below) are originally from Wandel, Peterson & Malkan (1999), as listed in Kaspi et al. (2000). Data for NGC 3783 are from Onken & Peterson (2002). Data for NGC 3227, NGC 3516, & NGC 4593 are from Onken et al. (2003). Data for NGC 4051 are from Peterson et al. (2000).

[b] Assuming $H_0 = 75$ km s^{-1} Mpc^{-1}.

[c] Hβ was not observed, therefore the rms velocity refers to Hα.

1.6.1 Observational Support as to the Reliability of Reverberation Mapping Masses

Of the assumptions made in deriving SMBH masses from reverberation mapping studies, perhaps the most critical are those regarding the velocity and geometry of the BLR.

The virial hypothesis is, of course, at the heart of reverberation mapping as a mass estimator. If the hypothesis is invalidated (for instance, if the BLR is in a radial flow), then the whole method is undermined, at least as far as mass estimates are concerned. Reassuringly, the virial hypothesis has recently received strong observational support from the work of Peterson & Wandel (2000) and Onken & Peterson (2002). If the motion of the gas is gravitational, using the lags derived for different emission lines in the same AGN must produce the same mass measurement. NGC5548 was the first galaxy for which this was indeed verified: the highest ionization lines are observed to have the shortest time lag, so that the virial product rv^2 remains constant. The same has now been proven in three additional galaxies, NGC7469, NGC3783, and 3C390.3. Although there exist non-equilibrium kinematical configurations that can mimic a virial relationship (e.g., cloud outflows or disk winds could also explain the observations; Blumenthal & Mathews 1975; Murray et al. 1995; Emmering, Blandford, & Shlosman 1992; Chiang & Murray 1996; Bottorff et al. 1997), these alternative explanations need to be very fine-tuned to account for the mounting quantity of observational evidence and are therefore becoming more and more untenable. The Hβ line in NGC5548 also maintains a virial relationship as a function of time: as the continuum brightens, the lags become larger and the lines becomes narrower (Peterson et al. 2002).

The geometry of the BLR is, unfortunately, unconstrained. For lack of evidence that favors one model over another, most studies (Wandel, Peterson, & Malkan 1999; Kaspi et al. 2000; Onken et al. 2002; cf., Sergeev et al. 2002; Vestergaard, Wilkes, & Barthel 2000) assume that the BLR is spherical and characterized by an isotropic velocity distribution.[7] Until 2D reverberation mapping becomes a reality, the most compelling evidence as to the reliability of the current mass estimates is shown in Figure 1.7. For quiescent galaxies, the SMBH mass measured through resolved stellar or gas dynamics is tightly connected to the velocity dispersion of the host bulge (Ferrarese & Merritt 2000; Gebhardt et al. 2000b; see §1.7). Reverberation mapping masses are found to agree remarkably well with the values expected in quiescent galaxies with comparable bulge velocity dispersions (Ferrarese et al. 2001), indicating that, as simplistic as it might appear, the picture adopted for the BLR might not be too far off.

---

[7] Different assumptions are sometimes made. For instance, McLure & Dunlop (2000) assume a thin disk geometry, leading to velocities 1.7 times and black hole masses 3 times greater than in the spherical, isotropic case.

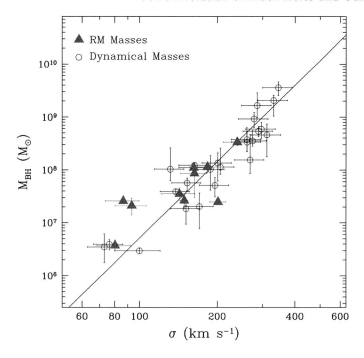

Figure 1.7. Location of reverberation mapped AGN *(triangles)* in the $M_\bullet - \sigma$ plane. Circles denote quiescent galaxies, which define the relation. Figure adapted from Ferrarese et al. (2001).

1.6.2   Secondary Mass Estimators Based on Reverberation Mapping

Since the first monitoring studies, it became apparent that the BLR radius and the central mass both correlate with continuum luminosity (Koratkar & Gaskell 1991; Wandel et al. 1999; Kaspi et al. 2000; Peterson et al. 2000). The most recent characterization of the $R_{BLR} - L_\lambda$ relation (Peterson 2002), uses time delays from the Hβ broad lines and monochromatic continuum luminosity measured at 5100 Å. An analogous relation, but using UV data (CIVλ1549 emission and 1350 Å continuum luminosity) has the tremendous advantage of being applicable to high redshift AGN (Vestergaard 2002).

Besides holding interesting information about the physical nature of the ionization process (Koratkar & Gaskell 1991; Kaspi et al. 2000), the $R_{BLR} - L_\lambda$ relation has tremendous practical appeal. It allows an estimate of the BLR size to be made from a quick, simple measurement of the continuum luminosity, bypassing the need for lengthy monitoring programs. Although the scatter is large (Vestergaard 2002 estimates an uncertainty of a factor of 2.5 in $M_\bullet$, although a smaller scatter has been claimed if the continuum luminosity is measured at 3000 Å instead; McLure & Jarvis 2002), the errors can be beaten down for large samples of objects (e.g., Bian & Zhao 2003; Shields et al. 2003, Netzer 2003; Woo & Urry 2002a, b; Oshlack et al. 2002; Wandel 2002; Boroson 2002; McLure & Dunlop 2001).

1.7 Scaling Relations for SMBHs

Proving the existence of SMBHs is only the first step in a long journey. In this section, a second step will be taken: establishing empirical correlations between the masses of SMBHs and the overall properties of the host galaxies. Ultimately, we want to know how common SMBHs are, how they form, how they evolve, whether they coalesce as a consequence of galaxy mergers, and to what extent they interact with their galaxy hosts. SMBH scaling relations serve as the basis for every study of SMBH demographics, formation, and evolution. We discuss four relations in detail: those between SMBH mass and (1) the luminosity of the host galaxy (§1.7.1), (2) the bulge velocity dispersion (§1.7.2), (3) the concentration of bulge light (§1.7.3); and (4) the mass of the surrounding dark matter halo (§1.7.4).

1.7.1 The $M_\bullet - L_B$ Relation

Using the eight SMBH detections available at the time, Kormendy & Richstone (1995) noticed that supermassive black hole masses correlate with the blue luminosity of the surrounding hot stellar component, this being the bulge of spiral galaxies, or the entire galaxy in the case of ellipticals. Whether the observed correlation was simply the upper envelope of a distribution that extends to smaller masses was unclear: while there should be no observational bias against detecting large SMBHs in small bulges, a failed detection of a small SMBH in a large system could have gone unreported at the time (but none has been reported since). Kormendy & Richstone point out that the existence of the correlation indicates that SMBH and bulge formation are tightly connected or even, based on the claimed absence of a SMBH in the bulge-less spiral M33, that the presence of a bulge might be essential for SMBH formation.

Further SMBH detections have confirmed the existence of a correlation. The most up-to-date $M_\bullet - L_B$ relation, using all SMBH detections for which the sphere of influence is resolved by the observations, is shown in Figure 1.8. A best fit, accounting for errors in both coordinates as well as intrinsic scatter gives:

$$\log(M_\bullet) = (8.37 \pm 0.11) - (0.419 \pm 0.085)(B_T^0 + 20.0) \qquad (1.9)$$

The scatter in the $M_\bullet - L_B$ relation is the subject of some debate. Using the 12 reliable SMBH masses known at the time, Ferrarese & Merritt

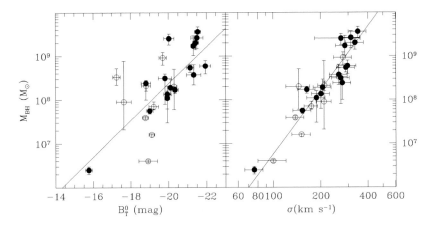

Figure 1.8. The $M_\bullet - L_B$ (left) and $M_\bullet - \sigma$ relations for all SMBH detections for which $r_h/r_{res} > 1.0$. Filled symbols show elliptical galaxies, while open symbols show spiral galaxies and lenticulars. The solid lines are the best fits to the data, accounting for errors in both coordinates as well as intrinsic scatter.

(2000) reported a reduced $\chi_r^2 = 23$ when the magnitudes are measured in the B-band, and a scatter of 0.6 dex in $M_\bullet$ at any given luminosity. This result has not changed significantly in the past four years. The expanded sample shown in Figure 1.8 has a scatter around the best fit line of 0.79 dex in $M_\bullet$. McLure & Dunlop (2001) were the first to point out that if the sample is restricted to elliptical galaxies, the scatter is reduced significantly, down to 0.33 dex. This is comparable to the scatter in the $M_\bullet - \sigma$ relation to be discussed in the next section. Elliptical galaxies are distinguished from lenticulars and spirals in Figure 1.8; they indeed define a relation with significantly reduced scatter, 0.40 dex in $M_\bullet$ in the updated sample presented in this review. This is only slightly larger than calculated by McLure & Dunlop, due to the addition of Cyg A (Tadhunter et al. 2003) and NGC 5845 (Gebhardt et al. 2003), both of which deviate (significantly in the case of Cyg A) from the best fit relation.

A natural question is whether the Hubble dependence of the scatter (and possibly characterization) of the $M_\bullet - L_B$ relation is indicative of a SMBH formation and/or evolution history which differs in elliptical and spiral galaxies. This seems unlikely for several reasons. First and foremost is the fact that the characterization of the $M_\bullet - \sigma$ relation, to be discussed in the next section, does not seem to depend on Hubble type. McLure & Dunlop note that disentangling the bulge from the disk light, especially in late type spirals, and especially in the B-band, is a difficult and uncertain process. For instance, Ferrarese & Merritt (2000) used the Simien & De Vaucouleurs relation (that is known to have large scatter) to estimate the fraction of total light contained in the bulge as a function of Hubble type. It is quite possible, therefore, that the overall larger scatter exhibited by spiral galaxies is simply due to an inaccurate determination of B_T^0 for the bulge component.

This conclusion is supported by the recent study of Marconi & Hunt (2003). The authors used K-band images from the recently released 2MASS database, and performed an accurate bulge/disk decomposition for the nearby quiescent spiral galaxies with dynamically measured masses. When only SMBH masses based on data that resolved the sphere of influence are used, Marconi & Hunt find that the scatter in the K-band $M_\bullet - L_B$ relation is very small (0.31 dex), comparable to the scatter in the $M_\bullet - \sigma$ relation, no matter whether spiral galaxies are included. This is not altogether surprising. Both the $M_\bullet - \sigma$ and $M_\bullet - L_B$ relations betray the existence of a correlation between the mass of supermassive black holes, and the mass of the host bulge. Near IR magnitudes are a better tracer of mass than B-band magnitudes (and are also less sensitive to the disk component in spiral galaxies). If mass is the underlying fundamental parameter as Marconi & Hunt suggest, the scatter in the $M_\bullet - L_B$ relation should depend on the photometric band in which the magnitudes are measured.[8]Marconi & Hunt reach a second notable conclusion, namely that the scatter in the $M_\bullet - L_K$ relation depends on the sample of galaxies used, increasing significantly (from 0.31 dex to 0.51 dex) when including galaxies for which the data do not resolve the SMBH sphere of influence. This is a strong endorsement of the argument, first proposed by Ferrarese & Merritt (2000), that resolving the sphere of influence is a necessary condition for a reliable mass determination to be made.

1.7.2   The $M_\bullet - \sigma$ Relation

Bulge magnitudes and velocity dispersions are correlated through the Faber–Jackson relation. The $M_\bullet - L_B$ relation, therefore, immediately entails a correlation between $M_\bullet$ and σ. In spite of the ample attention devoted to the $M_\bullet - L_B$ relation since 1995, five years went by before the first $M_\bullet - \sigma$ relation was published (Ferrarese & Merritt 2000; Gebhardt et al. 2000b). The reason for the long delay is easily understood. Figure 1.9 (bottom panels) shows the $M_\bullet - L_B$ and $M_\bullet - \sigma$ correlations using all SMBH masses, regardless of their accuracy, available in 2000. Both relations have large intrinsic scatter, and there doesn't appear to be any obvious advantage in preferring one to the other. The breakthrough came with the realization that the scatter in the $M_\bullet - \sigma$ relation is significantly dependent on sample selection. The upper panels of Figure 1.9 only show SMBH masses (as were available in 2000) derived from data that resolved the sphere of influence.

---

[8] It must however be pointed out that Marconi & Hunt measure a smaller scatter for the B-band $M_\bullet - L_B$ relation than previously reported.

While no significant changes (besides the obvious decrease in sample size) are noticeable in the $M_\bullet - L_B$ relation, the scatter in the $M_\bullet - \sigma$ relation decreases significantly when the restricted sample is used. Based on these findings, Ferrarese & Merritt (2000) concluded that the reliability of the SMBH mass depends critically on the spatial resolution of the data, and that the $M_\bullet - \sigma$ relation is tighter, and therefore more fundamental, than either the Faber-Jackson or the $M_\bullet - L_B$ relation.

The latter conclusion was also reached by Gebhardt et al. (2000b), who more than doubled the sample used by Ferrarese & Merritt (2000) by including SMBH masses for 13 additional galaxies. There are some noticeable differences between the Ferrarese & Merritt and Gebhardt et al. study, besides the size of the sample. Gebhardt et al. use "luminosity-weighted line-of-sight [velocity] dispersions inside a radius R", while Ferrarese & Merritt used central velocity dispersions normalized to an aperture of radius equal to 1/8 of the galaxy effective radius (Jorgensen et al. 1995; the same definition is used in studies of the fundamental plane of elliptical galaxies). Based on the fact that the σ measured by Gebhardt et al. (2000b) and those used by Ferrarese & Merritt (2000) differ slightly but systematically from each other, Tremaine et al. (2002) argue that the latter are flawed.

The original slope measured by Gebhardt et al. for the $M_\bullet - \sigma$ relation (3.75 ± 0.3) was considerably flatter than the one measured by Ferrarese & Merritt (4.8 ± 0.5). The reasons were identified by Merritt & Ferrarese (2001b) in the different algorithm used to fit the data, the σ value adopted for the Milky Way, and the inclusion in the Gebhardt et al. sample of SMBH masses for which the data did not resolve the sphere of influence. The first two issues were recognized and corrected in Tremaine et al. (2002), who unfortunately never addressed the third issue. Tremaine et al. (2002) propose a slope for the $M_\bullet - \sigma$ relation of 4.02 ± 0.32, and trace the reason for the difference between their value and that of Ferrarese & Merritt to the aforementioned discrepancies in the adopted definition of the velocity dispersion. To date, there has been no reconciliation of the issue. Independent investigation of the $M_\bullet - L_B$ (§1.7.1, Marconi & Hunt 2003) and $M_\bullet - C$ relations (S 1.7.3, Graham et al.2001) have led to the conclusion that resolving the SMBH sphere of influence significantly affects the characterization and scatter of those relations, arguing that masses based on data with poor spatial sampling (several of which are included in the Tremaine et al. study) should not be used. On the other hand, Tremaine et al.'s claim needs to be pursued, if nothing else because central velocity dispersions are commonly used in studies of the fundamental plane of elliptical galaxies.

Figure 1.8 shows the most current version of the $M_\bullet - \sigma$ relation, including only SMBH for which the data resolved the sphere of influence (from Table 1.3, which now includes more galaxies than used in any previously published study of the $M_\bullet - \sigma$ relation), using published central values of the velocity dispersions, and the fitting routine from Akritas &

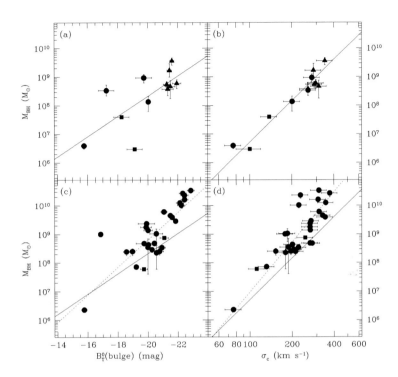

Figure 1.9. The $M_\bullet - L_B$ (left panels) and $M_\bullet - \sigma$ (right panels) relations for all SMBH masses (as available in 2000) for which the sphere of influence was (top panels) or was not (bottom panel) resolved by the data. The solid and dotted lines are the best-fit regression lines to the galaxies for which the SMBH sphere of influence was and was not resolved respectively. From Ferrarese & Merritt (2000).

Bershady (1996):

$$\frac{M_\bullet}{10^8 \, M_\odot} = (1.66 \pm 0.24) \left(\frac{\sigma}{200 \text{ km s}^{-1}}\right)^{4.86 \pm 0.43} \tag{1.10}$$

(using the fitting algorithm used in Tremaine et al. (2002) actually produces a slightly steeper slope, 5.1 ± 0.4). The reduced χ^2 is 0.880, indicating that the intrinsic scatter of the relation is negligible. The scatter around the mean is only 0.34 dex in $M_\bullet$. Neither the scatter, slope, or zero point of the relation depend on the Hubble type of the galaxies considered, as can be judged qualitatively from Figure 1.8.

It is, of course, a trivial exercise for anyone to produce his or her own fit to the preferred sample. Whatever the slope of the $M_\bullet - \sigma$ relation might turn out to be, there is one point on which there seems to be universal agreement: because of its negligible scatter, the $M_\bullet - \sigma$ relation is of fundamental relevance for many issues related to the studies of SMBHs. In particular:

- The $M_\bullet - \sigma$ relation allows one to infer SMBH masses with 30% accuracy from a single measurement of the large scale bulge velocity dispersion. Through σ, it has therefore become possible to explore the role played by the SMBH mass in driving the character of the nuclear activity, not only in individual galaxies (Barth et al. 2002), but also in different classes of AGNs (Barth et al. 2003; Falomo et al. 2003).
- Studies of SMBH demographics (§1.8), both in quiescent (Merritt & Ferrarese 2001a; Ferrarese 2002a; Yu & Tremaine 2003; Aller & Richstone 2003; Whythe & Loeb 2002, 2003) and active galaxies (Ferrarese et al. 2001) have relied heavily on the relation.
- The $M_\bullet - \sigma$ relation has become the litmus test of models of SMBH formation and evolution. Reproducing its slope, normalization, and, above all, scatter, and maintaining it in spite of the merger events that inevitably take place during galaxy evolution, is currently the biggest challenge faced by the models (Adams, Graff, & Richstone 2000; Monaco, Salucci & Danese 2000; Haehnelt, Natarajan, & Rees 1998; Silk & Rees 1998; Haehnelt & Kauffmann 2000; Cattaneo, Haehnelt & Rees 1999; Loeb & Rasio 1994).

Marconi & Hunt (2003) suggest that the $M_\bullet - \sigma$ relation is the reflection of a more fundamental relation between black hole and bulge mass. Such a relation is shown in Figure 1.10 , where the bulge mass has been derived following Marconi & Hunt as $M_{bulge} \sim 3R_e\sigma/G$, with R_e representing the galaxies' effective radii.

Finally, Figures 1.11 and 1.12 are included as a way to assess, qualitatively, the scatter and character of the $M_\bullet - L_B$ and $M_\bullet - \sigma$ relations as a function of Hubble Type and the ability of the data to spatially resolve the sphere of influence. The figures support the following conclusions, which can be rigorously proven by fitting the data given in Table 1.3: 1) the scatter in the B-band $M_\bullet - L_B$ relation decreases when bulge, instead of total, magnitudes are used, as pointed out by Kormendy & Gebhardt (2001). 2) The scatter in the $M_\bullet - L_B$ relation decreases when the sample is restricted to elliptical galaxies only, as first pointed out by McLure & Dunlop (2000). 3) The scatter further decreases if the sample is restricted to galaxies for which the sphere of influence is well resolved, independently on the Hubble Type of the galaxy considered. 4) For all samples, the scatter in the B-band $M_\bullet - L_B$ relation is larger than in the $M_\bullet - \sigma$ relation (c.f. Marconi & Hunt 2003).

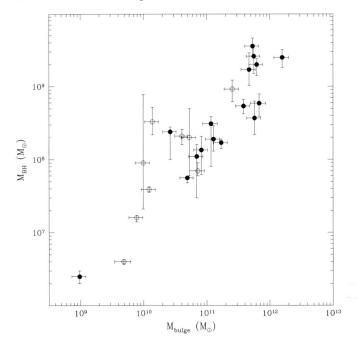

Figure 1.10. The relation between SMBH and virial bulge masses (following Marconi & Hunt 2003).

1.7.3 Black Hole Masses and Light Concentration

Graham et al. (2001) found a remarkably tight correlation between SMBH masses and the concentration of bulge light, defined as the ratio of flux inside one-third of the half-light radius to the one within the entire half-light radius. The existence of the correlation is not entirely surprising. At least in ellipticals (which comprise most of the sample) the shape of the brightness profile correlates with galaxy luminosity (Ferrarese et al. 1994; Lauer et al. 1995; Rest et al. 2001). Larger and more luminous galaxies have shallower brightness profiles (larger radii enclosing 1/3 of the light and thus less light concentration) and host more massive black holes. As is the case for the $M_\bullet - \sigma$ relation, however, what is surprising is the scatter, which appears to be negligible. As noticed by Ferrarese & Merritt (2000) for the $M_\bullet - \sigma$ relation and, later, by Marconi & Hunt for the $M_\bullet - L_B$ relation, Graham et al. find that the scatter in the $M_\bullet - C$ relation decreases significantly when only SMBH masses derived from data that resolve the sphere of influence are used.

The $M_\bullet - C$ and $M_\bullet - L_B$ relations have the practical advantage of needing only imaging data, generally more readily available than the spec-

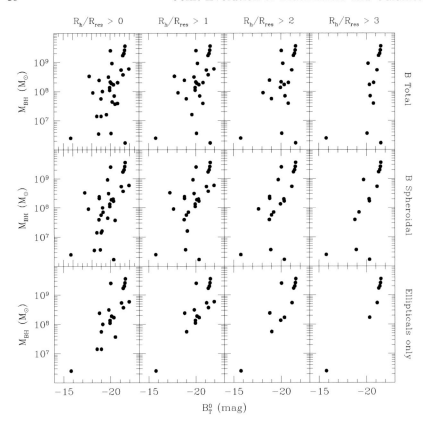

Figure 1.11. Top panels: The $M_\bullet - L_B$ relation obtained when total B-band magnitude is used, regardless of the Hubble Type of the host galaxy. Middle panels: as for the upper panel, except bulge, rather than total magnitudes are used. Bottom panels: The $M_\bullet - L_B$ relation for elliptical galaxies only. From left to right, the sample is increasingly restricted in terms of how well the data resolve the sphere of influence of the measured SMBH, as indicated by the labels at the top.

troscopic data necessary to measure σ. The $M_\bullet - C$ relation is, however, dependent on a parametric characterization of the light profile (Graham et al. use a modified Sersic law) which might not prove to be a good fit for some galaxies. Graham et al. point out that cD and merging/interacting galaxies might be particularly problematic, and exclude NGC 6251 and (judging from their Figure 2) M87 and NGC 4374 from their analysis. It is interesting to notice that while outliers in the $M_\bullet - C$ relation, these galaxies fit the $M_\bullet - \sigma$ relation.

Regrettably, there have not yet been theoretical studies targeting the

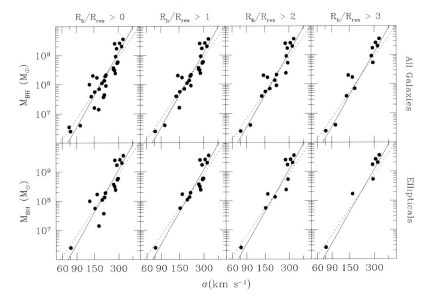

Figure 1.12. The $M_\bullet - \sigma$ relation obtained regardless of the Hubble Type of the host galaxy (top panels) and for elliptical galaxies only (bottom panel). From left to right, the sample is increasingly restricted in terms of how well the data resolve the sphere of influence of the measured SMBH, as indicated by the labels at the top. The two lines represent the fit from Tremaine et al. (2002, dashed line) and this Chpater (Equation 1.20, solid line)

$M_\bullet - C$ relation specifically, although a number of authors have studied the effect of a SMBH (or a SMBH binary) on the density profile of the surrounding stellar population (Young 1980; Quinlan, Hernquist & Sigurdsson 1995; Lee & Goodman 1989; Cipollina & Bertin 1994; Sigurdsson, Hernquist & Quinlan 1995; Quinlan & Hernquist 1997; Nakano & Makino 1999, van der Marel 1999; Milosavljevic & Merritt 2001).

1.7.4 Black Hole Masses and Dark Matter Halos

Kormendy & Richstone (1995) first pointed out that the existence of the $M_\bullet - L_B$ relation indicates that SMBH and bulge formation are tightly connected or even that the presence of a bulge might be a necessary condition for SMBH formation. Six years later, based on the observation that the scatter in the $M_\bullet - L_B$ relation increases mildly when total magnitude is substituted for bulge magnitude, Kormendy & Gebhardt (2001) argued that SMBH masses do not correlate with total mass. These conclusions, however, can be questioned on several grounds. B-band observations are a poor tracer of mass, AGNs in bulge-less galaxies do exist (e.g., Filippenko &

Ho 2003), and most self-regulating theoretical models of SMBH formation predict the fundamental connection to be between $M_\bullet$ and the total gravitational mass of the host galaxy, rather than the bulge mass (Adams, Graff, & Richstone 2000; Monaco, Salucci, & Danese 2000; Haehnelt, Natarajan & Rees 1998; Silk & Rees 1998; Haehnelt & Kauffmann 2000; Cattaneo, Haehnelt & Rees 1999; Loeb & Rasio 1994). Unfortunately, finding observational support for the existence of a relation between SMBHs and the total gravitational mass of the host galaxy is very difficult: measuring dark matter halo masses makes the measurement of $M_\bullet$ look almost trivial!

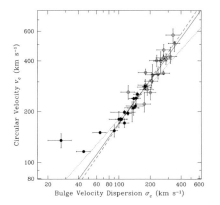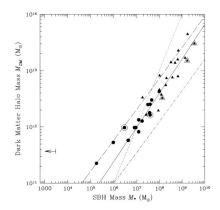

Figure 1.13. Left: Correlation between bulge velocity dispersion σ_c and disk circular velocity v_c for a sample of 20 elliptical galaxies (from Kronawitter et al. 2000, open symbols) and 16 spiral galaxies with H I rotation curves extending beyond the galaxies' optical radius (filled symbols). The galaxy to the far left, with the smallest value of σ_c, is NGC 598 (M33). The dotted line is a fit to all spiral galaxies with the exception of NGC 598, while the solid line is the fit to all spirals with $\sigma_c > 70$ km s^{-1}. The dashed line is a fit to the entire sample of ellipticals plus spirals with $\sigma_c > 70$ km s^{-1}. Right: σ_c has been transformed into SMBH mass using the $M_\bullet - \sigma$ relation, and v_c has been transformed into dark matter halo mass following the prescriptions of Bullock et al. (2001). Filled circles are spirals; filled triangles are ellipticals. Symbols accentuated by a larger open contour identify galaxies having a dynamical estimate of $M_\bullet$, which was used in the plot. The upper limit on the SMBH mass in NGC 598 is marked by the arrow. The dotted line represents a constant ratio $M_{DM}/M_\bullet = 10^5$. The solid line corresponds to the best fit to the data. The dashed and dot-dashed line show the best fits that would be obtained if different prescriptions to relate v_c and M_{halo} were used.

Ferrarese (2002b) provided the first — albeit indirect — observational evidence that SMBH and dark matter halos might be tightly connected. Figure 1.13 shows that in spiral galaxies as well as in ellipticals, the bulge velocity dispersion, which is typically measured within a radius of 0.5 kpc, and the large-scale circular velocity, measured at radii which range from 10 to 50 kpc, are tightly connected. For the spirals, the disk circular velocity is measured directly from HI observations that extend beyond the optical radius of the galaxy, well within the region where the rotation velocity stabilizes into a flat rotation curve. For the elliptical galaxies, the circular velocity is derived from dynamical models applied to the stellar kinematics (Gerhard et al. 2001; Kronawitter et al. 2000). The best fit $v_c - \sigma_c$ relation gives:

$$\log v_c = (0.84 \pm 0.09) \log \sigma_c + (0.55 \pm 0.29) \tag{1.11}$$

The existence of the $v_c - \sigma_c$ relation, with virtually unchanged characterization, has since been confirmed using a larger sample of galaxies by Baes et al. (2003).

The interest of the $v_c - \sigma_c$ relation reaches beyond the realm of SMBHs. For instance, characterizing the slope and normalization of the relation might help in constraining theoretical models and numerical simulations following the formation and evolution of galaxies (e.g., Steinmetz & Muller 1995). For the purpose of this review, however, the $v_c - \sigma_c$ relation clearly betrays a correlation between the masses of SMBHs and those of the surrounding dark matter halos (Figure 1.13), although the exact characterization of such a relation remains to be explored. The velocity dispersion is easily transformed to $M_\bullet$ using the $M_\bullet - \sigma$ relation but, unfortunately, the relation between circular velocity and dark matter halo mass is more uncertain. In a CDM-dominated universe, the dark matter halo mass is uniquely determined by the halo velocity measured at the virial radius. The latter is defined as the radius at which the mean density exceeds the mean universal density by a constant factor, generally referred to as the "virial overdensity". Based on this definition, and by virtue of the virial theorem, it immediately follows that the halo mass must be proportional to the third power of the virial velocity, but unfortunately the constant of proportionality depends on the adopted cosmology and may vary with time (e.g., Bryan & Norman 1998; Navarro & Steinmetz 2000). To convert the v_c relation to a $M_\bullet - M_{DM}$ relation, the virial velocity must also be related to the measured circular velocity, a step that entails further assumptions on the value and mass dependence of the halo concentration parameter. Using the $M_\bullet - \sigma$ relation, and the ΛCDM cosmological simulations of Bullock et al. (2002), Ferrarese (2002b) derives

$$\frac{M_\bullet}{10^8 \ M_\odot} \sim 0.10 \left(\frac{M_{DM}}{10^{12} \ M_\odot} \right)^{1.65} \tag{1.12}$$

The dependence of $M_\bullet$ on M_{DM} is not linear, a result that seems quite robust no matter what the details of the adopted cosmological model are. The ratio between SMBH and halo mass decreases from $\sim 2 \times 10^{-4}$ at $M_{DM} = 10^{14}$ M$_\odot$ to $\sim 10^{-5}$ at $M_{DM} = 10^{12}$ M$_\odot$, meaning that less massive halos seem less efficient in forming SMBHs. Ferrarese (2002b) notes that this tendency becomes more pronounced for halos with $M_{DM} < 5 \times 10^{11}$ M$_\odot$, and further note that, since there is no direct evidence that SMBH of masses smaller than 10^6 M$_\odot$ exist, SMBH formation might only proceed in halos with a virial velocity larger than ~ 200 km s^{-1}, in qualitative agreement with the formation scenario envisioned by Haehnelt et al. (1998) and Silk & Rees (1998).

Although there is no doubt that the $v_c - \sigma_c$ relation implies that the formation of SMBHs is controlled, perhaps indirectly, by the properties of the dark matter halos in which they reside, it is unclear at this time whether the connection between SMBHs and dark matter halos is of a more fundamental nature than the one between SMBHs and bulges, reflected in the $M_\bullet - \sigma$ relation. Ferrarese (2002b) notes that the scatter in the $M_\bullet - M_{DM}$ relation could be as small as or perhaps even smaller than the one in the $M_\bullet - \sigma$ relation for a given choice of the dark matter density profile and cosmological parameters, although "secure conclusions will have to await an empirical characterization of the $M_\bullet - M_{DM}$ relation, with both $M_\bullet$ and M_{DM} determined directly from observations."

1.8   Black Hole Demographics

Tracing the mass function of SMBHs from the quasar era to the present day can yield important clues about the formation and growth of SMBHs. It is, however, a road that can prove treacherous, unless we are fully aware of the uncertainties associated with this process. For instance, in 1998, a widely cited study by Magorrian et al. led to a cumulative mass density recovered in local SMBHs in excess, by a factor five, relative to the one required to power quasars during the optically bright phase ($z \sim 2-3$). Based on these findings, Richstone et al. (1998) suggested that "a large fraction of black hole growth may occur at radiative efficiencies significantly less than 0.1." This possibility was also put forth by Haehnelt, Natarayan, & Rees (1998), although the authors do mention alternative explanations, including the possibility that the Magorrian et al. (1998) local SMBH mass density could have been "strongly overestimated." The latter was shown to be the case by Ferrarese & Merritt (2000). Merritt & Ferrarese's (2001a) reassessment of the local SMBH mass density gave values in acceptable agreement with the SMBH mass density at high redshifts. Today, the emerging picture is one in which the total cumulative SMBH mass density does not seem to depend on look-back time, although the SMBH mass function (the mass density per unit SMBH mass interval) *might*. Again, much has been read into this (e.g., Yu & Tremaine 2002) although, as we will see, firm conclusions are still premature. In the following we will briefly discuss how SMBH mass functions are derived, paying particular attention to potential systematic errors that might affect the results.

Lacking a dynamical way of measuring SMBH masses in high redshift quasars, these have been inferred directly from quasar counts in the optical (Soltan 1982; Small & Blandford 1992; Chokshi & Turner 1992) and, most recently, x-rays. Andrzej Soltan (1982) first pointed out that under the assumption that the QSO luminosity is produced by gas accretion onto the central SMBH, i.e., $L_{bol} = \epsilon \dot{M}_{acc} c^2$, the total cumulative SMBH mass density can be derived directly from the mean comoving energy density in QSO light. Once a bolometric correction and a conversion efficiency ϵ of mass into energy are assumed, the latter is simply the integral, over luminosity, of the optical quasar luminosity function multiplied by luminosity. Based on these arguments, Soltan concluded that the SMBHs powering high redshift ($z > 0.3$) quasars comprise a total mass density of $\sim 5 \times 10^4$ $M_\odot$ Mpc^{-3}, each SMBH having a mass in the $10^8 - 10^9$ $M_\odot$ range. It was Soltan's argument that first led to the inescapable conclusion that most, if not all, nearby galaxies must host dormant black holes in their nuclei. This realization has been the main driver for SMBH searches in nearby quiescent galaxies and has kindled interest in the accretion crisis in nearby galactic nuclei (Fabian & Canizares 1988), ultimately leading to the revival of accretion mechanisms with low radiative efficiencies (Rees et al. 1982; Narayan & Yi 1995).

Small & Blandford (1992) proposed a different formalism to describe the time evolution of the SMBH mass function, based on a continuity equation of the type:

$$\frac{\partial N}{\partial t} + \frac{\partial}{\partial M}(N < \dot{M} >) = S(M,t) \qquad (1.13)$$

where $N(M,t)$ is the number density of SMBHs per unit comoving volume and unit mass, $< \dot{M}(M,t) >$ is the mean growth rate of a SMBH of mass M, and $S(M,t)$ is a source function. Under the assumption that during the optically bright phase quasars and bright AGNs accrete at an essentially constant $\dot{M}/M$, the mass accretion rate $\dot{M}$ is related to the luminosity as $L = \lambda M c^2/t_S = \epsilon \dot{M}/(1-\epsilon)c^2$, where λ is the fraction of the Eddington rate at which the SMBH is accreting, and t_S is the Salpeter time (Equation 1.2). $N(M,t)$ is related to the optical luminosity function $\Phi(L,t)$ as $\Phi(L,t) = N(M,t)\delta(M,t)M$. $\delta(M,t)$ is the QSO "duty cycle", i.e., the fraction of the time that the QSO spends accreting from the surrounding medium, so that $< \dot{M}(M,t) >= \delta(M,t)\dot{M}(M,t)$. Under the assumption that SMBHs are originally in place with a minimum mass, and that subsequently no SMBHs are formed or destroyed (for instance through merging processes), Equation (1.13) can then be written as

$$\frac{\partial N}{\partial t} = -\left(\frac{\lambda c}{t_E}\right)^2 \frac{1}{\epsilon} \frac{\partial \Phi}{\partial L}, \qquad (1.14)$$

which can be integrated for a given set of initial conditions (Small & Bland-ford 1992; Yu & Tremaine 2002; Ferrarese 2002a; Marconi & Salvati 2002; Marconi et al. 2004; Shankar et al. 2004).

The most recent application of these arguments is by Shankar et al. (2004). Starting from the 2dF QSO Survey from Croom et al. (2003), the authors find a total cumulative mass density in SMBH accreted during the bright AGN phase of $\sim 1.4 \times 10^5$ $M_\odot$ Mpc^{-3} for a radiative efficiency $\epsilon = 0.1$. This value is consistent with several other estimates in the literature (see Table 1.8), although the exact value is sensitive to the bolometric correction that needs to be applied to the optical fluxes (uncertain by at least 30% and perhaps as much as a factor of two; Elvis et al. 1994, Vestergaard 2003) and, to a lesser extent, the luminosity function used in the analysis. In this regard, it is worth pointing out that the magnitude limits of the 2dF ($0.3 < z < 2.3$; Boyle et al. 2000, Croom et al. 2003) and Sloan QSO surveys ($3.0 < z < 5.0$; Fan et al. 2001) correspond to Eddington limits on the SMBHs masses of 4.5×10^7 $M_\odot$ and 7.3×10^8 $M_\odot$ respectively. This implies that, at $z > 3$ in particular, QSO surveys probe only the most massive supermassive black holes, in sharp contrast with the local SMBH sample (e.g., Figure 1.9), requiring a not necessarily trivial extrapolation to smaller masses. Furthermore, the QSO luminosity function is not sampled in the $2.3 < z < 3.0$ range. Extrapolating within this range is reasonable but not altogether satisfactory, and could produce an overestimate of the cumulative SMBH mass density below 10^8 of a factor of a few (Ferrarese 2002a).[9]

SMBH mass densities derived from optical surveys fail to account for the contribution from obscured populations of AGNs that are known to exist from x-ray observations. Early studies relied on spectral synthesis models of the x-ray background, the general consensus being that lumi-nous x-ray absorbed AGNs, possibly showing a fast redshift evolution, are necessary to reproduce the 2–10 keV integrated source spectrum (Fabian & Iwasawa 1999; Salucci et al. 1999; Gilli, Salvati, & Hasinger 2001; Elvis et al. 2002). Without redshift measurements, estimates of the contribution of these obscured sources to the SMBH mass density were necessarily based on the assumption that they share the same redshift evolution as their better studied, unobscured counterparts. Such estimates generally yielded values in excess, by a factor of several, of the mass density in SMBH recovered

[9] While the extrapolation to $z = 2.3$ of the mass density of the SDSS QSOs joins smoothly with those measured from the 2dF survey at the same redshift for $M_\bullet \lesssim 10^8$ $M_\odot$, this is not true for smaller masses, where the SDSS mass density extrapolated to $z = 2.3$ overpredicts the QSO mass density (per unit redshift) derived from the 2dF data by an order of magnitude.

locally in quiescent or weakly active galaxies (see below). For example, Barger et al. (2001), integrated the accretion rate density inferred from the bolometric luminosity of a sample of 69 hard x-ray sources to obtain a total cumulative mass density of SMBHs of $\sim 2 \times 10^6$ M$_\odot$Mpc^{-3} (for $\epsilon = 0.1$), a factor $5 - 10$ larger than recovered locally (Yu & Tremaine 2002; Ferrarese 2002a; Marconi & Salvati 2001).

The launch of the Chandra and the XMM-Newton satellites has made it possible to resolve most of the x-ray background into individual sources; optical followups revealed that the bulk of the sources are at a redshift $z \sim 0.7$, which is significantly lower than the one that characterizes the optically bright phase of the quasars (Alexander et al. 2001; Barger et al. 2002; Hasinger 2002; Rosati et al. 2002; Cowie et al. 2003). The redshift distribution of the obscured sources affects both their intrinsic luminosity and the bolometric correction to the observed x-ray fluxes (which differs for high luminosity and low luminosity AGNs). Although accounting for this has led to a downward revision in the cumulative SMBH mass density in obscured AGNs ($\sim 4.1 \times 10^5$ M$_\odot$ Mpc^{-3} according to Shankar et al. 2004; see also Table 1.8), the mass accreted could be enough to account for all SMBH found locally if x-ray selected AGNs accrete with $\sim 10\%$ radiative efficiency and radiate at $\sim 30\%$ of the Eddington luminosity (Shankar et al. 2004). The current scenario is therefore of a population of unobscured, high luminosity quasars that dominate the accretion at high redshift, and a lower redshift population of obscured AGNs accreting at much lower rates,[10] but possibly accounting for the bulk of the accreted mass.

The SMBH mass function in local ($z < 0.1$), optically selected AGNs has received considerably less attention for several reasons. Compared to QSOs, lower-luminosity AGNs have a small ratio of nuclear to stellar luminosity, making it difficult to assess what fraction of the total luminosity is due to accretion onto the central black hole. Furthermore, the past accretion history is not known, and it is quite likely that a significant fraction of the SMBH mass predates the onset of the present nuclear activity. Finally, it is unlikely that lower luminosity AGNs are accreting at a constant fraction of the Eddington rate. Padovani, Burg, & Edelson (1990) estimated a total cumulative mass density in local Seyfert 1 galaxies of ~ 600 M$_\odot$ Mpc^{-3}, based on the photoionization arguments applied to the CfA magnitude limited sample. A recent revision of this result (Ferrarese 2002a), led to an estimate a factor ~ 8 larger. This notwithstanding, the main conclusion reached by Padovani et al. still holds: "the bulk of the mass related to the accretion processes connected with past QSO activity does not reside in Seyfert 1 nuclei. Instead, the remnants of past activity must be present in a much larger number of galaxies." This is true even

---

[10] For instance, Cowie et al. estimate that obscured AGNs accrete at no more than a few 10^{-5} M$_\odot$ yr^{-1} Mpc^{-3}.

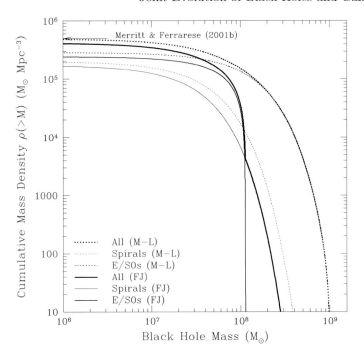

Figure 1.14. The mass function in local black holes for spirals (thin line), E/S0 (thicker line) and a complete sample of galaxies (thickest line). Dotted lines are derived from the $M_\bullet - L_B$ relation, solid lines from the $M_\bullet - \sigma$ combined with the Faber–Jackson relation (as described in the text).

after correcting for the contribution of AGNs other than Seyfert 1 galaxies (in the local universe, the ratio of Seyfert 2 to Seyfert 1 galaxies is $\sim$ 4, while LINERs are a factor of a few more numerous than Seyferts; e.g., Maiolino & Rieke 1995; Vila-Vilaro 2000).

Finally, in local quiescent galaxies — the necessary repository of the SMBHs that powered quasar activity at high redshifts — the SMBH mass function can be calculated from the $M_\bullet - L_B$ or, preferably due to its smaller scatter, the $M_\bullet - \sigma$ relation. If the mean ratio between the mass of the SMBH and that of the host bulge is known, the observed mass density of spheroids (e.g., Fukugita et al. 1998) can be easily transformed into a SMBH mass density. This approach was adopted by Merritt & Ferrarese (2001a). The $M_\bullet - \sigma$ relation was used to estimate $M_\bullet$ for a sample of 32 early-type galaxies with a dynamical measurement of the total mass (from Magorrian et al. 1998). For this sample, the frequency function $N[\log(M_\bullet/M_{\text{bulge}})]$ is approximated by a Gaussian with $\langle \log(M_\bullet/M_{\text{bulge}}) \rangle \sim$ -2.90 and standard deviation ~ 0.45. This implies $M_\bullet/M_{\text{bulge}} \sim 1.3 \times 10^{-3}$

Summary of SMBH Mass Densities

$\rho_\bullet$ ($M_\odot\,\mathrm{Mpc}^{-3}$)	Reference
Local Quiescent Galaxies ($z < 0.025$)	
$2.3^{+4.0}_{-1.5} \times 10^5$	Wyithe & Loeb (2003)
$2.4 \pm 0.8 \times 10^5$	Aller & Richstone (2002)
$\sim 2.5 \times 10^5$	Yu & Tremaine (2002)
$2.8 \pm 0.4 \times 10^5$	McLure & Dunlop (2004)
$4.2 \pm 1.0 \times 10^5$	Shankar et al. (2004)
$\sim 4.5 \times 10^5$	Ferrarese (2002a)
$4.6^{+1.9}_{-1.4} \times 10^5$	Marconi et al. (2004)
$\sim 5 \times 10^5$	Merritt & Ferrarese (2001a)
$\sim 5.8 \times 10^5$	Yu & Tremaine (2002)
QSOs Optical Counts ($0.3 < z < 5.0$)	
$\sim 1.4 \times 10^5$	Shankar et al. (2004)
$\sim 2 \times 10^5$	Fabian (2003)
$\sim 2.1 \times 10^5$	Yu & Tremaine (2004)
$\sim 2.2 \times 10^5$	Marconi et al. (2004)
$2 - 4 \times 10^5$	Ferrarese (2002a)
AGN X-Ray Counts ($z(peak) \sim 0.7$)	
$\sim 2 \times 10^5$	Fabian (2003)
$\sim 4.1 \times 10^5$	Shankar et al. (2004)
$4.7 - 10.6 \times 10^5$	Marconi et al. (2004)

or, when combined with the mass density in local spheroids from Fukugita et al. (1998), $\rho_\bullet \sim 5 \times 10^5\ M_\odot\,\mathrm{Mpc}^{-3}$.

Alternatively, $\rho_\bullet$ can be derived by combining the $M_\bullet - \sigma$ relation with the velocity function of galaxies. Compared to the previous method, this has the added bonus of producing an analytical representation of the cumulative SMBH mass density as a function of $M_\bullet$. The process, however, can be rather involved. The velocity function is generally constructed starting from a galaxy luminosity function (e.g., Marzke et al. 1998; Bernardi et al. 2003). This needs to be converted to a bulge luminosity function, which entails the adoption of a ratio between total and bulge luminosity (e.g., from Fukugita et al. 1998). Both steps need to be carried out separately for galaxies of different Hubble types. The derived bulge luminosity function can be combined directly with the $M_\bullet - L_B$ relation to give a SMBH mass function, or it can first be transformed to a velocity dispersion function (through the Faber-Jackson relation, which again depends on the Hubble type), and then combined with the $M_\bullet - \sigma$ relation. Both approaches have drawbacks. The $M_\bullet - L_B$ relation has significant (and possibly Hubble-type dependent) scatter, while the Faber–Jackson relation has large scatter and is ill defined, especially for bulges. The approach above was followed by Salucci et al. (1999), Marconi & Salvati (2001), Ferrarese (2002a), Yu & Tremaine (2002), Aller & Richstone (2002), Marconi et al. (2004), and

Shankar et al. (2004). Figure 1.14 shows the cumulative SMBH mass functions separately for the E/S0 and spiral populations, derived by Ferrarese (2002a) from the $M_\bullet - L_B$ relation and the $M_\bullet - \sigma$ relation combined with the Faber–Jackson relations for ellipticals and spirals. While the two distributions differ in the details, there is little difference in the total mass density, which falls in the range $(4 - 5) \times 10^5$ $M_\odot$ Mpc^{-3}.

Wyithe & Loeb (2003) and Shankar et al. (2004) used yet another, more direct approach, and employed the velocity dispersion function directly measured by Sheth et al. (2003) from SDSS data. The only disadvantage of this approach is that the SDSS velocity dispersion function is defined only for early-type galaxies, and therefore incomplete below 200 km s^{-1}, i.e., in the regime populated by spiral galaxies ($M_\bullet \lesssim 1.5 \times 10^8$ $M_\odot$). As pointed out by the authors, the mass function differs from the one derived using the methods described in the preceding paragraphs (e.g., Yu & Tremaine 2002), declining more gradually at the high mass end (where the velocity dispersion function is measured directly). By comparing the local mass function with that derived by combining the Press–Schechter (1974) halo mass function with the $M_\bullet - M_{DM}$ relation, Wyithe & Loeb find that SMBHs with $M_\bullet \lesssim 10^8$ $M_\odot$ formed during the peak of quasar activity ($z \sim 1 - 3$), while more massive black holes were already in place at higher redshifts, as high as $z \sim 6$ for SMBHs of a few $\times 10^9$ $M_\odot$. Integrating over $M_\bullet$, the total cumulative mass density is estimated to be $\rho_\bullet = (2.3^{+4.0}_{-1.5}) \times 10^5$ $M_\odot$ Mpc^{-3} (see also Shakar et al. 2004).

These results indicate that, given the current uncertainties, the total cumulative mass function in SMBHs hosted by local galaxies is of the correct order of magnitude to account for the AGNs energetics, although the relative contribution of high redshift optical QSOs and local obscured AGNs remains somewhat uncertain (Marconi et al. 2004; Shankar et al. 2004; Table 1.8). The fine details remain to be worked out. For instance, Yu & Tremaine (2002) claimed a larger fraction of very massive black holes, $M > 10^9$, in high redshift QSOs than have been recovered locally, although this result is not supported by more recent studies (Marconi et al. 2004; Shankar et al. 2004). The case is complicated by the fact that, as noted by Wyithe & Loeb (2003) and Ferrarese (2002a), the local SMBH mass function is *not* defined in the high ($\geq 10^{10}$ $M_\odot$) mass range. In this context, targeting SMBHs in brightest cluster galaxies (§1.9), will be particularly illuminating.

1.9 The Future

The various methods of SMBH detection, scaling relations, and secondary mass estimators based on those relations are schematically represented in Figure 1.15. None of the information shown was available before 1995, and most of it has surfaced only in the past 2 to 3 years. Still, we are only at the beginning of what is sure to be a long, fascinating journey. We have yet to find out, for instance, if SMBHs spin and how fast, whether binary SMBHs exist and how quickly they coalesce following galaxy mergers, what are the detailed modalities of accretion, and what controls the character and level of nuclear activity. We have not yet probed the morphology and kinematics of the gas within a few thousands of Schwarzschild radii from the central SMBH. On a grander scale, we do not know how SMBHs form. For the first time, it appears realistic that answers will be found in the near future.

On the observational front, the field is ready for the next technological leap. HST is mostly responsible for the progress made in the past decade, but it is inadequate to meet the challenges we are now facing. For instance, even a cursory look at Table 1.3 will reveal that the sample of galaxies targeted so far is remarkably homogeneous. Most are early-type galaxies. Most host SMBHs in the $10^8 \lesssim M_\bullet \lesssim 10^9$ M$_\odot$ range. With the exception of M33 (see below), no detections have (or can) be attempted below 10^6 M$_\odot$, and the $10^6 \lesssim M_\bullet \lesssim 10^7$ M$_\odot$ range is very poorly sampled. Most galaxies are in small clusters—M87 is the only cD. Only two galaxies are beyond 30 Mpc. HST stellar dynamical studies are prohibitive in giant ellipticals, due to the low central surface brightness that characterizes these objects (e.g., Ferrarese et al. 1994; Lauer et al. 1995; Rest et al. 2001).

Similarly, the vast majority of dwarf elliptical galaxies are beyond HST capabilities. Assuming that the $M_\bullet - \sigma$ relation holds in the $\sigma \sim$ a few $\times 10$ km s^{-1} regime that characterizes these objects, the sphere of influence of the putative SMBH at their centers is accessible only in the most nearby systems; in these cases, the stellar population is resolved into individual stars, each too faint to be handled by HST's small mirror.

The situation for late-type spirals is no better. The upper limit on $M_\bullet$ in M33, the closest Sc galaxy, puts the sphere of influence of the putative BH a factor of ~ 20 below the resolution capabilities of HST. Bulge less and low surface brightness galaxies, both of which are able to host AGN, have not been targeted.

Figure 1.16 shows distance against SMBH mass (calculated from the $M_\bullet - L_B$ relation given by Ferrarese & Merritt 2000) for all galaxies in the Center for Astrophysics (CfA) redshift sample (Huchra et al. 1990). The resolution limits for HST and for 8 m and 30 m diffraction-limited telescopes are shown by the diagonal lines. If only resolution constraints are considered (exposure time requirements would further exacerbate the problem), most late-type spirals (Sb–Sc) are expected to host SMBHs too small to be resolved by HST. Only a handful of Sa galaxies are within reach. It is only with an 8 m class telescope that a complete sample of galaxies spanning the whole Hubble sequence can be collected. Even then, little would be gained below 10^6 M$_\odot$. Pushing this limit down by an order of mag-

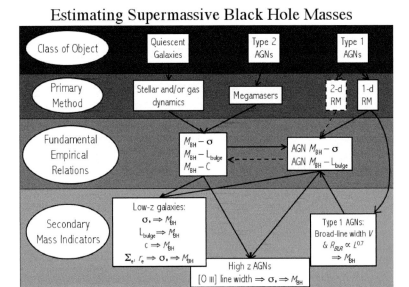

Figure 1.15. *(Upper two panels)* Primary methods for determining SMBH masses, and the galaxy classes to which they are most easily applied. "2D RM" and "1D RM" refer to two and one-dimensional reverberation mapping (§1.6), respectively. Only the latter has been addressed by current monitoring programs, while the former, which would lead to a complete morphological and kinematical picture of the BLR, awaits a future dedicated space mission. *(Third panel)* Fundamental relations discussed in §1.7. Reverberation mapping can be used to build the $M_\bullet - L$ and $M_\bullet - \sigma$ relations for AGN; by comparing them with the same relations defined by quiescent galaxies, constraints can be set on the reliability of reverberation mapping studies. Dashed lines represent "future" connections: if 2D reverberation mapping becomes a reality, the AGN $M_\bullet - L$ and $M_\bullet - \sigma$ relations can be compared directly to the relations observed in quiescent galaxies to evaluate the role played by $M_\bullet$ in driving the character of the nuclear activity. *(Bottom panel)* Secondary mass estimators. In low redshift galaxies, both quiescent and active, SMBH masses can be estimated through the $M_\bullet - L$, $M_\bullet - \sigma$, or $M_\bullet - C$ (where C is the concentration parameter for the bulge; Graham et al. 2001) relations, provided that L_{bulge}, σ, or C can be measured. For large samples of early-type galaxies, if σ is not directly observed, it can be derived using the fundamental plane, if the surface brightness profile and effective radius are known. In high redshift AGN, the bulge luminosity and velocity dispersion are difficult to measure due to contamination from the active nucleus. A method that has been applied in these cases is to use the width of the [OIII]λ5007 Å emission as an estimate of stellar velocity dispersion (Nelson & Whittle 1995; but see also Boroson 2003); $M_\bullet$ can then be derived from the $M_\bullet - \sigma$ relation. In type I AGN, the black hole mass can be derived from the virial approximation, if the AGN luminosity, which correlates with BLR size (§1.6.2), is known. Courtesy of Bradley Peterson.

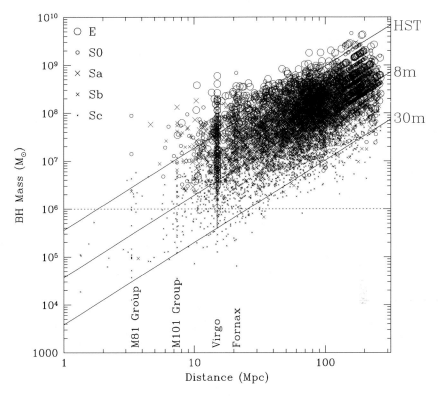

Figure 1.16. SMBH mass vs. distance for all galaxies in the CfA Redshift Sample (Huchra et al. 1990). $M_\bullet$ is calculated from the $M_\bullet - L_B$ relation, after a correction suitable to each Hubble type is applied to convert total luminosity to bulge luminosity (Fukugita, Hogan, & Peebles 1998). Only for the galaxies that lie above the solid lines can the sphere of influence of the putative nuclear SMBH be resolved by HST's STIS or 8 m and 30 m diffraction-limited telescopes. A few nearby groups and clusters are marked. Figure from Ferrarese (2003; that paper's Figure 1).

nitude requires a 30 m diffraction-limited telescope—unless improvements in radio detectors suddenly unveil a significant number of H_2O masers in these galaxies.

A 30 m diffraction-limited telescope is also necessary to probe the Coma cluster and the most nearby rich Abell clusters at $d \sim 100$ Mpc, and to study the influence of environment on the formation and evolution of SMBHs. Resolved dynamical studies must be abandoned altogether if one wishes to push even further and tackle the redshift evolution of SMBH scaling relations. In this case, reverberation mapping studies (or use of the $R_{BLR} - L_\lambda$ relation, which is based on these studies) appear to be the only viable alternative. To understand the uncertainties and systematics affecting reverberation mapping and any of the secondary methods based on it, knowledge of the morphology and kinematics of the BLR is a must. *Kronos*, a proposed optical/UV/x-ray NASA Medium Explorer mission, is designed to meet this goal in a selected sample of nearby AGN (Peterson & Horne 2003).

Although the issues listed above are certain to be key topics of future research, it is almost as certain that the field will take many unanticipated and serendipitous turns, and it is perhaps these turns that we look forward to most.

References

Adams F C, Graff, D S and Richstone D 2001 *ApJ* **551** L31

Akritas M G and Bershady M A 1996 *ApJ* **470** 706

Alexander D M, et al 2001 *ApJ* **122** 2156

Aller M C and Richstone D 2002 *AJ* **124** 3035

Antonucci R R J and Miller J S 1985 *ApJ* **297** 621

Arav N, Barlow T A, Laor A and Blandford R D 1997 *MNRAS* **288** 1015

Arav N, Barlow T A, Laor A, Sargent W L W and Blandford R D 1998 *MNRAS* **297** 990

Baade W and Minkowski R 1954 *ApJ* **119** 206

Bacon R, Emsellem E, Combes F, Copin Y, Monnet G and Martin P 2001 *A&A* **371** 409

Baes M, Buyle P, Hau G K T and Dejonghe H 2003 *MNRAS* **341** L44

Balick B and Brown R L 1974 *ApJ* **194** 264

Barger A J et al. 2001 *AJ* **122** 2177

Barger A J et al. 2002 *AJ* **124** 1839

Barth A, Ho L C and Sargent W L W 2003 *ApJ* **583** 134

Barth A J, Sarzi M, Rix H -W, Ho L C, Filippenko A V and Sargent W L W 2001 *ApJ* **555** 685

Begelman M C 2001 *ApJ* **551** 897

Bernardi M et al. 2003 *AJ* **125** 1817

Bian W -H and Zhao Y -H 2003 *PASJ* **55** 143

Binney J and Mamon G A 1982 *MNRAS* **200** 361

Binney J and Tremaine S 1987 *Galactic Dynamics* (Princeton: Princeton University Press)

Blandford R D and Begelman M C 1999 *MNRAS* **303** L1

Blandford R D and McKee C F 1982 *ApJ* **255** 419

Blumenthal G R and Mathews W G 1975 *ApJ* **198** 517

Bolton C T 1972 *Nature* **235** 271

Boroson T A 2002 *ApJ* **565** 78

Boroson T A 2003 *ApJ* **585** 647

Bottorff M, Korista K T, Shlosman I and Blandford R D 1997 *ApJ* **479** 200

Bower G A et al. 1998 *ApJ* **492** L111

Boyle B J et al.2000 *MNRAS* **317** 1014

Braatz J A, Wilson A S and Henkel C 1994 *ApJ* **437** L99

Braatz J A, Wilson A S and Henkel C 1996 *ApJS* **106** 51

Brucato R J and Kristian J 1972 *ApJ* **173** 105

Bryan G L and Norman M L 1998 *ApJ* **495** 80

Bullock J S, Kolatt T S, Sigad Y, Somerville R S, Kravtsov A V, Klypin A A, Primack J R and Dekel A 2001 *MNRAS* **321** 559

Cappellari M, Verolme E K, van der Marel R P, Verdoes Kleijn G A, Illingworth G D, Franx M , Carollo C M and de Zeeuw P T 2002 *ApJ* **578** 787

Cattaneo A, Haehnelt M G and Rees M J 1999 *MNRAS* **308** 77

Chandrasekhar S 1931 *ApJ* **74** 81

Chandrasekhar S 1935 *MNRAS* **95** 226

Cheung A C et al. 1969 *Nature* **221** 626

Chiang J and Murray N 1996 *ApJ* **466** 704

Chokshi A and Turner E L 1992 *MNRAS* **259** 421

Cipollina M and Bertin G 1994 *A&A* **288** 43

Claussen M J and Lo K Y 1986 *ApJ* **308** 592

Claussen M J, Hellingman G M and Lo K Y 1984 *Nature* **310** 298

Collin-Souffrin S, Dyson J E, McDowell J C and Perry, J J 1988 *MNRAS* **232** 539

Cowie L L et al. 2003 *ApJ* **584** L57

Crane P et al. 1993 *AJ* **106** 1371

Crawford M K, Genzel R, Harris A I, Jaffe D T, Lugten J B, Serabyn E, Townes C H and Lacy J H 1985 *Nature* **315** 467

Cretton N and van den Bosch F C 1999 *ApJ* **514** 704

Croom S M et al. 2003 Astroph/0403040

de Vaucouleurs G, de Vaucouleurs A, Corwin H G Jr, Buta R J, Paturel G and Fouqué P 1991 *Third Reference Catalogue of Bright Galaxies* (New York: Springer-Verlag)

Devereux N, Ford H C, Tsvetanov Z and Jacoby G 2003 *AJ* **125** 1226

Dietrich M, Wagner S J, Courvoisier T J -L, Bock H and North P 1999 *A&A* **351** 31

Di Matteo T, Fabian A C, Rees M J, Carilli C L and Ivison R J 1999 *MNRAS* **305** 492

Doeleman S S et al. 2001 *AJ* **121** 2610

Done C, Madejski G M and Zycki P T 2000 *ApJ* **536** 213

dos Santos P M and Lepine J R D 1979 *Nature* **278** 34

Eckart A, Genzel R, Hofmann R, Sams B J and Tacconi-Garman L E 1993 *ApJ* **407** L77

Einstein A 1915 Sitzungsberichte der Deutschen Akademie der Wissenschaften zu Berlin Klasse fur Mathematik Physik, und Technik 844

Eisenhauer F, Schödel R, Genzel R, Ott T, Tecza M, Abuter R, Eckart A and Alexander T 2003 *ApJ* **597** L121

Ekers R D, Goss W M, Schwarz U J, Downes D and Rogstad D H 1975 *A&A* **43** 159

Elvis M, Risaliti G and Zamorani G 2002 *ApJ* **565** 75

Elvis M et al. 1994 *ApJS* **95** 1

Emmering R T, Blandford R D and Schlosman I 1992, *ApJ* **385** 460

Emsellem E, Dejonghe H and Bacon R 1999 *MNRAS* **303** 495

Fabian A C 2003 *Astron Nach* **324** 4

Fabian A C and Canizares C A 1988 *Nature* **333** 829

Fabian A C and Iwasawa K 1999 *MNRAS* **303** L34

Falcke H, Henkel Chr Peck A B, Hagiwara Y, Almudena Prieto M and Gallimore J F 2000 *A&A* **358** L17

Fan X et al. 2001 *AJ* **121** 54

Ferland G J, Peterson B M, Horne K, Welsh W F and Nahar S N 1992 *ApJ* **387** 95

Ferrarese L 2002a in *Current High-Energy Emission around Black Holes* Eds C -H Lee & H -Y Chang (Singapore: World Scientific Publishing) p3

Ferrarese L 2002b *ApJ* **578** 90

Ferrarese L 2003 in *Hubble's Science Legacy: Future Optical/Ultraviolet Astronomy from Space* Eds K R Sembach, J C Blades G D Illingworth and R C Kennicutt (San Francisco: ASP Conference Series) **291** 196

Ferrarese L and Ford H C 1999 *ApJ* **515** 583

Ferrarese L and Merritt D 2000 *ApJ* **539** L9

Ferrarese L, Ford H C and Jaffe W 1996 *ApJ* **470** 444

Ferrarese L, Pogge R W, Peterson B M, Merritt D, Wandel A and Joseph C M 2001 *ApJ* **555** L79

Ferrarese L, van den Bosch F C, Ford H C, Jaffe W and O'Connell R W 1994 *AJ* **108** 1598

Filippenko A V and Ho L C 2003 *ApJ* **588** L13

Ford H C et al. 1994 *ApJ* **435** L27

Fukugita M, Hogan C J and Peebles P J E 1998 *ApJ* **503** 518

Gardner F F and Whiteoak J B 1982 *MNRAS* **201** 13

Gebhardt K et al. 2000a *AJ* **119** 1157

Gebhardt K et al. 2000b *ApJ* **539** L13

Gebhardt K et al. 2003 *ApJ* **583** 92

Genzel R and Townes C H 1987 AR*A&A* **25** 377

Gerhard O 1993 *MNRAS* **265** 213

Gerhard O, Kronawitter A, Saglia R P and Bender R 2001 *AJ* **121** 1936

Ghez A M, Klein B L, Morris M and Becklin E E 1998 *ApJ* **509** 678

Ghez A M, Morris M, Becklin E E, Tanner A and Kremenek T 2000 *Nature* **407** 349

Ghez A M et al. 2003 *ApJ* **586** L127

Gilli G M, Salviati M and Hasinger G 2001 *A&A* **366** 407

Gondoin P, Barr P, Lumb D, Oosterbroek T, Orr A and Parmar A N 2001a *A&A* **378** 806

Gondoin P, Lumb D, Siddiqui H, Guainazzi M and Schartel N 2001b *A&A* **373** 805

Graham A W, Erwin P, Caon N and Trujillo I 2001 *ApJ* **563** L11

Greenhill L J, Gwinn C R, Antonucci R and Barvainis R 1996 *ApJ* **472** L21

Greenhill L J, Henkel C, Becker R, Wilson T L and Wouterloot J G A 1995a *A&A* **304** 21

Greenhill L J, Jiang D R, Moran J M, Reid M J, Lo K Y and Claussen M J 1995b *ApJ* **440** 619

Greenhill L J, Kondratko P T, Lovell J E J, Kuiper T B H, Moran J M, Jauncey D L and Baines G P 2003 *ApJ* **582** L11

Greenhill L J, Moran J M and Herrnstein J R 1997 *ApJ* **481** L23

Greenhill L J et al. 1997 *ApJ* **486** L15

Greenhill L J et al. 2002 *ApJ* **565** 836

Haehnelt M G and Kauffmann G 2000 *MNRAS* **318** L35

Haehnelt M G, Natarajan P and Rees M J 1998 *MNRAS* **300** 817

Hagiwara Y, Diamond P J and Miyoshi M 2002 *A&A* **383** 65

Hagiwara Y, Diamond P J, Nakai N and Kawabe R 2001a *ApJ* **560** 119

Hagiwara Y, Henkel C, Menten K M and Nakai N 2001b *ApJ* **560** L37

Harms R J et al. 1994 *ApJ* **435** L35

Haschick A D Baan W A and Peng E W 1994 *ApJ* **437** L35

Hasinger G 2002 in *x-ray astronomy in the new millennium* R D Blandford A C
    Fabian and K Pounds eds Roy Soc of London Phil Tr A vol 360 Issue 1798
    p 2077

Henkel C, Braatz J A, Greenhill L J and Wilson A S 2002 *A&A* **394** L23

Henkel C, Guesten R, Downes D, Thum C, Wilson T L and Biermann P 1984
    A&A **141** L1

Herrnstein J R, Greenhill L J and Moran J M 1996 *ApJ* **468** L17

Herrnstein J R et al. 1999 *Nature* **400** 539

Huchra J P, Geller M J de Lapparent V and Corwin H G Jr 1990 *ApJS* **72** 433

Ishihara Y, Nakai N, Iyomoto N, Makishima K, Diamond P and Hall P 2001
    PASJ **53** 215

Jaffe W, Ford H C, Ferrarese L, van den Bosch F and O'Connell R W 1993 *Nature*
    364 213

Jennison R C and Das Gupta M K 1953 *Nature* **172** 996

Jones D L, Wehrle A E, Meier D L and Piner B G 2000 *ApJ* **534** 165

Jorgensen I, Franx M and Kjaergaard P 1995 *MNRAS* **276** 1341

Kaspi S, Smith P S, Netzer H, Maoz D Jannuzi, B T and Giveon U 2000 *ApJ*
    533 631

Koratkar A P and Gaskell C M 1991 *ApJ* **370** L71

Kormendy J 1988 *ApJ* **335** 40

Kormendy J and Gebhardt K 2001 in *20th Texas Symposium on Relativistic As-
    trophysics* Eds J C Wheeler & H Martel (Melville New York: AIP Conference
    Proceedings) **586** p363

Kormendy J and Richstone D 1992 *ApJ* **393** 559

Kormendy J and Richstone D 1995 *ARA&A* **33** 581

Krolik J H 2001 *ApJ* **551** 72

Kronawitter A, Saglia R P, Gerhard O and Bender R 2000 *A&AS* **144** 53

Lacy J H, Townes C H and Hollenbach D J 1982 *ApJ* **262** 120

Lacy J H, Townes C H, Geballe T R and Hollenbach D J 1980 *ApJ* **241** 132

Landau L D 1932 *Phys Z Sowjetunion* **1** 285

Lauer T R et al. 1995 *AJ* **110** 2622

Lee M /H and Goodman J 1989 *ApJ* **343** 594

Loeb A and Rasio F 1994 *ApJ* **432** L52

Lynden-Bell D 1969 *Nature* **223** 690

Lyutyi V M, Syunyaev R A and Cherepashchuk A M 1973 *Soviet Astronomy* **17**
    1

Macchetto F, Marconi A, Axon D J, Capetti A, Sparks W and Crane P 1997
    ApJ **489** 579

Maciejewski W and Binney J 2001 *MNRAS* **323** 831

Magorrian J et al. 1998 *AJ* **115** 2285

Maiolino R and Rieke G H 1995 *ApJ* **454** 95

Maoz E 1998 *ApJL* **494** 181

Marconi A and Hunt L K 2003 *ApJ* **589** L21

Marconi A and Salvati M 2002 in *Issues in Unification of AGNs* R Maiolino A Marconi and N Nagar eds ASP Conf Series

Marconi A, Capetti A, Axon D J, Koekemoer A, Macchetto D and Schreier E J 2001 *ApJ* **549** 915

Marconi A, Risaliti G, Gilli R, Hunt L K, Maiolino R and Salvati M 2004 *MNRAS* **351** 169

Marconi A et al. 2003 *ApJ* **586** 868

Marzke R O et al. 1998 *ApJ* **503** 617

Mauder H 1973 *A&A* **28** 473

McHardy I 1988 *MmSAI* **59** 239

McLure R J and Dunlop J S 2000 *MNRAS* **317** 249

McLure R J and Dunlop J S 2001 *MNRAS* **327** 199

McLure R J and Jarvis M J 2002 *MNRAS* **337** 109

Melia F and Falcke H 2001 AR*A&A* **39** 309

Merritt D and Ferrarese L 2001a *MNRAS* **320** L30

Merritt D and Ferrarese L 2001b *ApJ* **547** 140

Merritt D and Ferrarese L 2001c in *The Central Kiloparsec of Starbursts and AGN: The La Palma Connection* Eds J H Knapen J E Beckman I Shlosman and T J Mahoney (San Francisco: ASP Conference Series) 249 p335

Mezger P G and Wink J E 1986 *A&A* **157** 252

Milosavljevic M and Merritt D 2001 *ApJ* **563** 34

Miyoshi M, Moran J, Herrnstein J, Greenhill L, Nakai N, Diamond P and Inoue M 1995 *Nature* **373** 127

Monaco P, Salucci P and Danese L 2000 *MNRAS* **311** 279

Moran J, Greenhill L, Herrnstein J, Diamond P, Miyoshi M, Nakai N and Inoue 1996 in *Quasars and AGN: High Resolution Imaging* Proceedings of the National Academy of Sciences, Vol 92 Issue 25 p11427

Murray N and Chiang J 1997 *ApJ* **474** 91

Murray N, Chiang J, Grossman S A and Voit G M 1995, *ApJ* **451** 498

Nakai N, Inoue M and Miyoshi M 1993 *Nature* **361** 45

Nakano T and Makino J 1999 *ApJ* **510** 155

Nandra K, George I M, Mushotzky R F, Turner T J and Yaqoob T 1997 *ApJ* **477** 602

Nandra K, George I M, Mushotzky R F, Turner T J and Yaqoob T 1999 *ApJ* **523** L17

Narayan R and Yi I 1994 *ApJ* **428** L13

Navarro J F and Steinmetz M 2000 *ApJ* **538** 477

Nelson C H and Whittle M 1996 *ApJ* **465** 96

Netzer H 2003 *ApJ* **583** L5

Neufeld D A and Maloney P R 1995 *ApJ* **447** L17

Neufeld D A, Maloney P R and Conger S 1994 *ApJ* **436** L127

Onken C A and Peterson B M 2002 *ApJ* **572** 746

Onken C A, Peterson B M, Dietrich M, Robinson A and Salamanca I M 2003 *ApJ* **585** 121

Oppenheimer J R and Snyder H 1939 *Phys Rev* **46** 455

Oppenheimer J R and Volkoff G M 1938 *Phys Rev* **55** 374

Oshlack A Y K N, Webster R L and Whiting M T 2002, *ApJ* **576** 81

Padovani P, Burg R and Edelson R A 1990 *ApJ* **353** 438

Peebles P J E 1972 *Gen Relativ Gravitation* **3** 63

Pelletier G and Pudritz R E 1992 *ApJ* **394** 117

Peterson B M 1993 *PASP* **105** 247

Peterson B M 2002 in *Advanced Lectures on the Starburst-AGN Connection* (Singapore: World Scientific) p3

Peterson B M and Horne K 2003 in *Astrotomography*, 25th meeting of the IAU Joint Discussion 9

Peterson B M and Wandel A 2000 *ApJ* **540** L13

Peterson B M et al. 2000 *ApJ* **542** 161

Peterson B M et al. 2002 *ApJ* **581** 197

Pogge R W 1989 *ApJS* **71** 433

Press W H and Schechter P 1974 *ApJ* **187** 425

Quinlan G D and Hernquist L 1997 *NewA* **2** 533

Quinlan G D, Hernquist L and Sigurdsson S 1995 *ApJ* **440** 554

Rees M J 1984 *ARA&A* **22** 471

Rees M J, Phinney E S, Begelman M C and Blandford R D 1982 *Nature* **295** 17

Reid M J, Menten K M, Genzel R, Ott T, Schödel R and Eckart A 2003 *ApJ* **587** 208

Rest A, van den Bosch F C, Jaffe W, Tran H, Tsvetanov Z, Ford H C, Davies J and Schafer J 2001 *AJ* **121** 2431

Reunanen J, Kotilainen J K and Prieto M A 2003 *MNRAS* **343** 192

Reynolds C S 1997 *MNRAS* **286** 513

Reynolds C S and Nowak M A 2003 *Phys Rept* **377** 389

Rhoades C E and Ruffini R 1974 *Phys Rev Lett* **32** 324

Richstone D O and Tremaine S 1985 *ApJ* **286** 370

Richstone D O et al. 1998 *Nature* **395** A14

Robinson I, Schild A and Schucking E L 1965, in *1st Texas Symposium on Relativistic Astrophysics*, Eds I Robinson A Schild and E L Schucking (Chicago: University of Chicago Press)

Rosati P et al. 2002 *ApJ* **566** 667

Salpeter E E 1964 *ApJ* **140** 796

Salucci P et al. 1999 *MNRAS* **307** 637

Sargent W L W, Young P J, Lynds C R, Boksenberg A, Shortridge K and Hartwick F D A 1978 *ApJ* **221** 731

Sarzi M, Rix H -W, Shields J C, Rudnick G, Ho L C, McIntosh D H, Filippenko A V and Sargent W L W 2001 *ApJ* **550** 65

Schlegel D J, Finkbeiner D P and Davis M 1998 *ApJ* **500** 525

Schödel R, Ott T, Genzel R, Eckart A, Mouawad N and Alexander T 2003 *ApJ* **596** 1015

Schwarzschild K (1916a) 1916 Sitzungsberichte der Deutschen Akademie der Wissenschaften zu Berlin Klasse fur Mathematik Physik und Technik 189

Schwarzschild K (1916b) 1916 Sitzungsberichte der Deutschen Akademie der Wissenschaften zu Berlin Klasse fur Mathematik Physik und Technik 424

Scorza C and Bender R 1995 *A&A* **293** 20

Serabyn E and Lacy J H 1985 *ApJ* **293** 445

Sergeev S G, Pronik V I, Peterson B M, Sergeeva E A and Zheng W 2002 *ApJ*

576 660

Seyfert C K 1943 *ApJ* **97** 28

Shankar F, Salucci P, Granato G L, De Zotti G and Danese L 2004 *MNRAS* in press

Sheth R K et al. 2003 *ApJ* **594** 225

Shields G A, Gebhardt K, Salviander S, Wills B J, Xie B, Brotherton M S, Yuan J and Dietrich M 2003 *ApJ* **583** 124

Sigurdsson S, Hernquist L and Quinlan G D 1995 *ApJ* **446** 75

Silk J and Rees M J 1998 *A&A* **331** L1

Simien F and de Vaucouleurs G 1986 *ApJ* **302** 564

Small T and Blandford R D 1992 *MNRAS* **259** 725

Smith M D and Raine D J 1985 *MNRAS* **212** 425

Smith M D and Raine D J 1988 *MNRAS* **234** 297

Soltan A 1982 *MNRAS* **200** 115

Steinmetz M and Muller E 1995 *MNRAS* **276** 549

Tadhunter C, Marconi A, Axon D, Wills K, Robinson T G and Jackson N 2003 *MNRAS* **342** 861

Takahashi K, Inoue H and Dotani T 2002 *PASJ* **54** 373

Tonry J L, Dressler A, Blakeslee J P, Ajhar E A, Fletcher A B, Luppino G A, Metzger M R and Moore C B 2001 *ApJ* **546** 681

Tran H D, Tsvetanov Z, Ford H C, Davies J, Jaffe W, van den Bosch F C and Rest A 2001 *AJ* **121** 2928

Tremaine S et al. 2002 *ApJ* **574** 740

Trotter A S, Greenhill L J, Moran J M, Reid M J, Irwin J A and Lo K -Y 1998 *ApJ* **495** 740

Ulrich M -H and Horne K 1996 *MNRAS* **283** 748

Urry C M and Padovani P 1995 *PASP* **107** 803

Valluri M, Merritt D and Emsellem E 2004 *ApJ* **602** 66

van de Ven G, Hunter C, Verolme E K and de Zeeuw P T 2003 *MNRAS* **342** 1056

van den Bosch F C and de Zeeuw P T 1996 *MNRAS* **283** 381

van den Bosch F C Jaffe W and van der Marel R P 1998 *MNRAS* **293** 343

van der Marel R P 1994 *MNRAS* **270** 271

van der Marel R P 1999 *AJ* **117** 744

van der Marel R P and Franx M 1993 *ApJ* **407** 525

van der Marel R P and van den Bosch F C 1998 *AJ* **116** 2220

Verdoes Kleijn G A van der Marel R P Carollo C M and de Zeeuw P T 2000 *AJ* **120** 1221

Verolme E K et al. 2002 *MNRAS* **335** 517

Vestergaard M 2002 *ApJ* **571** 733

Vestergaard M, Wilkes B J and Barthel P D 2000 *ApJ* **538** L103

Vila-Vilaro B 2000 *PASPJ* **52** 305

Wandel A 2002 *ApJ* **565** 762

Wandel A Peterson B M and Malkan M A 1999 *ApJ* **526** 579

Wang J -X, Wang T -G and Zhou Y -Y 2001 *ApJ* **549** 891

Watanabe M et al. 2003 *ApJ* **591** 714

Watson W D and Wallin B K 1994 *ApJ* **432** L35

Wollman E R, Geballe T R, Lacy J H, Townes C H and Rank D M 1977 *ApJ*

218 L103

Woo J -H and Urry C M 2002a *ApJ* **579** 530

Woo J -H and Urry C M 2002b *ApJ* **581** L5

Wyithe J S B and Loeb A 2002 *ApJ* **581** 886

Wyithe J S B and Loeb A 2003 *ApJ* **595** 614

Young P J 1980 *ApJ* **242** 1232

Yu Q and Tremaine S 2002 *MNRAS* **335** 965

Zel'dovich Y B and Novikov I D 1964 *Sov. Phys. Dokl.* **158** 811

Chapter 2

Joint Evolution of Black Holes and Galaxies: Observational Issues

Alberto Franceschini
Department of Astronomy
University of Padova
Padova, Italy

New crucial information on the relationship between nuclear black-holes and the host galaxies has been obtained from high spatial resolution spectroscopy with the Hubble Space Telescope on local objects, but also from direct inspection of the high-redshift universe with newly implemented powerful telescopes from radio to x-rays. By these means, the most active phases in galaxy evolution at high-redshifts, involving both stellar formation and black-hole gravitational accretion, have been probed in-situ.

We review in this lecture some of the most recent observations concerning local and high-redshift Active Galaxies and quasars. Such observational properties of cosmic sources are then compared with the local remnants of the past activity, including the old stellar populations in galaxies, the distribution of nuclear supermassive black-holes, and the cosmological background radiations (particularly those in x-rays, optical-UV, and the far-infrared).

From all these recent facts, a consistent picture seems to emerge, requiring progressive formation and assembly of galaxies and AGNs taking place over a very extended interval of cosmic time (from $z > 6$ to $z < 1$), well consistent with the time-honored hierarchical picture of structure formation. Many important details of this scheme are however still unclear or missing, and deserve much further inspection by forthcoming or future instrumentation and theoretical insight.

2.1   Galaxy Activity: Generalities

The activity of galaxies concerns transient phenomena occurring during the galaxy lifetime, the consequence of a rapid gas inflow into the nucleus. This gas is either processed in stars, or it is accreted by the strong gravitational field of a supermassive Black Hole (BH), or both. Large amounts of energy are then produced over periods of $10^7 - 10^8$ yrs, in the form of radiation, ultra-relativistic particles, and high-velocity plasma outflows. We also expect intense emissions of gravitational waves from black-hole merging and neutrinos of stellar origin.

This phenomenon has an important relationship and consequences for cosmology; it likely follows important structural and morphological transformations, like the generation of stellar bulges and of a nuclear supermassive BH, it is responsible for the background radiations in the gamma, x-ray, IR and radio bands, and it provides luminous "lighthouses" essential for the exploration of the remote universe.

Galaxy activity is also deeply related to high energy astrophysics and fundamental physics, to which it provides a cosmic laboratory of otherwise untestable physical phenomena, and provides important tests of special and general relativity. In particular, the putative supermassive BH may on one side explain the highly efficient energy production. On the other side it provides capabilities of large-scale spatial organization of e.m. fields and of the gaseous flow, producing ultra-energetic particles, relativistic bulk motions of plasma clouds, high-energy (e.g., TeV) photons, e+e- pairs, photon–photon interactions, etc.

Altogether, the phenomenon of *galaxy activity*, including both the starburst phase and the AGN proper (henceforth the term will be used in such general sense), concerns the most relevant evolutionary phases of galaxies. We summarize in this introduction some of the main points that will be discussed in the chapter.

2.2   Local Evidence on the Interplay Between the Stellar and Gravitational Origin of the AGN Activity

2.2.1   The Starburst-AGN Connection

Several multi-band observations have found evidence for a connection between nuclear activity and star formation in the host galaxy or in the circumnuclear region. This connection appears stronger for obscured AGNs. This issue is currently highly debated due to the possible implications for the cosmic co-evolution of AGN and galaxies. Starbursts and AGNs are probably only indirectly connected: non-axisymmetric morphologies both boost the star formation activity and drive gas into the nuclear region to obscure and feed the AGN (Maiolino 2002). The obscuring torus invoked by the unified model (§ 2.6.1) should also be the site of vigorous star formation: according to some recent models, the energy released by the circumnuclear starburst might yield to a thickening of the gaseous medium that would increase the obscuration of the AGN. In the context of the starburst–AGN connection, a hotly debated issue is the relative contribution of these two components to the bolometric luminosity of the cosmic sources (most of which is emitted in the IR). Various studies, based on indirect estimators of the primary radiation spectrum, have provided sometimes contradictory results.

2.2.2 The Missing Type-II AGN Population and Ultraluminous Infrared Galaxies

A general problem comes from the fact that type-II AGNs, those in which the broad emission lines and UV continuum are suppressed by circumnuclear dust, are difficult to identify (§ 2.6.1). Luminous and Ultra-luminous Infrared Galaxies (ULIRGs) are often invoked to hide this missing population, and deserve a dedicated discussion (§ 2.6.2). These systems are characterized by bolometric luminosities similar to QSOs (mostly irradiated in the IR) and are commonly identified with strongly interacting or merging galaxies. Star formation is certainly very strong. However, several ULIRGs also show evidence for the presence of a hidden AGN. These extreme properties make ULIRGs a promising population to study the connection among AGNs, star formation and morphology of the host system (Sects. 2.6.4 and 2.6.5). For these objects, in particular, the relative contribution of AGN and starbursts to the bolometric luminosity has been a debated issue in the past 10 years. X-ray observations have found hidden powerful AGNs in some ULIRGs, which are the best examples of type 2 QSOs. On the other hand, mid-IR spectroscopic observations have shown that the bolometric luminosities in most ULIRGs are powered by star formation. Surveys at other wavelengths (radio, optical, and near-IR) have occasionally given contradictory results. Forthcoming x-ray and IR facilities will provide new important information to tackle this issue.

Further interest in the ultraluminous IR galaxies comes from the consideration, developed in § 2.6.5, that this population may trace a critical event in galaxy evolution leading, after the exhaustion of the circumnuclear dusty envelope, to the optical quasar phase and later to the emergence of a passively evolving galaxy spheroid.

2.3 The Cosmic History of Galaxy Activity

Thanks to the continuous improvements of observational techniques, a progressively increasing effort is being dedicated to establish how galaxy activity has evolved, at which cosmic epochs it has originated, and to understand its relationship to the global process of structure formation.

Given their high intrinsic luminosity and point-like morphology, active galaxies, quasars, and radio galaxies (Sects. 2.7.1 and 2.7.2) have always been used as the ideal lighthouses to explore the universe at its observable edge (the most distant known objects are the $z > 6$ quasars from the SLOAN Survey, currently providing critical information on the re-ionization epoch). These properties of radio galaxies have offered during the fifties, in particular, the first incontrovertible evidence in favor of an evolutionary universe, soon supported by the discovery of the Cosmic microwave background. For a long time active galaxies and quasars have been used to attempt to constrain the geometrical parameters of the universe, until it was understood that evolution dominates over the geometrical effects.

The description of the cosmological evolution of active galaxies has been among the main motivations and has provided the most important results of various generations of space observatories (such as those in the x-ray band, UHURU, the Ariel series, the NASA HEAO-1 including the Einstein satellite, EXOSAT, ASCA, BeppoSAX, Chandra and XMM), and of the IR space observatories, IRAS, ISO, and soon SIRTF. Also the main optical and radio telescopes have long been used to select, identify, and analyze high-z active galaxies.

A strong increase of the volume emissivity (interpreted as either luminosity or density evolution, or both) has been found as a common property of most kinds of active galaxies selected in all wavebands (§ 2.7). The evolution rate may depend on the AGN type (e.g., the BLAZARs seem to evolve very little) and observing waveband. A similar evolutionary phenomenon seems to equally characterize active galaxies dominated by star formation, and selected in the optical (CFRS survey), in the IR (ISO), and mm (JCMT/SCUBA).

Following the HST discovery that galaxy activity is a phenomenon occurring in all massive galaxies (supermassive black holes, considered as the fossil records of an ancient quasar phase, have been identified in all galaxies with appreciable bulge components), interest is currently focused on understanding how galaxy activity is related to the formation of structures. The proportionality relation between BH mass and host mass found by HST indicate that the two processes are intimately connected. Other interesting constraints have come from studies of high-redshift radio galaxies (shown to trace the formation of high mass spheroidal galaxies), of radio-optical alignments, of the IR-mm continuum emissions from dusty ISMs observed with large mm telescopes and with ISO, and of the clustering analyses of optical quasars. The primary cause of galaxy activity is, according to the current understanding, the galaxy interactions and merging, which destroy the rotational equilibrium of the gas and produce a gas inflow into the galaxy core, with a consequent trigger of a starburst and with a fraction of the gas going to accrete onto the nuclear BH (§ 2.6). If this is the case, we also expect that the cosmic evolutionary timescales for starbursts and AGNs are the same: this is indeed what the most recent observations by Chandra and XMM in x-rays and ISO in the IR seem to indicate.

2.4 Constraints on the Cosmic Energy Budget: Active Galaxies, Background Radiation and Local Remnants

2.4.1 Obscured AGNs and Origin of the X-Ray Background

The presence of dust in the circumnuclear gas flow onto the central BH has important effects on the observability of radio-quiet AGNs (the most numerous, by a factor of ~ 10), depending on the fact that the line-of-sight to the nucleus might or might not intersect the accreting dusty structure: this unified model of AGNs had great success in explaining the physical and statistical properties of type 1 (unobscured) and 2 (obscured) AGNs.

The cosmic x-ray background (XRB), the first to be discovered in 1962, provides relevant constraints on the evolution of various AGN kinds. Though the XRB for a long time has eluded the various attempts to interpret it, eventually deeper and deeper x-ray surveys have found decisive proof that XRB is due to obscured AGNs at high redshifts, as an effect of photoelectric absorption and scattering by a circumnuclear medium on an intrinsic steep primary spectrum. The XRB then contains integrated information on the whole history of gravitational accretion. It has been suggested that this might be a proof of the fact that the validity of the unified AGN model holds also during past cosmic epochs. Recent observations by Chandra and XMM seem however to require some significant complication of this scheme. In any case, this interpretation of the XRB implies that it contains only a small fraction ($\sim 10\%$) of the energy emitted by AGNs, most of their emission being absorbed and re-emitted at longer wavelengths.

2.4.2 The Cosmic IR Background: Contributions by AGNs and Starburst Galaxies

It is expected that 90% of the x-ray emission by AGN was degraded in energy and emerges in the IR mainly between 1 and 50 microns as thermal dust emission. This adds to the intense UV emission of type 2 AGNs, also reprocessed by dust at long wavelengths. Hence we expect that the mid- and far-IR contain a dominant fraction of AGN emission, currently observable in the form of a cosmic background radiation peaking at 100 mm recently discovered by COBE (the CIRB). The CIRB has been partially resolved into sources at its long and short wavelength edges (by ISO in the mid-IR and JCMT and IRAM in the sub-mm): indeed detailed spectroscopic analyses have shown that a significant fraction of the CIRB is due to obscured AGNs. The bulk of it ($\sim$80–90%) is instead due to luminous dusty starbursts (LIRGs and ULIRGs) with a peak activity at $z \sim 1$.

As summarized in § 2.8.1, the two main diffuse radiative components, the CIRB and XRB, are the fossil records of past galaxy activity, the former mostly due to stellar activity, the latter to gravitational accretion. Given the similar evolutionary timescales, it is plausible that this ratio 8:2 between the two contributions may simply reflect the cosmic average of the stellar vs. AGN duty cycle. It will be very informative to compare these energetics with the local densities of the baryonic remnants in low-mass stars and nuclear supermassive black holes (§ 2.8.2).

2.5 Current Observational Programs and Future Perspectives

The relationship between galaxy evolution and quasar formation is a branch of astrophysics that has experienced a dramatic evolution during the last few decades, thanks in particular to new observational techniques, both from space and from ground, progressively widening the e.m. spectral domain accessible to observation. Given the enormous complexity and unpredictability of the involved phenomena, observational discoveries more than theoretical elaboration have led the development of this field.

We will eventually mention the most relevant current observational programs and sketch future perspectives to exploit forthcoming instrumentation, such as the Spitzer Space Telescope, ASTRO-F, SOFIA, the Herschel Space Observatory, ALMA, and JWST in the next 10 years (§ 2.8.5).

2.6 Current Issues on Local Active Galaxies

Before coming to a detailed investigation of evolutionary properties of active galaxies, it is essential to discuss a local census of the phenomenon.

2.6.1 A Census of Local Active Galaxies

Active galaxies in the local universe are characterized by an enormous variety of phenomena, which however involve a minority fraction of normal galaxies at the present time.

The fundamental physical differentiation concerns objects dominated by stellar activity and those whose bolometric luminosity is mostly due to gravitational accretion in a compact nuclear source. Among the latter, which we will refer to in the following as Active Galactic Nuclei (AGN), two main classes of objects have been identified from optical spectroscopic

observations: the type-I AGNs, whose optical spectra show broad permit-
ted (as well as narrow forbidden) lines, and type-II AGNs showing high
ionization narrow emission lines (Figure 2.1).

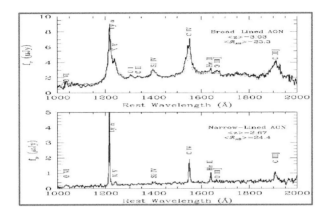

Figure 2.1. Optical spectra of the two main AGN kinds: that characteristic
of broad-lined type-I AGNs (top), and of narrow emission lined type-II AGNs
(bottom).

Among the AGN-dominated, two main sub-classes are also defined
(Figure 2.2): the radio-bright and the radio-quiet objects. A very rough
(local) statistics based on optical, IR, and radio observations indicates that:
∼10% of local normal galaxies are starbursts; ∼10% of the starbursts (1% of
galaxies) are type-I AGNs; and finally ∼10% of local AGNs are radio-loud
(0.1% of the local normal galaxies).

Most AGN classes show evidence of cosmological evolution (but with
a remarkable difference for the BLAZAR class).

The basic differentiation between type-I and -II AGNs is currently
interpreted as induced by the presence of obscuring material in the nuclear environment, likely associated with the gas flow fueling the AGN.
This concept has been formalized as the AGN unification model in which
this material acquires a torus shape by the presence of residual angular
momentum in the accreting gas, such that a narrow-line type-II object is
interpreted as one in which the line-of-sight to an observer does intersect
the torus and the emission by the inner broad-line region (within 0.1 pc of
the nucleus) is obscured, while for a type-I object the observer line-of-sight
keeps external to the torus and the inner broad-line region becomes visible
(see illustration in Figure 2.3).

The relevance of the orientation effect has been proven by a variety
of observational tests, including: (a) the linear polarization of residual
scattered broad-line emission in type-II AGNs (e.g., Antonucci & Miller

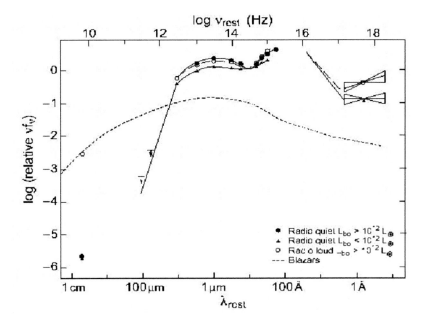

Figure 2.2. Broad band Spectral Energy Distributions (SEDs) of radio-quiet, radio-loud AGNs and Blazars.

1985), (b) the ionization cones in the narrow-line components, and, (c) the missing continuum flux in type-II AGNs inferred from spectroscopic studies, all emphasizing the anisotropic directionality of the ionizing nuclear continuum, exactly as predicted by the model.

Eventually, the unification scheme was successfully used as a paradigm to provide a solution to the old problem of the origin of the X-Ray Background (XRB), based on the assumption of a random sky distribution of AGN inclinations. Contributions to the XRB of 20% by type-I and 80% by type-II to the XRB have been claimed to be roughly consistent with the expected sky-covering fraction of the nucleus by the surrounding torus (Madau et al. 1994; Comastri et al. 1995; Gilli et al. 2001; see Sects. 2.7.3 and 2.8.1).

However, in spite of the great success of such a unification scheme (involving not only optical AGNs, but also radio galaxies and AGNs selected in the IR and x-rays), many details are still missing 20 years after its proposition. First, the statistics about the incidence of the obscured relative to unobscured AGNs is still unclear and several conflicting results have been reported. While Maiolino & Rieke (1995) infer a canonical 4:1 ratio of the two classes (consistent with a torus opening angle of 75°), based

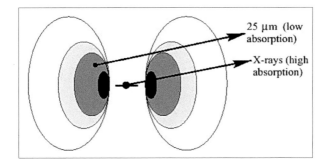

Figure 2.3. Scheme of a cut through a toroidal structure surrounding the nuclear ionizing continuum source. A polar view and an equatorial one would produce a type-I and a type-II AGN out of the same source, respectively, simply due to orientation. The scheme also illustrates potential differences that may occur when observing an AGN in the far-IR or in x-rays: if the dust particles are destroyed in the inner dark region, then the optical depth inferred from x-rays and in the optical-IR may be quite different and larger in x-rays (taken from Risaliti et al. 2000).

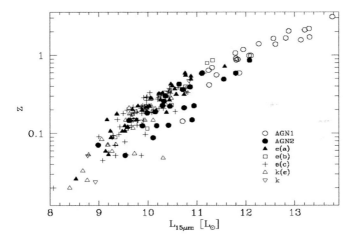

Figure 2.4. Luminosity at 15 μm of the ELAIS-S1 spectroscopically identified sources as a function of redshift (from La Franca et al. 2004). The various source categories are indicated in the figure label.

on [OIII]5007 observations of a sample of 91 Seyfert galaxies from the optically selected RSA catalogue, most of the other analyses have emphasized a lack of the type-II class. A first such indication has come from the review

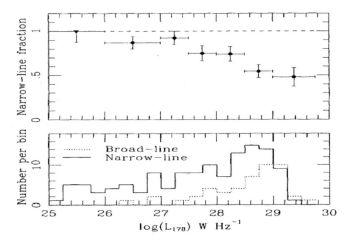

Figure 2.5. Dependence of properties of radio-selected AGNs on radio luminosity. In the top panel the narrow-line fraction appears to be sensibly dependent on the luminosity. The number distribution of narrow- and broad-line AGNs is reported in the lower panel. (From Lawrence 1991.)

by Lawrence (1991) and from the Seyfert-1 and Seyfert-2 statistics in the CfA galaxy redshift survey, but the most remarkable constraints have come from far-IR selected samples.

While the optical selection of AGNs is subject to the selection effects by the obscuring medium, a far-IR selection is expected to provide a largely unbiased view based on bolometric luminosity (Spinoglio & Malkan 1989). Rush, Malkan & Spinoglio (1993) find evidence for a roughly similar number of type-I and type-II AGNs in a rich flux-limited ($S_{12} > 400$ mJy) local sample selected from the 12 μm IRAS survey. Remarkably, these statistics were completely confirmed at much fainter flux levels by the European Large Area ISO Survey (ELAIS) using the Infrared Space Observatory (ISO): in a sample of 406 15 μm sources with $S_{15} > 0.5$ mJy, 332 ($\sim$82%) sources were optically identified by La Franca et al. (2004), of which 25% are AGNs. Among these AGNs, 25 are type-I (line FWHM >1200 Km/s) and 23 type-II, hence again very close to the one-to-one ratio (Figure 2.4).

The obvious difficulty is to disentangle the latter from starburst galaxies. La Franca et al. use classical diagnostic line ratios based on [OIII]λ5007/ Hβ and [NII]λ6583/Hα. Of the starburst galaxy population, which includes $\sim$50% of the sample, 37 have spectra of dust-enshrouded starbursts [spectral class e(a)], 11 [e(b)], 68 are normal spirals [e(c)], 32 have passive elliptical-like spectra with tiny emission lines [k(e)] and only 3 are spectra of old stellar populations [k].

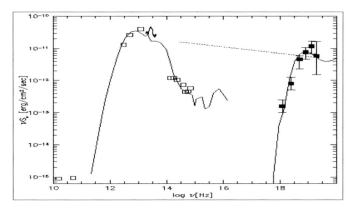

Figure 2.6. Observed SED of the type-II quasar IRAS 09104+4109 over 10 decades in photon energy. The IR and hard x-ray spectra are fitted with the same self-consistent AGN model (Franceschini et al. 2000), including a dusty torus responsible for the IR emission with the associated gas photoelectrically absorbing the x-ray photons. The dotted line shows a schematic fit to the average x-ray to IR spectrum of type-I AGNs (Barcons et al. 1995). X-ray data points are from BeppoSAX observations.

Another difficulty for the AGN Unification Scheme is illustrated in Figure 2.5, resulting from an analysis of a radio-selected sample by Lawrence (1991) and showing a strong dependence of the narrow-line fraction on luminosity: the most luminous objects tend to be type-I (typically blue, luminous, broad-emission line objects like the optical quasars), while very luminous type-II AGNs seem to be missing. This seems to keep true also for hard x-ray selected AGN samples down to moderate x-ray fluxes.

2.6.2   The Missing Type-II AGN population

Altogether, the type-II AGN population is posing a challenge to our understanding: there is a global lack of this class of sources compared with the ratio of ~ 4–5 to 1 expected on the basis of physical models of the obscuring torus. In addition, there seems to be a luminosity-dependent effect in this shortage, such that at the highest luminosities the type-II class completely disappears.

Perhaps the first luminous type-II quasar was indeed discovered in the infrared luminous galaxy IRAS 09104+4109, the central luminous galaxy of a cluster at $z = 0.44$: Beppo-SAX observations by Franceschini et al. (2000) have shown this to host a hard x-ray object with unabsorbed broadband 2–100 keV emission with bolometric L $\simeq 2.5 \ 10^{46}$ erg/s, well within the range of quasar luminosities (Figure 2.6). The object shows very high gas column density of $N_H \geq 5 \ 10^{24}$ cm^{-2}. The association of this quasar with a huge cooling-flow of ~ 1000 M$_\odot$ in the cluster, also indicated by the x-ray data, might suggest that such condition of extremely fast mass accumulation could favor the survival of a thick obscuring envelope, which would otherwise be quickly destroyed by the very luminous central source.

The discovery of IRAS 09104+4109, and a few more, all found with deep x-ray observations (e.g., Mkn 231, $L_x \geq 10^{44}$ erg/s, Braito et al. 2004), illustrates the exceptionality of these objects. For some time this rarity of luminous type-II quasars, and AGNs in general, was ascribed to the heavy selection operating in the optical, although not even the IR selection apparently revealed many of them at first sight.

At closer inspection, however, the all-sky far-IR view of the local universe by the IRAS mission (1984) may offer a way to solve the problem. Probably the major discovery of the mission was the identification of luminous, ultraluminous, and hyperluminous phases in galaxies (respectively LIRGs, with $L > 10^{11}L_\odot$; ULIRGs, $L > 10^{12}L_\odot$, HyLIRGs, $L > 10^{13}L_\odot$), among which the elusive type-II AGNs and quasars may hide. This is supported by the fact that the few above-mentioned cases all belong to these IR categories. Together with the optical quasars, these are the most luminous objects in the local universe.

After many years from their discovery, the nature of the LIRGs and ULIRGs still remains rather enigmatic: large gas and dust column densities in the galaxy cores, responsible for the IR emission, prevent a direct observation of the primary energy source and the IR spectral shapes are highly degenerate with respect to the illuminating spectrum. Both starburst and AGN activity may be responsible for the observed luminosities (Genzel et al. 1998; Veilleux, Kim & Sanders 1999; Risaliti et al. 2000; Franceschini et al. 2003). Conflicting evidence has been reported about the relative contributions of the two energy sources, stellar and gravitational, in ULIRGs.

2.6.3   IR Spectroscopy of Obscured Sources

IR/sub-mm spectroscopy offers unique opportunities to probe the physical conditions ($n[atoms]$, P, T, extinction, ionization state) in obscured active galaxies. The widespread presence of dust makes optical-UV-NIR line diagnostics completely unreliable; and most of the line emission by molecular clouds, where stars form, is extinguished and does not appear in the optical; several fundamental cooling lines of gas and lines from an extremely wide range of ionization states are observable in the IR.

In the *cold molecular gas*, the most abundant molecule (H$_2$) is not easily observed directly. It is seen in absorption in UV, or in the NIR roto-vibrational transitions at 2.121 and 2.247 μm. Only with mid-IR spectroscopy by ISO was it possible to observe the fundamental rotational lines at 17 μm (S[1]), 28.2 μm (S[0]), and 12.3 μm (S[2]) in NGC6946, Arp220, Circinus, NGC3256, NGC4038/39.

Table 2.1 The most important IR fine-structure lines used to characterize the nature of obscured active galaxies.

Species	Excitation potential	λ (μm)	n_{crit} cm^{-3}	F/F[CII][a]
OI	-	63.18	$5\ 10^5$	1.4
OI	-	145.5	$5\ 10^5$	0.06
FeII	7.87	25.99	$2\ 10^6$	
SiII	8.15	34.81	$3\ 10^5$	2.6
CII	11.26	157.7	$3\ 10^2$	1
NII	14.53	121.9	$3\ 10^2$	0.37
NII	14.53	203.5	$5\ 10^1$	0.11
ArII	15.76	6.99	$2\ 10^5$	0.11
NeII	21.56	12.81	$5\ 10^5$	2.1
SIII	23.33	18.71	$2\ 10^4$	0.68
SIII	23.33	33.48	$2\ 10^3$	1.1
ArIII	27.63	8.99	$3\ 10^5$	0.23
NIII	29.60	57.32	$3\ 10^3$	0.31
OIII	35.12	51.82	$5\ 10^2$	0.74
OIII	35.12	88.36	$4\ 10^3$	0.66
NeIII	40.96	15.55	$3\ 10^5$	0.16
OIV	54.93	25.87	10^4	–

[a] Line intensity compared with the observed [CII]158 μm for the prototypical starburst M82, when available, or predicted by Spinoglio & Malkan (1992) from a model reproducing the physical conditions in M82.

Because of the difficulty of a direct measure, the amount of molecular gas (H_2) is often inferred from easier measurement of CO emission lines, assuming an H_2/CO conversion. CO rotational transitions allow excellent probes of cold ISM in galaxies: the CO brightness temperature ($\propto$ line intensity) is almost independent on z at $z = 1$ to 5, due to the additional $(1+z)^2$ factor with respect to the usual scaling with the luminosity distance (Scoville et al. 1996). CO line measurements have been performed for all IRAS sources in the Bright Galaxy Sample; the majority have been detected with single-dish telescopes. Typically 50% or more of the mass in molecular gas is found within the inner kpc from the nucleus, substantially contributing to the total dynamical mass ($> 50\%$ of M_{dyn}).

The *diffuse neutral ISM* is commonly traced by the HI 21 cm line with radio observations. HI cooling, which is essential to achieve temperatures and densities needed to trigger SF, depends mainly on emission by the 158 μm [CII] line, the 21 cm line, and the 63 μm [OI] line. The 158 μm [CII] line is a major coolant for the diffuse neutral gas and a fundamental cooling channel for the photo-dissociation regions (PDRs), the dense phase interfacing cold molecular clouds, where stars form, with the HII or HI lower-density gas. Carbon is the most abundant element with ionization potential (11.3 eV) below the H limit (13.6 eV): CII atoms are then present in massive amounts in neutral atomic clouds. The two levels in the ground state of CII responsible for the $\lambda = 158$ μm transition correspond to a relatively low critical density $n_{crit} \simeq 300$ cm^{-3} (the density at which collisional excitation balances radiative de-excitation): CII is excited by electrons and protons and cools down by emitting a FIR photon. The [OI]145 μm and 63 μm lines are also coolants, though less efficient.

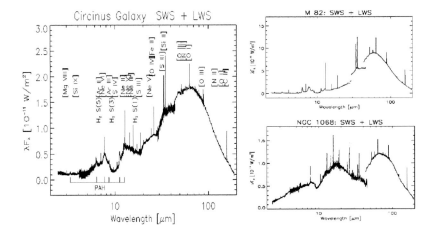

Figure 2.7. ISO spectra of prototypical IR-bright active galaxies: the starburst galaxy M82, the close-by type-II AGN NGC 1068, and the heavy extinguished AGN, the *Circinus* galaxy. A number of coronal emission lines, PAH emission bundles at 6–10 μm and many other spectral features are visible. This figure (taken from Genzel & Cesarsky 2000) illustrates the enormous amount of information made available for the first time by the ISO observatory. Spectra from the LWS and SWS long-λ and short-λ ISO spectrographs are visible.

A number of lines from ionized species, covering an extremely wide range of ionization conditions, are observable in the far-IR. For a detailed physical investigation, line ratios sensitive to either gas temperature T or density n are used. To estimate electron density n one can use the strong dependence of the fine-structure line intensities for doublets of the same ion on n: one example is the [OIII] lines at 5007 Å, 52 μm and 88 μm. Similarly one can estimate T and the shape of the ionizing continuum. Particularly relevant to test the spectral shape of the ionizing continuum are the *fine-structure lines from photo-ionized gas*, which allow us to discriminate spectra of stellar and quasar origin. Low-ionization transitions typically strong in starbursts are [OIII]52 and 88, [SiII]34, [NeII]12.8, [NeIII]15.6, [SIII]18.7 and 33.4, while higher ionization lines in AGNs are [OIV]25.9 and [NeV]24. Table 2.1 reports a few of the most important IR ionic lines.

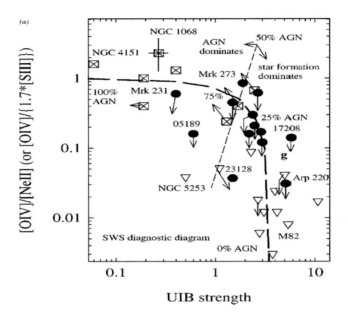

Figure 2.8. ISO IR diagnostic diagram (from Genzel et al. 1998). The vertical axis measures the flux ratio of high excitation to low excitation mid-IR emission lines, and the horizontal axis measures the strength (i.e., feature to continuum ratio) of the 7.7 μm PAH feature. The latter is the continuum-subtracted 7.7 μm peak line divided by the underlying continuum, where the continuum is computed from a linear interpolation of the flux densities measured at 5.9 and 11.2 μm. AGN templates are marked as rectangles with crosses, starburst templates as open triangles, and UILRGs as filled circles. The long dashes are a mixing curve from 0% to 100% AGN. The short-dash line effectively discriminates the AGN- from starburst-dominated galaxies.

Clearly, IR spectroscopy is essential for studies of galaxy activity, but requires a continuous coverage of the IR spectrum which has been made possible by the Infrared Space Observatory (see Figure 2.7), allowing us to probe for the first time the inner optically thick nuclei.

In the spectra of Figure 2.7 we can recognize low-excitation atomic/ionic fine-structure lines ([FeII], [SiII], [OI], [CII]) sampling photodissociation regions, the ionic lines with excitation potential <50 eV ([ArIII], [NeII], [NeIII], [SIII], [OIII], [NII]) sampling HII regions photoionized by OB stars, and lines from species with excitation potentials up to 300 eV (e.g., [OIV], [NeV], [NeVI], [SiIX]) probing highly ionized coronal gas and requiring very hard radiation fields (as in AGNs). These line ratios give information about the physical characteristics of the emitting gas. The extinction corrections are small at these long wavelengths $[A(\lambda)/A(V) \simeq 0.1$ to 0.01 in the 2 to 40 μm region].

Using diagnostics based on these coronal line intensities and PAH line-to-continuum ratios, Lutz et al. (1996), Genzel et al. (1998), and Rigopoulou et al. (1999) have performed a systematic investigation of the physical conditions in the heavily extinguished nuclei of LIRGs and ULIRGs. As an example, the right-end panels of Figure 2.7 show the IR spectra of a prototype starburst (M82) and type-II AGN (NGC 1068): the two are remarkably different, the first one showing prominent Polycyclic Aromatic Hydrocarbon (PAH) emission bands between 6 and 10 μm, low-excitation fine-structure lines, and very faint $\lambda < 10\,mu$m continuum. NGC 1068 shows instead the opposite behavior: strong continuum, no PAHs, and high excitation lines.

Figure 2.8 shows results of the diagnostic study by Genzel et al. (1998) using a combination of the 25.9 μm[OIV] to 12.8 μm[NeII] line flux ratio (or [OIV]/[SIII], or [NeV]/[NeII]) on one axis, and of the PAH strength on the other axis. This diagram clearly separates known star-forming galaxies from AGNs. The diagram is insensitive to dust extinction if the numerator and denominator in both ratios are affected by a similar amount of extinction. A complication here may arise from the fact that low-level [OIV] emission may be due to ionizing shocks in galactic super winds (Lutz et al. 1998b).

Laurent et al. (2000) proposed another diagnostic diagram based on the 15 μm to 6 μm continuum ratio versus PAH to 6 μm continuum ratio, again fairly well effective in distinguishing AGNs from starbursts.

The result of all these analyses was that in the majority of LIRGs and ULIRGs high-ionization lines are missing (filled dots with upper limits in Figure 2.8); hence they are mostly powered by star formation.

It is interesting to note that, while ISO allowed us to investigate spectroscopically nearby IR active galaxies, future missions (the Spitzer Telescope, the Herschel Space Observatory, JWST) will make possible similar studies for galaxies at any redshifts.

2.6.4   X-Ray Spectroscopy of ULIRGs

The results of IR spectroscopy, however, may not be conclusive. A significant fraction of ULIRGs exhibit nuclear optical emission line spectra characteristic of Seyfert galaxies (Veilleux, Sanders & Kim 1997), some contain compact central radio sources (Smith, Lonsdale & Lonsdale 1998) indicative of an AGN. Soifer et al. (2000) have found that several ULIRGs

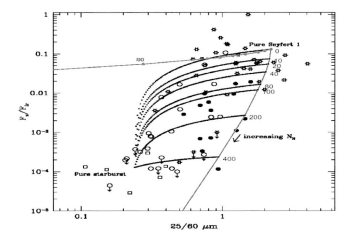

Figure 2.9. Comparison of far-IR colors and the hard x-ray to far-IR flux ratios in a sample of LIRGs and ULIRGs (filled dots and open circles). The lines on the right represents AGN-dominated sources with varying absorbing x-ray column density (in units of 10^{22} cm^{-2}). The dotted curves are mixing curves of AGN and starburst emission, with the latter increasing towards the left. The line on the top represents a model where the same material absorbs both the IR and the x-ray emission of an AGN, with galactic dust-to-gas ratio. (Taken from Risaliti et al. 2000.)

show very compact (100–300 pc or unresolved) structures dominating the mid-IR flux, a fact they interpret as favoring AGN-dominated emission.

A critical diagnostics for the presence of an AGN is provided by hard x-ray observations, only moderately affected by photoelectric absorption (an example is reported in Figure 2.6). Risaliti et al. (2000), from a comparison of the far-IR colors and the hard x-ray to far-IR flux ratios of a sample of LIRGs and ULIRGs, find them to be consistent with combined AGN and starburst emissions in the majority of the sources, both emissions being photoelectrically absorbed and extinguished: at least 50% of these sources appear to host a (weak) AGN.

A deep spectroscopic survey with XMM–Newton observations of a complete ULIRG sample is reported by Franceschini et al. (2003). Although all objects are detected in hard x-rays, their x-ray luminosities (in the 2–10 keV interval) are rather low, typically $L_{2-10} \simeq 10^{41-43}$ erg/s, whereas their bolometric IR luminosities are $> 10^{45}$ erg/s. In all sources there is evidence for thermal emission from hot plasma with a rather constant temperature kT $\simeq$ 0.7 keV, dominating the x-ray spectra below 1 keV, and likely associated with a nuclear or circumnuclear starburst: this

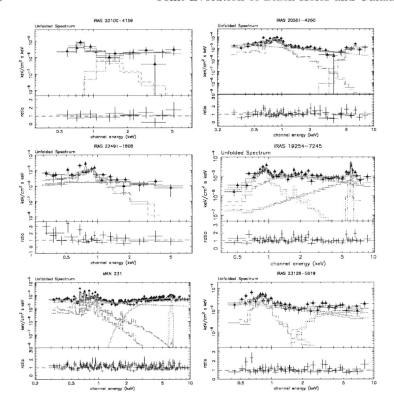

Figure 2.10. XMM–Newton spectra (in units of keV/cm^2/s/keV) for the sources IRAS20100-4156, IRAS 20551-4250, IRAS 22491-1808, IRAS 19254-7245, Mkn 231, and IRAS 23128-5919, from top left to bottom right. Panels below each spectrum show the ratios of data to the best-fit model. All spectra are fitted with a hot thermal plasma emission at low energies (in the case of MKN 231 there are 2 such components), plus a photoelectrically absorbed hard x-ray component. (From Franceschini et al. 2003.)

is well visible in the x-ray spectra of Figure 2.10 as an excess dominating the spectra below 1 keV. At higher energies, typically between 1 and 4 keV rest-frame, there is evidence for emissions by high-mass x-ray binaries, hence proportional to the number of newly formed stars and well correlated with the far-IR emission. Finally, in 50% of the sources there is evidence for photoelectrically absorbed hard x-ray components indicative of deeply buried AGNs (see Figure 2.12).

The composite nature of ULIRGs is then confirmed, with concomitant starburst and absorbed AGN emissions. At variance with the results of

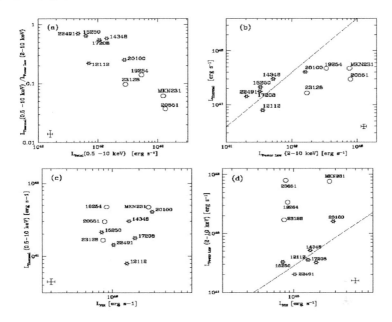

Figure 2.11. Luminosity–luminosity plots of various emission components identified in the XMM–Newton spectra of the ULIRG sample (from Franceschini et al. 2003). Starred symbols refer to sources that we classify as dominated by starbursts, open circles to AGN-dominated sources. Panel (a): luminosity ratio between the thermal and the power-law (2–10 keV) component after correction for intrinsic absorption, versus the bolometric x-ray luminosity measured in the total 0.5–10 keV band. Panel (b): luminosity–luminosity plot of the thermal versus PL components. Panels (c) and (d): luminosities of the thermal and PL components against the bolometric far-IR luminosities LFIR.

the mid-IR spectroscopy, these x-ray results often show the presence of buried AGNs (not otherwise evident in the optical or IR) in ultraluminous sources whose bolometric (mostly far-IR) luminosity is dominated by a circumnuclear starburst.

2.6.5   An Evolutionary Sequence for Galaxy Activity

2.6.5.1   Galaxy Merging as the Origin of Activity

In normal inactive spirals the rate of star formation is enhanced in spiral arms in correspondence with density waves compressing the gas. This favors the growth and collapse of moderately massive molecular clouds and eventually the formation of stars. This process is, however, slow and

Name (1)	Optical (2)	Mid-IR (3)	X-ray (4)
IRAS 12112+0305	L	SB	SB
Mkn 231	Sey1	AGN	AGN
IRAS 14348-1447	L	SB	SB
IRAS 15250+3609	L	SB	SB
IRAS 17208-0014	HII	SB	SB
IRAS 19254-7245	Sey2	SB	AGN
IRAS 20100-4156	HII	SB	SB/AGN
IRAS 20551-4250	HII	SB	AGN
IRAS 22491-1808	HII	SB	SB
IRAS 23128-5919	HII	SB	SB/AGN

Note. – Col.(1) Object name. Col.(2) Optical Classification (Lutz et al. 1999; Veilleux et al. 1999). Col.(3) Mid Infrared classification based on ISO spectroscopy (Genzel et al. 1998). Col.(4) X-ray Classification.

Figure 2.12. Classification of ULIRGs from Franceschini et al. (2003) based on optical, mid-IR spectroscopy, and hard x-ray spectroscopy.

inefficient in making stars (also because of the feedback reaction to gas compression produced by young stars): very long (several Gyrs) timescales are needed to convert the ISM into a significant stellar component.

On the contrary, because of the extremely high compression of molecular gas inferred from CO observations in the central regions of luminous starburst galaxies, SF can proceed there much more efficiently. Both on theoretical and observational grounds, it is now well established that the trigger of a powerful nuclear starburst is due to a galaxy–galaxy interaction or merger, driving a sustained inflow of gas into the nuclear region. This gas has a completely different behavior with respect to stars: it is extremely dissipative (gas clouds have a much larger cross-section and during cloud collisions gas efficiently radiates thermal energy generated by shocks). A strong dynamical interaction breaks the rotational symmetry and centrifugal support for gas, induces violent tidal forces producing very extended tails and bridges, and triggers central bars, which produce shocks in the leading front, and efficiently disperses the ordered motions and the gas angular momentum. The gas is then efficiently compressed in the nuclear region and allowed to form stars. High-resolution imaging of ULIRGs (e.g., Figure 2.13) provides strong support to these ideas.

These concepts are confirmed by numerical simulations of galaxy encounters, pioneered by Toomre (1977). Much more physically motivated and numerically detailed elaborations have more recently been published by Barnes and Hernquist (1992), who model the dynamics of the encounters between 2 gas-rich spirals including disk/halo components, using a combined N-body and gas-dynamical code based on the Smooth Particle

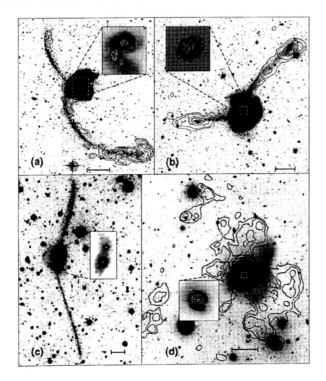

Figure 2.13. Well-studied IR-luminous mergers: (a) The Antennae (Arp 244); (b) (Arp 226); (c) IRAS 192547245 (Super-Antennae); (d) (Arp 220). All are R-band images. Inserts show expanded views in the K-band of the nuclear regions. The scale bars correspond to 20 kpc. (Taken from Sanders and Mirabel 1996).

Hydrodynamics (SPH). Violent tidal forces act on the disk producing extended tails and triggering central bars, which sweep the inner half of each disk and concentrate the gas into a single giant cloud. The final half-mass radii of gas are much less than those of stars: for an M^* galaxy of 10^{11} $M_\odot$, $\sim 10^9$ $M_\odot$ of gas are compressed within 100–200 pc, with a density of 10^3 $M_\odot/pc^3$ (Barnes & Hernquist 1996).

Various other simulations confirm these findings: SPH/N-body codes show in particular that the dynamical interaction in a merger has effects not only on the gas component, but also on the stellar one, where the stars re-distribute following the merging and violent relaxation of the potential.

2.6.5.2   From Starburst, to ULIRG, to Quasar

An evolutionary connection relating ULIRGs and quasars has been suggested by Sanders et al. (1988) based on an exhaustive morphological and spectroscopic analysis of a complete IRAS-selected ULIRG sample. The optical spectra showed a mixture of starburst and AGN energy sources, both appearing fueled by enormous reservoirs of molecular gas. It is proposed that ULIRGs represent an initial dust-enshrouded phase of quasars. Once the nuclei release their dusty and obscuring envelope, then the AGN start to dominate the fading starburst, finally becoming an optically selected quasar.

This evolutionary connection was essentially suggested by the analysis of the IR Spectral Energy Distributions (SED): Sanders et al. showed that a significant fraction of ULIRGs with warm infrared colors (with an excess at 25 μm over 60 μm) have SEDs with mid-infrared emission over an order of magnitude stronger than cooler ULIRGs and starburst galaxies in general (with IR spectra peaking at 100 μm). These warm galaxies comprise a wide variety of classes of extragalactic objects, including powerful radio galaxies, optically selected QSOs. Their suggestion was that the IR spectra of active galaxies evolve from the cool starburst IR spectral phase, to the warmer spectra of the most active ULIRGs, to the optically bright normal QSO phase.

This connection is further strengthened by IRAS data for QSOs (§ 2.7.1.1 and Figure 2.20 below), which show that the mean SED of optically selected QSOs is dominated by thermal emission from an infrared/submillimeter bump ($\sim 150\,\mu$m) in addition to the big blue bump (0.051 μm); the former is typically 30% as strong as the latter and is presumably thermal emission from dust in an extended circumnuclear disk surrounding the active nucleus. This small mid- and far-IR component in optical quasars is interpreted as the remnant of much more IR-prominent previous phases in the activity evolutionary sequence.

2.6.5.3   Origin of Elliptical Galaxies and Galaxy Spheroids

Important information on the mass, structure, and kinematics of the molecular gas in LIRGs and ULIRGs has been obtained from observations of millimeter-wave CO emission for large samples of IRAS galaxies and from analyses of the H_2 NIR vibrational lines (Sanders and Mirabel 1996). An important discovery has been that all appear to be extremely rich in molecular gas. These emissions are more peaked than stars, and located in between the merging nuclei, consistent with the fact that gas dissipates and concentrates more rapidly, while stars are expected to relax violently and follow on a longer timescale the new gravitational potential ensuing the merger.

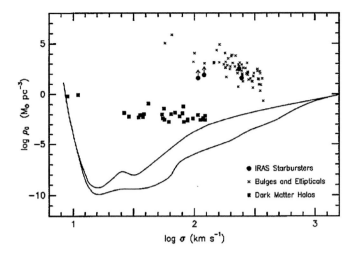

Figure 2.14. The "cooling diagram" (density versus temperature) for the cores of bulges, elliptical galaxies, and for the dark-matter halos characteristic of spirals. ULIRGs appear as filled circles close to the bulge region, and their velocity dispersion σ is measured as the width of the CO line. (Taken from Kormendy and Sanders 1992.)

A detailed study of Arp 220 by van der Werf (1996) has shown that most of the H_2 line emission, corresponding to $\sim 2 \; 10^{10} \; M_\odot$ of molecular gas, comes from a region of 460 pc diameter; the gas mass is shocked at a rate of $\sim 40 \; M_\odot/\text{yr}$. Compared with the bolometric luminosity of Arp 200, this requires a IMF during this bursting phase strongly at variance with respect to the Salpeter's one ($dN/dM \propto M^{-2.35}$) and either cut at $M_{\min} \gg 0.1 \; M_\odot$ or displaying a much flatter shape.

The basic observables to check this hypothesis are the "Kormendy diagram" central surface brightness versus effective radius, the stellar velocity dispersion versus core density. The two are "manifestations" of the fundamental plane of E/S0s. As pointed out for the first time by Kormendy and Sanders (1992; see Figure 2.14), the typical gas densities found by interferometric imaging of CO emission in ultra-luminous IR galaxies turn out to be very close to the high values of stellar densities in the cores of E/S0 galaxies. This is suggestive of the fact that the ULIRG phenomenon bears some relationships with the long-standing problem of the origin of early type galaxies and spheroids.

Originally suggested by Toomre (1977), the concept that E/S0 could form in mergers of disk galaxies faced the problem of explaining the dramatic difference in phase-space densities between the cores of E/S0 and

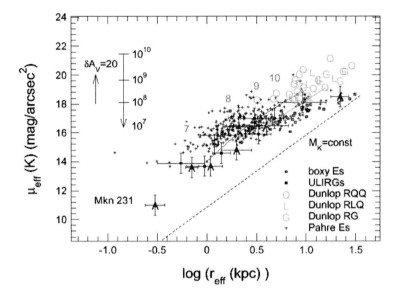

Figure 2.15. The Kormendy relation for ULIRGs (filled squares) compared with boxy ellipticals (open squares) and the host galaxies of radio-loud and radio-quiet quasars. (From Tacconi et al. 2002.)

those of spirals. Some efficient dissipation is required during the merger, which can be provided by the gas. Indeed, the CO line observations in ULIRGs, also combined with those of the stellar nuclear velocity dispersions and effective radii, show them to share the same region of the "cooling diagram" occupied by ellipticals (Figure 2.14).

Also proven by simulations, after the formation of massive nuclear star clusters from the amount of gas (up to 10^{10} $M_\odot$) collapsed in the inner Kpc, part of the stellar recycled gas has low momentum and further contracts into the dynamical center, eventually producing a supermassive black hole with the associated AGN or quasar activity (Norman & Scoville 1988; Sanders et al. 1988).

In conclusion, the line of thought started with the paper by Sanders et al. seems indeed to provide a very logical and convincing interpretative scheme of an important variety of phenomena concerning the activity phases in galaxies. Such a scheme foresees, as a consequence of a merger or strong interaction of gas-rich galaxies, an evolutionary sequence starting with an enhanced phase of star-formation leading to either a LIRG or ULIRG, at the end of which an optical QSO will shine. This process is finally expected to leave a spheroidal galaxy as a remnant.

One additional consideration concerns a particular set of AGNs, the radio galaxies. Simple arguments about the directionality of the radio emission and the usual lack of the big blue bump suggest that the supermassive black hole has to be maximally rotating and with minimal ongoing disk accretion, both conditions that may naturally happen at the end of the optical quasar phase (when the disk is fading away and the black hole has acquired rotational energy from the disk). Radio-loud AGNs might then constitute a link between the quasar phase and the passively evolving ensuing spheroid, a conclusion strongly supported by the association of radio galaxies with early-type galaxies.

2.6.5.4 Tests of the Active Galaxy Standard Evolutionary Model

A long-standing observational effort has been dedicated to test the Standard Evolutionary Model for Active Galaxies. We summarize here a few of the most important results.

In support of the idea that ellipticals may form through merging processes traced by the LIRG and ULIRG population there is the evidence coming from high-resolution near-IR imaging that the starlight distribution in ULIRGs and hyperluminous IR galaxies follows a de Vaucouleurs $r^{-1/4}$ law typical of E/S0 (Wright et al. 1990).

High-resolution near-infrared spectroscopy of ULIRG mergers is reported in Genzel et al. (2001) based on VLT and Keck observations. They confirm that ULIRG mergers are ellipticals-in-formation, in which random motions dominate their stellar dynamics, but significant rotation is also commonly detected. Gas and stellar dynamics appear decoupled in most systems, consistent with their very different reactions to the violent dynamical disturbance.

ULIRGs fall remarkably on or near the fundamental plane of spheroidal stellar systems, as shown in Figure 2.15. The ULIRG velocity dispersion distribution and their location in the fundamental plane closely resemble those of intermediate mass (L_*) elliptical galaxies with moderate rotation ("disky" ellipticals), but they typically do not resemble giant ellipticals with large cores and little rotation. So local ULIRGs, dissipative mergers of gas-rich disk galaxies, are related with disky ellipticals with compact cores, while giant ellipticals with large cores have a different formation history, and are likely confined to higher redshifts.

In a follow-up study by Tacconi et al. (2002), a comparison has been made between ULIRGs containing or not QSO-like active galactic nuclei. The main properties of the host galaxies of AGN-dominated and starburst-dominated ULIRGs are similar, fully supporting an evolutionary relation between the two. Both of them, however, have smaller effective radii and velocity dispersions than the local QSO/radio galaxy population. Their host masses and inferred black hole masses are correspondingly smaller, more similar to those of local Seyfert galaxies.

2.6.6   Local Active Galaxies, Conclusions

Galaxy activity involves extremely complex physics and astrophysics, with a variety of manifestations. In many cases AGN and starburst activity appear to happen concomitantly in the same objects, especially evident in LIRGs and ULIRGs.

An evolutionary scheme, that we named the Standard Evolutionary Model of active galaxies, seems very effective in explaining a variety of AGN phenomena in a time-sequential process triggered by galaxy interactions and mergers.

Formation of low-mass spheroidal galaxies and low-mass (Seyfert-like) black holes is still ongoing in the local universe.

2.7   Faint Active Galaxies at High Redshifts: Overview

Recent improvements in our observational capabilities have made possible a direct inspection of galaxy activity back at the cosmic time when the phenomenon was at its maximum. Since the early identification of optical quasars and radio galaxies in the '60s, and later from extensive spectroscopic surveys of galaxies in the '90s, it has become evident that essentially all populations of cosmic sources have been subject to fast cosmological evolution. The source volume emissivities measured in all wavebands and for all source classes turned out to be quite higher at cosmic times corresponding to redshift $z = 1$ to 2 than they are locally. Consequently, the epochs of major stellar formation and black hole accretion have to be dated back in the past.

It will be useful to review some results in this chapter, with emphasis on the most recent ones. We will mention in particular: high-z optical quasar searches and the evidence for a very early metal enrichment in their environments; high-redshift radio galaxies and the evidence that these trace the formation of the most massive spheroidal galaxies; we will review some recent evidence on the evolutionary patterns of active galaxies emerged from surveying the most natural AGN tracer, their x-ray emission; we will discuss recent outcomes of deep IR and mm surveys of the high-redshift Universe and the identification of the most luminous galaxies (LIRGs, ULIRGs, HyLIRGs) over a wide range of cosmic epochs; and finally will exploit broad-band diagnostics to disentangle starburst-dominated from AGN-dominated objects at high redshifts.

2.7.1 High-Redshift Optical Quasars

2.7.1.1 High-Redshift Optical Quasars and Early Metal Enrichment in the Universe

Quasars, since their discovery, are fundamental probes of the high-redshift universe. Their luminosities and compactness allow us to observe them to the highest accessible redshifts (currently $z > 6$; Fan et al. 2003). Figure 2.16, showing the average spectrum of a set of quasars at $z > 4$, illustrates the important feature in these objects: the presence of very strong broad emission lines by heavy elements (C, N, O, and heavier).

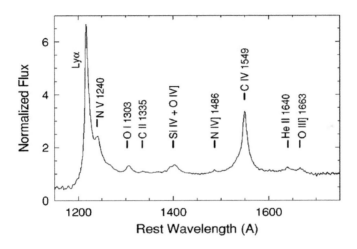

Figure 2.16. Mean spectrum of 13 QSOs at $z > 4$. (From Hamann & Ferland 1999.)

Such spectral features are surprisingly similar from one quasar to the other at all observed redshifts and over several orders of magnitude in luminosity. The presence of these metals can only be explained as due to early, fast, and efficient processing of primordial gas by massive stars (Burbidge & Burbidge 1967; Hamann & Ferland 1999). Considering the similarity of all QSO spectra in this respect, the star formation activity should predate the quasar phase. Quasar elemental abundances then provide unique probes of high redshift star formation and galaxy formation. The current evidence is that QSO environments have solar or higher metallicities out to redshifts > 6. There is also evidence for higher metal abundances in more luminous objects.

Quasars then identify sites in the high-redshift universe characterized by vigorous, high-redshift star formation. The high metallicity also requires that a substantial fraction of the local gas must be converted into stars.

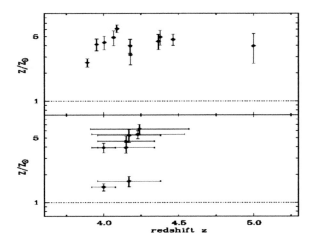

Figure 2.17. Metallicity estimates in high-redshift quasars compared with the solar value (dotted line). (From Dietrich et al. 2003.)

Dietrich et al. (2003; Figure 2.17), from observations of 11 high redshift quasars between $z = 4$ and 5, find average gas metallicities in the broad emission line region of $\sim$4 times solar: compared with nucleosynthetic stellar codes, this implies a redshift of $z = 6$ to 8 for the onset of the first major star formation episode. The short cosmic time available from the Big Bang at these high redshifts is plotted in Figure 2.18.

A further particularly strong constraint comes from the evidence of an enhanced Fe-abundance, which may require the operation of type-I supernovae from intermediate-mass stars, whose stellar evolutionary timescales are of the order of 1 Gyr. Freudling et al. (2003) have obtained low-resolution near-IR spectra of three QSOs at $5.7 < z < 6.3$ using the NIC-MOS instrument on the Hubble Space Telescope. The Fe II emission-line complex at 2500 Å is clearly detected in two of the objects, and possibly present in the third, and the inferred Fe metallicities are comparable to those measured for QSOs at lower redshifts. There thus seems to be no evolution of QSO metallicity to $z > 6$, which suggests that massive, chemically enriched galaxies formed within 1 Gyr of the Big Bang. If this chemical enrichment was produced by Type-Ia supernovae, then the progenitor stars may have formed at $z \sim 20 \pm 10$, in agreement with recent estimates based on the cosmic microwave background observations by WMAP (Spergel et al. 2003; Bennett et al. 2003).

Important information comes also from studies of the absorption lines in high-z quasar spectra due to gas clouds in the line-of-sight. Many published data on the average metallicities and abundance ratios for such

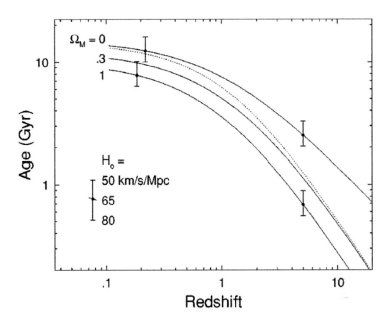

Figure 2.18. Relationship of cosmic time versus redshift in Big Bang cosmologies. The three solid curves correspond to $H_0=65$ km s^{-1} Mpc^{-1}; $\Omega_\Lambda = 0$, and $\Omega_M = 0$, 0.3, and 1. The dotted curve refers to the canonical parameter set $\Omega_\Lambda = 0.3$, and $\Omega_M = 0.3$.

absorption-line systems suggest that a dichotomy may exist between the chemical composition of damped $Ly\alpha$ systems (interpreted as *intervening* galaxies in the QSO line-of-sight) and absorption line systems *associated* with the quasar (those with $z_{abs} \simeq z_{em}$). Intervening systems have quite smaller than solar metallicities, whereas associated absorbers are solar or greater than solar (Petitjean et al., 1994); see Figure 2.19. It has been argued (Franceschini & Gratton 1997) that the latter may be explained by an early phase of efficient metal enrichment occurring only in the close environment of high-z QSOs, and characterized by an excess type-II supernova activity. This is reminiscent of the SNII phase required to explain the abundance ratios (favoring α- over Fe-group elements) observed in the IC medium of local galaxy clusters. Franceschini & Gratton discuss the following scenario: *(a)* well studied damped-Lyα, Lyα and metal lines in *intervening* systems trace only part of the history of metal production in the Universe – the one concerning slowly star-forming disks or dwarf irregulars; *(b)* the complementary class of early-type and bulge-dominated galaxies formed quickly (at $z \gtrsim 3$-5) through a huge episode of star for-

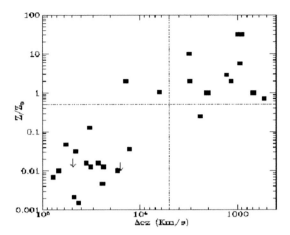

Figure 2.19. Plot of the average metallicity (in solar units) of the absorber versus escape velocity from the quasar. The metallicity refers mostly to C, Si, and O, or is an average estimated value for various elements. For negative velocities (infall) we plot the modulus in abscissa. The two dotted lines mark half-solar metallicity and an escape velocity of 5000 km/s, the usual value separating the *intervening* from the *associated* absorbers. (From Franceschini & Gratton 1997.)

mation favoring high mass stars; *(c)* the nucleus of the latter is the site of subsequent formation of a quasar, which hides from view the dimmer host galaxy; *(d)* the products of a galactic wind, following the violent episode of star formation in the host galaxy and polluting the IC medium of metals, are *directly* observable in the $z_{abs} \simeq z_{em}$ *associated* absorption systems on the QSO sight-line.

Consistent with all this are also the results of far-IR and mm observations of dust emission by very high-z quasars (Omont et al. 1996, 2001; Andreani et al. 1999). Figure 2.20 summarizes some of these results, based on modeling of IR-mm emission from circumnuclear dust torii around quasars. Under the assumption that the dust is heated by a point-like source with a power-law primary spectrum, as defined by the observed optical-UV continuum, the circumnuclear dust masses, temperature distributions, and torus sizes can be estimated by numerically solving the radiative transfer equation. These analyses constrain the properties of the enriched interstellar medium in the galaxies hosting the quasars. The dust abundance does not display appreciable trends as a function of redshift, from $z = 1$ to almost 5, and shows that dust and metals are at least as, and often more, abundant at these early epochs than they are in local galactic counterparts (Andreani et al. 1999). Also the inferred sizes of dust distributions around high-z

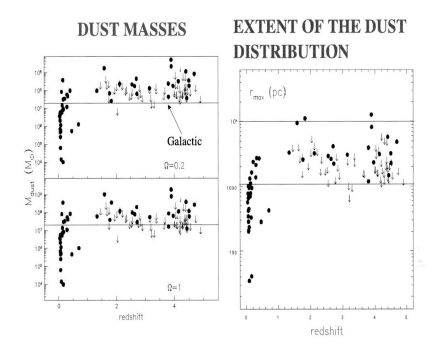

Figure 2.20. Left panel: circumnuclear dust mass in high-z quasars as a function of the redshift. Down arrows correspond to those sources undetected at FIR/sub-mm wavelengths. The dashed-line is the SED of a template nearby dusty spiral with a dust mass of 10^7 M$_\odot$. Right panel: Maximum radial size of the dust distribution surrounding the central source for the quasars. (Taken from Andreani, Franceschini, Granato 1999.)

QSOs are comparable with those of a large (several kpc) hosting galaxy.

All these results confirm active phases of star formation at early epochs in the QSO environments, probably related to the formation of massive spheroidal galaxies, and support models of an evolutionary link among star formation, nuclear activity, and the growth of supermassive black holes.

2.7.1.2   *The Optical Quasar Evolution*

Relevant progress on quasar evolution has been recently obtained from all-sky multi-color imaging data and improved selection techniques. The Sloan Digital Sky Survey (SDSS; York et al. 2000; Stoughton et al. 2002) provided high photometric quality, color information, and spectroscopic follow-up, able to effectively constrain the optical QSO luminosity density at very high redshifts, $z > 5$. Fan et al. (2003), in particular, report the discovery of quasars at $z > 6$, selected from 2870 deg^2 within the SDSS area and estimate the comoving density of luminous quasars at $z = 6$ to be a factor 30 lower than the maximum reached by luminous QSOs at $z = 2$ (see Figure 2.21).

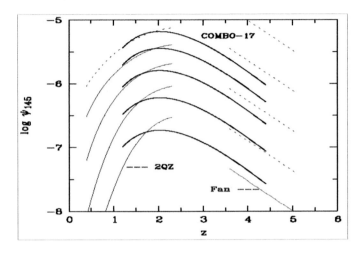

Figure 2.21. The evolution of quasar comoving spatial densities as a function of z. The various curves correspond to ranges in absolute magnitude at 1450 Åfor $M_{1450} < -24$ (top lines) down to $M_{1450} < -28$ (bottom lines). The lines at lower z are from the 2dF survey (Boyle et al. 2000), those at the highest z from Fan et al. (2003), and the intermediate ones from the COMBO-17 survey (Wolf et al. 2003).

Wolf et al. (2003) report on a large QSO optical survey (COMBO-17; see Figure 2.22) based on a narrow-band multi-color imaging in 17 filters between 350 nm and 930 nm, allowing us to simultaneously select quasars and determine their photometric redshifts with high accuracy ($\delta z < 0.03$) and obtain spectral energy distributions over a large redshift interval $1.2 < z < 4.8$ (Figure 2.22). The survey is maximally efficient at z where it was previously difficult to obtain information, close to the QSO LF's peak, and is able to detect moderate-luminosity AGNs as well. The results (summarized in Figure 2.21 as continuous curves at $z = 1$ to 4) are completely consistent with QSO luminosity function analyses based on more classical color selection (e.g., the 2DF QSO survey, Boyle et al. 2000, and SDSS).

Finally, Steidel et al. (2002) and Nandra et al. (2002) have analyzed

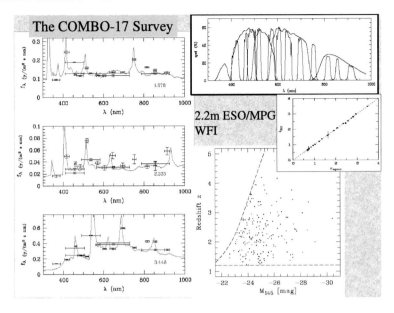

Figure 2.22. The COMBO-17 quasar survey. Top panel, right: response functions of the COMBO-17 multi-color imaging setup. Small insert: plot of spectroscopic and photometric redshifts. Lower panel: QSO distribution on the z-L plane. Left panel: comparison of spectral templates and observed flux densities.

large complete samples of $z \sim 3$ Lyman–Break galaxies and looked for faint AGNs in optical spectra and deep Chandra data. The former find among 1000 LBGs only 13 type-I and 16 type-II AGNs with high-ionization lines. The latter confirm such a low AGN fraction ($\sim 3\%$) in a sample of 148 LBGs, with the bulk of the faint x-ray signal to be attributed to starburst emission (as discussed in § 2.6.4).

2.7.2 High-Redshift Radio Galaxies

Powerful radio galaxies at low and moderate redshifts appear, for reasons not yet completely understood but for which a general interpretation has been suggested in § 2.6.5.3, exclusively associated with massive to very massive early-type galaxies, and often situated in moderately rich groups or clusters. Indeed radio galaxies at high redshifts have been found to reside in overdense environments, an evidence further emphasized by the discovery of extended H_α emitting regions around them that have been interpreted as plasma halos in forming clusters or groups (e.g., Reuland et al. 2003; Kurk et al. 2000). All this suggests that the hosts of high-redshift radio galaxies are the progenitors of present-day giant elliptical galaxies, a conclusion obviously supported by the locally established tight relationship between nuclear supermassive black holes and massive spheroidal galaxies.

While imaging of high-z radio galaxies in the optical shows the somehow enigmatic "alignment effect" with the radio axis, likely due to AGN scattered light and, in some cases, to jet-induced star formation (Best, Longair & Roettgering 1997; McCarthy et al. 1997), near-IR high-resolution studies reveal instead that the true nature of the host galaxy is a typically massive elliptical with very symmetric light profiles. Zirm, Dickinson & Dey (2003) report a detailed morphological investigation using NIC-MOS/HST for a sample of objects at $z = 0.8$ to 1.8, invariably showing well-defined $r^{1/4}$–law light profiles, large effective radii, and high values of the surface brightness (Figure 2.23). The light distributions of these host galaxies become consistent with the Kormendy relation of local galaxies if one assumes that these high-z counterparts are brightened by younger luminous stellar populations, whose redshift of formation is estimated at $z \simeq 2.5$. High-redshift radiosources are then rare, massive galaxies, with well-established early-type morphology.

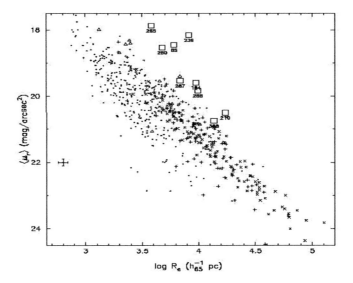

Figure 2.23. Kormendy relation for radio galaxies at $z = 0.8$ to 1.8, large squares (from NICMOS 1.6 μm observations by Zirm et al. 2003), compared with lower-z radio galaxy hosts (triangles), brightest cluster galaxies (crosses), and local E/S0 galaxies (dots and pluses). The high-z radio galaxy host becomes consistent with the local Kormendy relation if one assumes that they are brightened by the younger stellar populations and the latter are formed at redshift $z = 2.5$.

Similar conclusions hold also for the host galaxies of radio-quiet quasars, although in this case the population shows broader morphological properties and the lower-luminosity objects (in fact belonging to the Seyfert category) show late-type morphology (Dunlop 2003).

Radio samples also offer a good way for obtaining complete samples that are free of the optical selection effects, since their selection is relatively simple and well understood and unaffected by dust extinction. Surveys for high-z radio galaxies and quasars have long since indicated that the rapid increase in their numbers with redshift did not continue much beyond $z \simeq 2$. From a flux-limited sample of 878 flat-spectrum radio sources, of which 442 are radio-loud quasars with complete redshift information and no optical selection, Shaver et al. (1998) were able to map out the entire quasar epoch, unhindered by optical selection effects and any intervening dust: the rapid decline in true space density above $z \simeq 2$ indicated by previous studies of the high-z radio galaxy population (e.g., Dunlop & Peacock 1990) is fully confirmed. As shown in Figure 2.24, where some results on the cosmic evolution of the comoving emissivity of radio, optical, and x-ray selected AGN and QSO samples are summarized, this redshift cutoff looks similar to that obtained from optically selected samples (see also Figure 2.21), a result indicating that optical type-I QSO samples may not be too much affected by dust.

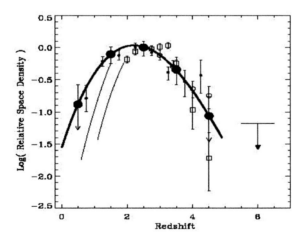

Figure 2.24. Comoving space densities as a function of redshift, normalized to $z = 2.5$ for various populations of active galaxies. Filled dots refer to flat-spectrum radio-loud quasars (from Shaver et al. 1998), open squares and circles are for optically selected quasars. (Taken from Shaver et al. 1998.)

2.7.3 The X-Ray View

Due to the so different background sky with respect to the optical deep images, the x-ray searches of AGNs offer important advantages to allow easier and more straightforward sampling of AGNs of any kind, given their dominance in x-rays (other prominent x-ray sources at bright fluxes, galaxy clusters, and groups, fade away quickly at faint limits given their shallow number counts). A particularly relevant advantage of the x-ray view is to allow effective sampling of the AGN low-luminosity end at various redshifts: a starburst- and an AGN-dominated galaxy are effectively discriminated in x-rays by a luminosity threshold of $L_{2-10} \simeq 10^{42.2}$ erg/s (§ 2.6.4). Flux-limited x-ray samples also provide simpler selection functions, while the optical ones are typically more complex and redshift-dependent. Finally, x-ray analyses of the AGN evolution can exploit the tight constraint set by the cosmic x-ray background (see Chapter 4).

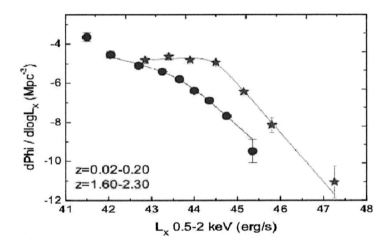

Figure 2.25. Luminosity function of type I AGN selected in the soft 0.5-2 keV x-ray band in two redshift intervals: $z = 0.02 - 0.2$ and $z = 1.6 - 2.3$. (From Hasinger 2003.)

Recent progress in this field has exploited the enormous improvement in sensitivity and spatial resolution allowed by the XMM–Newton and Chandra x-ray observatories. Important results have emerged from the variety of AGN surveys, now covering large portions of the luminosity functions at different redshifts.

Hasinger (2003) has merged the whole body of previously identified ROSAT, XMM, and Chandra AGN samples and computed the luminosity functions of type-I AGNs (for completeness and homogeneity) in the 0.5-2 keV band. The resulting sample spanned six orders of magnitude in flux and seven orders of magnitude in survey solid angle between the Bright ROSAT All-Sky and the Chandra Deep Surveys. The sources found in the very deep Chandra and XMM surveys are predominantly Seyfert galaxies at moderate redshifts. Figure 2.25 compares the AGN luminosity functions in two z–intervals at $z \sim 0.1$ and $z \sim 2$: the shapes of the two are significantly different, so that the cosmological evolution cannot be described by pure luminosity, nor by pure density evolution, and a more complex evolution pattern is indicated. While the low-z function appears rather steep at decreasing L, the high-redshift luminosity function is almost constant for luminosities below $10^{44.5}$ erg/s. So the classical pure–luminosity evolution pattern, which seemed to describe well the high-luminosity QSOs, breaks down completely for lower luminosity AGNs, for which the low-luminosity end shows much less or even negative density evolution.

We will discuss this important point further in the next chapter after having acquired more information from long-wavelength surveys of active galaxies and additional constraints by the diffuse background radiation.

2.7.4 The Long-Wavelength IR and Millimetric Surveys

High-redshift active galaxies and the generation of stars during the past cosmic history are usually investigated in the UV/optical/near-IR by exploiting very large telescopes on the ground and efficient photon detectors (e.g., Steidel et al. 1994; Madau et al. 1996). The outcomes of these *unbiased* optical surveys are either blue galaxies apparently characterized by moderate luminosities and modest activity of star formation (few $M_\odot$/yr on average), or passively evolving early-type galaxies (particularly evident in the near-IR) or finally type-I AGNs.

However, that this view could be rather incomplete is illustrated by the fact that *biased* optical surveys emphasize a quite more *active* universe, as revealed by the heavily metal-enriched environments around quasars and active galaxies at any redshifts discussed in Sect. 2.7.1; by the presence of populations of massive elliptical galaxies at $z > 1$, with dynamically relaxed profiles and complete exhaustion of the ISM (Sect. 2.7.3), expected to originate from violent starbursts; and by the intense activity of massive stars required to explain the metal-polluted hot plasmas present in galaxy clusters and groups (Mushotzky & Loewenstein, 1997). Optical surveys do not seem to reveal the expected cosmic sites where active transformations of baryons took place at rates high enough to explain the above findings.

Hints on such a possible missing link between the *active* and *quiescent* universe have emerged from long-λ observations, as mentioned in Sects. 2.6.2 to 2.6.5. The IRAS pioneering survey has revealed that in a small fraction of local massive galaxies (LIRGs and ULIRGs) star formation is taking place at very high rates. Interesting to note, the reddened optical spectra of these objects do not contain manifest signatures of the dramatic phenomena revealed by the far-IR observations (Poggianti & Wu 1998; Poggianti, Bressan, & Franceschini 2001). This emphasizes the role of extinction by dust, which is present wherever stars are forming, but is increasingly important in the most luminous objects.

It is only during the last several years that new powerful instrumentation has allowed us to start a systematic exploration of the distant universe at long wavelengths.

A first important discovery was made by COBE of a bright isotropic background in the far-IR/sub-mm, of likely extragalactic origin (CIRB) and interpreted as the integrated emission by dust present in distant and primeval galaxies and forming quasars (see Chapter 4).

The second important fact was the start of operation of imaging bolometer arrays (SCUBA and MAMBO) on large millimetric telescopes (JCMT and IRAM), able to resolve a substantial fraction of the CIRB background at long wavelengths into a population of very luminous IR galaxies at $z \sim 2$ or larger.

Finally the Infrared Space Observatory (ISO) allowed us to perform sensitive surveys of distant IR sources in the mid- and far-IR and to characterize the evolution of the IR emissivity of cosmic sources between $z = 0$ and 1. Mostly because of the different K-corrections, the sky explorations by ISO and the mm telescopes were nicely complementary in terms of the redshift/luminosity space surveyed, with the former mostly limited to moderate-L and $z < 1.3$ and the latter to very large-L and $z > 1$.

2.7.4.1 *Millimetric Searches of Very High-z Sources*

Surveys in the sub-millimeter offer a unique advantage to naturally generate volume-limited samples from flux-limited observations of sources down to a given luminosity, due to the peculiar shape of galaxy and AGN spectra rising very steeply from 1 mm to 200 μm [roughly $L(\nu) \propto \nu^{3.5}$]. Due to the strong K-correction almost completely counterbalancing the cosmological dimming of the flux, a sensitive sub-mm survey avoids local objects (stars and nearby galaxies) and preferentially selects sources at very high redshifts, a kind of direct image of the high-redshift universe impossible to obtain at other wavelengths.

Various groups have used SCUBA on JCMT for deep cosmological surveys, including surveys of distant galaxy clusters used as cosmic lenses (Smail et al. 1997, 1999) to overwhelm the source confusion limitation. Hughes et al. (1998) published a single very deep image of the HDF North to the SCUBA confusion limit $S_{850} \geq 2$ mJy. Lilly et al. (1999), Eales et al. (2000) and Blain et al. (2002), among others, have published samples of few tens of sources at similar depths. Source samples have also been obtained at 1.2 mm with MAMBO/IRAM (Bertoldi et al. 2001).

The extragalactic source counts at 850 and 1200 μm show evident departures from the Euclidean law ($dN \propto S^{-3} dS$ in the flux-density interval from 2 to 10 mJy), a clear signature of the strong evolution and high redshift of mm-selected sources.

Due to sensitivity limitations and small surveyed areas, mm surveys preferentially detect ULIRGs ($L_{bol} > 10^{12} L_\odot$) at $z > 1$ and up to $z = 4$. These sources have typically very faint and red optical counterparts, due to the high redshift and extinction. These extreme properties of the mm-selected sources entail a serious difficulty to identify them, also because of the large diffraction-limited beam.

An effort has been made in the last several years to identify the mm sources. Recently, improvements have been obtained by combining deep mm and radio observations to narrow down the source error box (55 radio-identified submm galaxies are reported by Chapman et al. 2003a). Another approach has been to exploit large-multiplexing spectrographs to follow-up spectroscopically all faint/red counterparts within a SCUBA errorbox, with some priors on the expected spectral features. The bulk of the source redshifts turn out to fall between $z = 1.5$ and 3.5, with essentially no sources identified above $z = 4$. In some cases these redshifts are confirmed by detection of the CO emission.

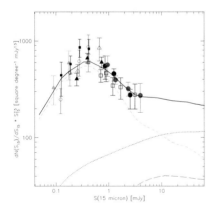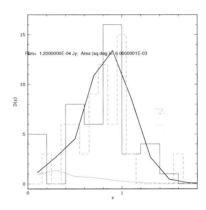

Figure 2.26. Left: differential counts at $\lambda_{\text{eff}} = 15\mu$m normalized to the Euclidean law ($N[S] \propto S^{-2.5}$). The data come from an analysis of the IGTES surveys by Elbaz et al. (1999). The dotted line corresponds to the expected counts for a population of non-evolving spirals. The dashed line comes from our modeled population of strongly evolving starburst galaxies. Right: redshift distribution of ISO sources in the HDF-North and CFRS. (From Franceschini et al. 2001.)

The corresponding bolometric luminosities are extremely high: together with quasars, these are the most luminous objects in the universe. However, it is debated how much this derives from ongoing star formation or from AGN accretion. Quite often the observed spectra are found showing type-II AGN features, like high ionization lines (CIV) and high $OIII/H_\beta$ line ratios, found in approximately 50% of the sources. If instead interpreted as due to young stars, the star formation rates inferred from the luminosity are several hundreds $M_\odot$ yr^{-1}.

As suggested by various authors, the similarity in properties (bolometric luminosities, SEDs) between this high-z population and local ultraluminous IR galaxies argues in favor of the idea that these represent the long-sought "primeval galaxies" originating the local massive elliptical and S0 galaxies. This is also supported by estimates of the volume density of these objects ($\sim 2 - 4 \times 10^{-4}$ Mpc^{-3}), high enough to allow most of the E/S0 to be formed in this way (Lilly et al. 1999). An excess of very luminous ($L \sim 10^{13} L_\odot$) sources at 850 μm around the $z = 3.8$ radio galaxy 4C41.17 (Ivison et al. 2000) may indicate the presence of a forming cluster surrounding the radio galaxy, where the SCUBA sources would represent the very luminous ongoing starbursts.

The morphologies of a subset of SCUBA galaxies have recently been studied with HST-STIS by Conselice et al. (2004 arXiv:astro-ph/0308198) and Chapman et al. (2003 0308197) on the radio-identified sample. The high-z millimetric galaxies often show faint, distorted blue/red composite structures to high-resolution HST observations, with evidence for multi-component, distorted galaxies reminiscent of mergers in progress. There is a remarkable lack of isolated systems, either compact or extended: this, together with the observed z-distribution peaking at $z \sim 2$ to 3, indicates that there is no evidence that spheroids do form in a monolithic collapse of primordial gas clouds at very high redshifts, as sometimes advocated.

Morphologies and z-distributions of SCUBA galaxies are possibly more consistent with hierarchical galaxy formation scenarios, in which the most intense activity occurs when gas-rich, high-redshift proto-galaxies collide and merge.

2.7.4.2   Mid- and Far-IR Surveys from Space

The Infrared Space Observatory provided unprecedented capabilities for deep cosmological surveys in the mid-IR and far-IR, although it was seriously confusion-limited at the longer wavelengths (due to the different spatial resolutions, $\sim$4.6 and $\sim$50 arcsec FWHM at 15 and 100 μm, ISO sensitivity limits ranged from $\sim$0.1 mJy in the mid-IR to $\sim$100 mJy at the long wavelengths). The most important ISO surveys have been performed with a wide-band filter at 12-18 μm (see results in Figure 2.26) and two far-IR ($\lambda = 90$ and 170 μm) wavebands. Among the former, those in the Hubble Deep Fields (Rowan-Robinson et al. 1997; Aussel et al. 1999;

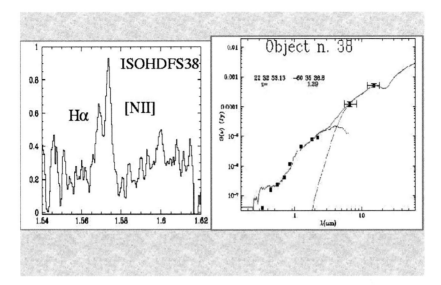

Figure 2.27. Left: the near-IR ISAAC spectrum of the source S40 in the Hubble Deep Field South detected by ISO at 15 μm at $z = 1.39$, showing $H_\alpha - NII$ emission lines (Franceschini et al. 2003). Note that the two line intensities are inverted with respect to a normal starburst, indicative of an obscured type-II AGN. Right: the optical-IR SED of the source S40 with red optical spectrum and relatively flat spectral shape between 7 and 15 μm, consistent with the type-II AGN nature.

Mann et al. 2003), the CFRS fields (Flores et al. 1999), Lockman Hole (Fadda et al. 2004; Rodighiero et al. 2004), Marano Southern area (Elbaz et al. 2005), and the large area ELAIS survey (Oliver et al. 2000) are to be mentioned. Far-IR surveys were published by Puget et al. (1999), Dole et al. (2001), and Rodighiero et al. (2003).

Faint IR-selected sources show very high rates of evolution with red-shift, exceeding those measured for galaxies at other wavelengths and comparable to or larger than those of quasars (Elbaz et al. 1999, 2002, Franceschini et al. 2001). This fast evolution implies that dust-obscuration, if moderately important in local galaxies, has instead strongly affected the past active phases of galaxies' life.

The faint ISO-selected sources are very luminous on average in the far-IR ($L_{bol} \geq 10^{11} L_\odot$), while they turn out unnoticeable in the optical. On the contrary, their areal density is much lower than found for optically-selected faint blue galaxies (Ellis 1997).

Franceschini et al. (2001, 2003), Flores et al. (1998, 2004) and Liang et al. (2004) have investigated the characters of IR emissions in ISO sources between $z \sim 0.2$ and 1.3, using in particular the mid-IR flux as a measure of the bolometric luminosity. Near-infrared and optical spectrographs on VLT have been used to analyze representative source samples. Fairly intense Hα+[NII] emission is detected in virtually all the observed sources. The comparison with the Hβ, Hγ and [OII], as well as the analysis of the spectral energy distributions, indicate typically high extinction values, between 1.5 and almost 3 magnitudes in V (see Figure 2.27), quite larger than found for local normal spirals (the de-reddened H$_\alpha$ flux is strong in these objects). Enhanced activity is also proven by the fact that the mid-IR flux is typically larger by factors >2-3 than expected for normal galaxies. Assuming that this flux is from young stars, typically inferred values of SFR range between $\sim$ 10 and 300 M$_\odot$/yr for these IR-selected galaxies, i.e., less extreme than those inferred for mm-selected galaxies. Remarkably, the $z \sim 1$ ISO galaxies show metal abundances of oxygen which are half of those in local spirals, implying that a large fraction of their metals and stellar masses have formed since $z = 1$ (Liang et al. 2004).

Launched in August 2003, the Spitzer Space Telescope has started a very systematic sky exploration in the mid-IR (4 bands between 3 and 10 μm) and far-IR(24, 70, and 160 μm), by dedicating a significant fraction of the mission's first year and a half to deep public surveys (e.g., Lonsdale et al. 2004; Dickinson 2004).

2.7.5   Physics of the IR-Submillimeter Galaxy Population

2.7.5.1   The AGN Contribution

The origin of the high bolometric luminosities of IR-mm source populations is a critical current-day issue, since these are among the most important contributors to the cosmic energy budget (see Chapter 4). An estimate of the global ratio of the energy produced by gravitational accretion in AGNs and by young stars then requires a detailed analysis of their physical properties.

The first SCUBA source unambiguously identified via optical/near-IR spectroscopy was found to contain an obscured AGN at $z = 2.81$. Interestingly, star formation and a hidden AGN seem to contribute $\sim$50% each of the bolometric luminosity (Frayer et al. 1999). Optical follow-up (Chapman et al. 2003) of other SCUBA sources has revealed further AGNs, although optical spectroscopy has clearly many limitations for AGN identification, particularly due to the faintness and high dust obscuration.

The presence of AGNs in faint ISO and sub-mm sources has also been investigated in the optical by looking at the broadness of the Balmer lines, and the low- to high-ionization line ratios. By combining these with the HST morphologies, the slopes of the mid-IR spectra, and the ratio of the radio to IR fluxes, Franceschini et al. (2003) found evidence for nuclear

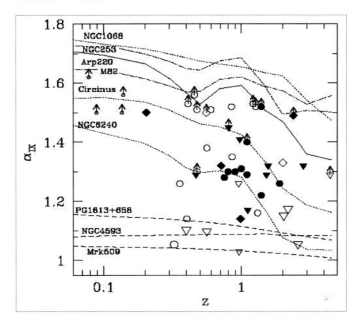

Figure 2.28. Distribution of the observed mid-IR to hard x-ray spectral indices $\alpha_{IX} = \log(L_{15\mu}/L_x)/(\nu_{15\mu}/\nu_x)$ as a function of redshift (Fadda et al. 2002). Triangles, diamonds, and circles represent type-I AGNs, type-II AGNs, and unknown types, respectively. Templates derived from various types of known active galaxies and AGNs are shown.

activity in 2 out of 21 ISO mid-IR galaxies (one example is in Figure 2.27), while for two others the presence of AGN was suspected.

Diagnostic diagrams of [OII]/Hβ) versus $\log$([OIII]/Hβ) have been used by Liang et al. (2004) to distinguish the HII-region objects from the LINERs and Seyferts among the ISO 15 μ population at $0.4 < z < 1.2$: the large majority turn out to be purely starburst galaxies, while 8/48 objects are identified to be AGNs, including five LINERs and three Seyfert 2 galaxies. AGN fractions around 10–20% are then indicated by these studies.

A $\sim$20% AGN fraction in sub-mm sources is also indicated by their unusually high brightness at radio wavelengths (Chapman et al. 2004), departing from the far-IR/radio distribution of low-z starbursts (Helou et al. 1985).

However, given the dust-polluted nature of the ISO and sub-mm galaxies, the best discriminator of AGN activity would instead be the detection of luminous hard x-ray emission, which is unaffected by photoelectric absorption up to very high gas column densities ($N_H > 10^{23.5}$ cm^{-2}). Also

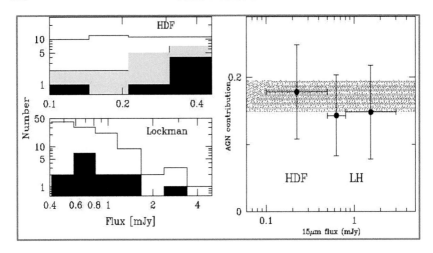

Figure 2.29. On the left: histogram of the 15 μm fluxes in the Lockman and HDF surveys. Sources detected in x-rays are shaded, while the black histogram shows the sources dominated by AGN emission. On the right: ratio of integrated 15 μm flux of AGN-dominated sources to that of all the mid-IR extragalactic sources as a function of limiting flux. (Fadda et al. 2002.)

the contribution to the x-ray flux from a starburst component is usually irrelevant compared with that from an AGN (e.g., Franceschini et al. 2003 XMM-ULIRG). As shown in § 2.6.4, hard x-ray observations can provide an AGN identification in sources where the optical signatures are weak or absent.

Due to their sustained high-energy responses, sensitivities, and positional accuracies, the Chandra and XMM–Newton observatories have recently allowed us to make important progress in this field. Fadda et al. (2002) discussed deep combined mid-IR and x-ray observations in HDF-North and the Lockman Hole. The HDF-North has been observed by ISO-CAM over a 24 square arcminutes area to a depth of 0.05 mJy (more than half of the mid-IR background is expected to be resolved at this limit) and by Chandra for 1 Msec. Almost 25% of the mid-IR sources are detected in the x-ray, while 30–40% of the x-ray sources show mid-IR emission. The fraction of mid-IR identifications rises to 63% for sources selected at higher energies, 5 to 10 keV. Under the assumption that AGN-dominated sources are luminous x-ray sources and have the same IR to x-ray SEDs typical of local AGNs, Fadda et al. find that (17±7)% of the mid-IR integrated flux from resolved sources in the HDF-North is due to AGN emission. Given its weak dependence on the IR flux, this figure may be taken as indicative of the likely AGN contribution to the extragalactic mid-IR background intensity.

Obviously, the sensitivity to the presence of any AGN components increases with the depth of the x-ray investigation. At the moment, up to ∼70% of the radio-identified SCUBA galaxies have x-ray detections in the deepest x-ray images. In particular, Alexander (2004) compared an ultradeep 2 Msec Chandra survey (Alexander et al. 2003) and a deep SCUBA sample with radio/optical identifications and spectroscopic follow-up in the Chandra Deep Field-North. He showed that a substantial fraction (∼40%) of SCUBA sources appear to host an AGN, some of which are luminous type-II QSOs (with $L_x > 3 \; 10^{44}$ erg/s). This however is not in contradiction with previous studies: from a comparison of the x-ray to bolometric luminosity ratio (see Figure 2.30), it is argued that in almost all cases these AGN components are not bolometrically important, the AGN contribution to the total flux being < 20%). Then the conclusion of these studies is that star formation appears to dominate the bolometric output of the majority of sub-mm sources, but many of them reveal the presence of minor AGN components.

In summary, AGNs may be frequently found to be hidden in the IR/sub-mm source samples (as expected from the concomitant evolution of the BH and bulge implied by the tight correlation between their masses; Magorrian et al. 1998), but only rarely appear to dominate their bolometric energy.

Another very interesting outcome of ultradeep x-ray surveys (Alexander et al. 2003) is that a substantial fraction (∼30%) of bright SCUBA galaxies show evidence for bimodal x-ray emission, indicating binary AGN activity. Since these x-ray pairs (separated by 2–3″, corresponding to 10–20 kpc at the source) are barely resolved by Chandra, we expect that many more have bimodal emission with lower separation, hence are unresolvable. Together with the outcomes of optical/near-IR imaging, showing that these systems are interacting and merging, this is remarkable evidence for the sub-mm sources to be major mergers in which enhanced star formation is accompanied by AGN activity and during which the supermassive black holes of the parent galaxies coalesce.

2.7.5.2   *Timescales of the Activity*

Franceschini et al. (2003) have attempted to estimate the timescales of star-formation activity in IR-selected galaxies by comparing the mass in stars M_{star} with the star formation rate. M_{star} was estimated by fitting the photometric SED data with a combination of stellar populations with different ages and extinction (solid thin line in Figure 2.27). This analysis showed that the faint ISO sources with fluxes $S_{15} > 100 \; \mu$Jy are hosted by massive galaxies ($M \simeq 10^{11}$ M$_\odot$). The timescale for the formation of stars, t_{SF}, was estimated for each galaxy as the ratio of the stellar mass

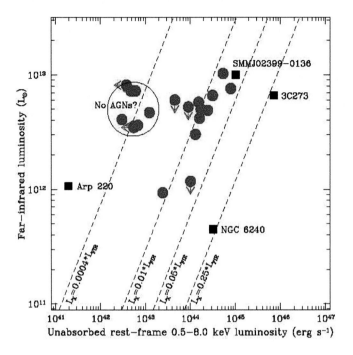

Figure 2.30. Bolometric far-IR versus 0.58.0 keV x-ray luminosities (after absorption correction) for x-ray detected SCUBA galaxies with spectroscopic redshifts (from Alexander 2004); (far-IR luminosities are also constrained by the 1.4 GHz radio flux). The line indicating the smallest L_X/L_{FIR} ratio corresponds to the average found for starburst galaxies. Assuming that the right dashed line fitting NGC 6240, and 3C273 corresponds to objects dominated by AGN emission, the two intermediate lines fitting the SCUBA galaxies correspond to an AGN contribution between 4 and 20%.

M_{star} to the rate of star formation deduced from the mid-IR flux. t_{SF} has been found to span a very wide range of values between 0.2 and 10 Gyrs or more: when compared with the typical starburst duration ($< 10^8$ yrs), this implies that the ongoing event of star formation can generate only a fraction of the stellar content in these galaxies, several of such repeated episodes during a protracted SF history being required for the whole galactic build-up. A trend towards a reduced level of star-formation activity in galaxies at decreasing redshifts was also apparent in the data.

These results were confirmed by Liang et al. (2004), who found that relatively small amounts of stellar mass and metals are formed during the ongoing starburst event in faint ISO galaxies. Indeed, several consecutive bursts are predicted by merging simulations, several of which may occur in a Hubble time. Detailed studies of the Balmer absorption lines of these LIRGs and ULIRGs will be needed to quantify in more detail the mass fraction of stars born during the starbursts, as well as the duration of these bursts.

The duration of the activity of the very luminous SCUBA galaxies has been sparsely studied with high signal-to-noise optical spectra compared with starburst synthesis models, to yield a timescale for the starburst activity visible at ultraviolet (UV) wavelengths. For one of their brightest SCUBA sources, Smail et al. (2003) found a best fit with a 5 Myr instantaneous burst of star formation, with an implied huge star-formation rate of ~ 1000 M$_\odot$/yr, consistent with that inferred directly from the sub-mm and radio fluxes. This is likely the paradigm for a significant fraction of SCUBA galaxies (Chapman et al. 2004).

Also for quasars, and AGNs in general, the activity pattern indicated by the data is that of a set of recurrent, short-lived events (with duration $\sim 10^7 - 10^8$ yrs) occurring in all massive galaxies, rather than a continuous long-lived phenomenon. The latter would imply a local mass function of BH remnants and Eddington ratios in high-z quasars entirely inconsistent with the observations (Cavaliere & Padovani 1989; Franceschini, Vercellone & Fabian 1998).

2.7.6 High-Redshift Active Galaxies: Conclusions

Together with relevant constraints and new interesting facts emerged from current deep optical, soft x-rays and radio surveys (e.g., the unexpected evolutionary trend for type-I AGNs as a function of luminosity), important information on the relationship of star formation and AGN activity in the past epochs has been provided by the IR/sub-mm selection.

IR/sub-mm sources appear to emphasize sites of strongly enhanced activity of star formation inside massive galaxies, which are typically the brightest members of galaxy groups. AGN (or even type-II quasar) components are also often concomitantly observed in these sources, but appear to be bolometrically important only for a minority of the sources. From the analogy with local ULIRGs, it seems at least plausible that the ULIRGs found by SCUBA mostly at $1 < z < 3$ may indeed be forming ellipticals, while the lower-z sources detected by ISO are lower-luminosity analogues and may be related with lower-mass bulges of present-day spiral galaxies.

These sources seem to trace evolutionary phases, involving strong dynamical interactions and mergers, leading to the formation of massive current-day galaxies and their nuclear supermassive black holes. The activity pattern derived from a comparison of the total cumulative mass in stars and in the supermassive BH with the ongoing SFR and AGN bolometric luminosity, is composed by a set of repeated "activations", triggered by mergers–interactions, during a substantial fraction of the Hubble time and peaking at $z \simeq 0.7$ to $z \simeq 3$, depending on luminosity.

Finally, the results of recent observations suggest that the AGN phenomenon, even during the past cosmic epochs, involves a minority of high-z (as well as low-z) active galaxies at any given time. This may be simply interpreted as a different duration of the starburst and AGN phases.

2.8   Radiative and Baryonic Remnants of Past Activity

Having discussed in chapter 3 direct observational evidence of galaxy activity at high-redshifts, we dedicate this Chapter to a comparison of these results with constraints set by a variety of remnants in the local universe of past activity.

One of these is the low-mass stellar populations in local galaxies, which are the record of past star formation. These are related to the massive luminous stars dominating the emissions in high-z galaxies through the stellar initial mass function.

A local record of the quasar and AGN phase has been recently identified and investigated (mostly through HST high-resolution spectroscopy) in the form of nuclear supermassive (10^8 to $> 10^9$ M$_\odot$) collapsed objects. Supermassive black holes and gas accretion in their strong gravitational field are the paradigm for the interpretation of AGN activity. The nuclear collapsed objects in local galaxies are then presumed to constitute the remnants of an ancient quasar phase characterizing all the most massive galaxies.

Eventually, both processes of baryon transformation leave imprints in the cosmic background radiation. We will attempt to match these three fossil records of past activity with the observed properties of the high-z sources presumed to originate them, to verify if we may have a globally acceptable picture, or if some fundamental problems come to light in this exercise.

2.8.1   The Cosmological Background Radiation

The cosmological background radiation is the integrated mean surface brightness due to resolved and unresolved extragalactic sources in a given waveband. It is a fundamental cosmological observable, able to constrain the total radiant energy produced at any cosmic time by sources emitting close to the observed wavelength in that particular sky area. We defer to Franceschini (2000) for a discussion about the relationship of this with other fundamental observables, such as the source number counts at the observed wavelength, the angular correlation function, and the galaxy luminosity functions.

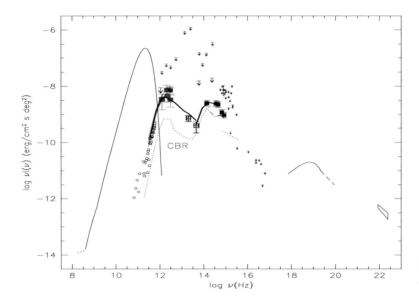

Figure 2.31. Cosmological background radiation spectral intensity, from radio (dotted at the extreme left) to the gamma-ray (extreme right). The thin continuous line marked CBR is the microwave background. The infrared, optical, and x-ray backgrounds are the components peaking at $\nu \sim 10^{12}$, $\nu \sim 10^{14}$, and $\nu \sim 10^{19}$ Hz, respectively.

A sketch of the current knowledge of the cosmic background radiation is reported in Figure 2.31. We will concentrate in the following on three components of the background radiation, the optical, the far-IR, and the x-ray backgrounds. We leave obviously aside the microwave background, originating from the diffuse primeval plasma at $z > 1000$. We also do not consider the radio centimetric and gamma-ray backgrounds, due to photon-production processes in high-redshift sources providing negligible contributions to the global energetics: the two respective emission mechanisms, synchrotron and inverse Compton, concern only marginal fractions of the energetic particle content of cosmic sources.

In conclusion, we expect that a largely dominant fraction of the integrated radiant energy emitted by active galaxies is currently observable in the three wavelength domains, the far-IR, the optical/near-IR, and the x-rays. Currently available limits in the difficult-to-test far-UV (Figure 2.31) seem to exclude that we are missing significant contributions there.

2.8.1.1   Do We Safely Constrain the Optical Background?

An extragalactic background light at optical wavelengths (say 2500 Å to 1 μm) is expected to be produced by stellar nucleosynthesis at redshifts $z < 7$, and by optically bright type-I AGNs.

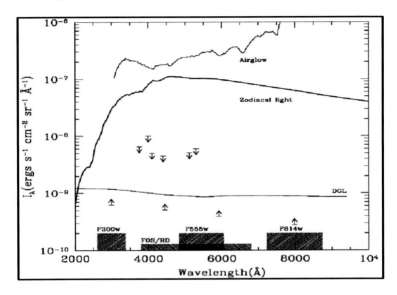

Figure 2.32. Spectral intensities of foreground radiation in the optical. The upper line is the terrestrial airglow spectrum. The zodiacal light is the sun scattered spectrum by the interplanetary particles. DGL is the Diffuse Galactic Light by galactic stars. The bandpasses WFPC-2 HST are indicated at the bottom of the plot; the corresponding lower limits are based on galaxy counts in HDF. (Taken from Bernstein et al. 2002.)

Figure 2.32 illustrates the difficulty in obtaining the true extragalactic background from subtraction of the much more intense foreground emissions. A variety of different approaches has been attempted to measure the optical background, including a pioneering one by Mattila (1976) trying to isolate it by the difference of the signals in and out the lines-of-sight to high galactic latitude dark clouds acting as blank screens, spatially isolating all foreground contributions to the background radiation. Toller (1983) later attempted to avoid both atmospheric and zodiacal foregrounds by using data taken with the Pioneer 10 spacecraft at 3 AU from the sun, beyond the zodiacal dust cloud. All such attempts provided us with only upper limits to the optical background. Among the fundamental difficulties was the estimation and subtraction of the galactic stellar light reflected by high latitude dust (the IRAS "cirrus").

To overcome these difficulties, an alternative approach has been followed by Madau & Pozzetti (2000) among others to exploit the convergence of deep galaxy counts in the HDF in all optical bandpasses UBVIJHK and to estimate a lower limit to the background by integrating the counts over the observed wide magnitude intervals. These data are reported as lower limits in Figure 2.32, and correspond to an integrated optical background at the level of around 10 nWatt/m^2/sr.

More recently, Bernstein et al. (2002) have re-attempted to estimate the optical background by subtracting from the total signal measured by HST the contributions of the zodiacal light and the diffuse galactic light. They arrive in such a way to background levels which are typically factors $\sim$3 larger than those estimated from the direct HST counts, which would mean that either a truly diffuse background (of unknown origin) exists, or the HST counts miss 80% of the light from faint galaxies, which may be surprising considering the depth of the HST HDF images on which the Madau estimate is based.

Some difficulties and inconsistencies in the Bernstein et al. (2002) measurement are discussed by Mattila (2003), particularly concerning the modeling of the diffuse galactic light and the evaluation of the terrestrial airglow for ground-based observations. For these reasons and following the analysis of Mattila (2003), we will more conservatively refer in the following to the determination based on the deep HST counts as the optical/near-IR background.

2.8.1.2 A New Entry: The Cosmic Infrared Background

The effects of the cosmological redshift and absorption of short-wavelength radiation by dust and re-emission at long wavelengths are expected to shift a significant fraction of the energy radiated by active and primeval galaxies into the near- and far-IR. Although the relevance of both effects is evident from our discussions in the previous chapters and a long-wavelength background was expected (e.g., Franceschini et al. 1994), its existence has remained for long time untestable. Eventually a dedicated space mission, the Cosmic Background Explorer, detected this extragalactic signal as an isotropic excess on top of the foreground emissions, produced by the interplanetary dust and galactic cirrus emissions (see Hauser & Dwek 2001 for a thorough review).

The COBE mission included two instruments designed to make absolute sky brightness measurements: the broad-band absolute photometer DIRBE from 1.25 to 240 μm, and the Fourier transform spectrometer FIRAS from 125 μm to the millimeter. Puget et al. (1996) reported the first tentative detection of the Cosmic Infrared Background (CIRB) using FIRAS all-sky maps, by subtracting the zodiacal and galactic foregrounds from the intercepts of the all-sky signal versus the column densities of the

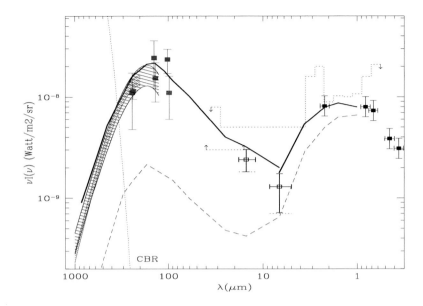

Figure 2.33. The Cosmic Infrared Background (CIRB) spectrum as measured by independent groups in the all-sky COBE maps (e.g., Hauser et al. 1998), compared with estimates of the optical extragalactic background based on ultradeep optical integrations by the HST in the HDF (Madau & Pozzetti 2000). The three lower datapoints in the far-IR are from DIRBE data by Lagache et al. (1999), and the shaded areas from Fixsen et al. (1998) and Lagache et al. The two mid-IR points are the resolved fraction of the CIRB by the deep ISO surveys IGTES (Elbaz et al. 2002), while the dashed histograms are limits set by TeV cosmic opacity measurements (Stanev & Franceschini 1998). The lower dashed line is the expected intensity based on the assumption that the IR emissivity of galaxies does not change with cosmic time. The thick line is the predicted CIRB spectrum of the presently discussed reference model for IR galaxy evolution. (Taken from Franceschini et al. 2001.)

respective tracers in ecliptic and galactic coordinates. The resulting CIRB spectral intensity was essentially confirmed by later analyses by Fixsen et al. (1998) and Lagache et al. (1999).

The existence and intensity of the CIRB at far-IR (140 and 240 μm) wavelengths were further confirmed by Hauser et al. (1998) and Schlegel et al. (1998) based on refined calibration of the DIRBE maps and detailed spatial models of the foregrounds exploiting their spatial variations. The residual, supposed extragalactic, signal was found to be consistent with isotropy.

There were claims of detections of a diffuse extragalactic background in the near-IR (2.2 and 3.5 μm) at a level of ~ 20 and $12\,\mathrm{nWatt/m^2/sr}$ after subtraction of the stellar contribution (Gorjian et al. 2000, Wright & Reese 2000), but with substantial uncertainty due to the uncertain contribution of the zodiacal light, and very limited possibility to check its isotropy (Hauser & Dwek 2001). The existence of a bright extragalactic background in the near-IR was also inferred from the analysis of data by the IRTS Japanese mission (Matsumoto 2001), with values at 2.2 and 3.5 μm even higher than reported by COBE. At shorter wavelengths, the Matsumoto (2001) and Xu et al. (2002) results continue to rise steeply to $\sim 65\,\mathrm{nWatt/m^2/sr}$, that is above the 95% confidence level COBE upper limits, with a spectrum very reminiscent of that of the sun-scattered zodiacal light. In the following we will adopt for the optical/near-IR background the values derived from direct integration of the ultradeep HST counts, as a safer reference.

Data on the CIRB and optical backgrounds are reported in Figure 2.33. The lower limits based on integrations of the galaxy number counts in the optical/near-IR appear to be already close to the upper limits (dotted line histogram) set by TeV-energy observations of blazars. These latter are obtained from the attenuation effect of the e^+e^- pair production due to collision of a very high energy photon (typically emitted by a low-redshift blazar) with the CIRB photons (Stecker, de Jager, & Salomon 1992). The corresponding absorption cross-section of γ–rays of energy E_γ [TeV] has a maximum for IR photons with energies obeying the condition:

$$\epsilon_{\mathrm{max}} = 2(m_e c^2)^2/E_\gamma, \qquad (2.1)$$

which implies

$$\lambda_{\mathrm{peak}} \simeq 1.24 \pm 0.6(E_\gamma[TeV])\,\mu\mathrm{m}. \qquad (2.2)$$

This technique was mostly used to constrain the CIRB spectrum in the wavelength range 5 to 25 μm range where its direct detection is currently impossible given the overwhelming dominance of the interplanetary dust emission, and where upper limits of 5 to 10 $\mathrm{nW/m^2/sr}$ have been obtained. Stanev & Franceschini (1998) have exploited the attenuation analysis on Mrk 501, using high-quality TeV data during a source outburst in 1997, to attempt constraining in particular the optical background between 2 and 0.5 μm. Taken literally, this analysis is consistent only with an optical/near-IR maximum background intensity of 10 $\mathrm{nW/m^2/sr}$, and may rule out background values much higher than the Madau & Pozzetti estimate.

Altogether, the integrated CIRB intensity between 100 and 1000 μm, where the present estimates show a small scatter, is $\sim (30 \pm 5)\,10^{-9}$ $\mathrm{nWatt/m^2/sr}$. The addition of the presently un-measurable fraction between 100 and 10 μm using the constraints summarized in Figure 2.33

brings the total energy density in the CIRB between 7 and 1000 μm to the value: $\nu I(\nu)|_{FIR} \simeq 40 \ 10^{-9}$ nWatt/m^2/sr. This flux is larger than the integrated bolometric emission by distant galaxies between 0.1 and 7 μm (the "optical background"), for which we adopt the value estimated by Madau & Pozzetti (2000) using the HST integrations in the Hubble Deep Field: $\nu I(\nu)|_{opt} \simeq (17 \pm 3) \ 10^{-9}$ nWatt/m^2/sr. This sets a relevant constraint on the evolution of cosmic sources, if we consider that for local galaxies only 30% on average of the bolometric flux is re-processed by dust into the far-IR.

Preliminary attempts to resolve the CIRB into sources have been mostly attempted in the mid-IR and in the sub-mm, while at the moment only a very small fraction of the CIRB is resolvable close to its peak-emission wavelength, due to the limitation of present-day far-IR imagers (§ 2.7.4.2). A major fraction of the sub-mm background has been resolved with the deepest SCUBA maps, exploiting a sensitivity boost from gravitational lenses (Chapman et al. 2003), but it is expected that a small fraction of the total CIRB is due to such luminous, rare, and high-z sources. From the observed correlations of mid-IR, far-IR, and radio luminosities, Elbaz et al. (2002) argue that the galaxies detected in the ISOCAM deepest 15 μm surveys may be responsible for about two-thirds of the integrated intensity of the CIRB at the peak wavelength of 140 μm.

Most recently, deep integrations with the Spitzer Space Telescope have resolved significant fractions of the near-IR extragalactic background. Fazio et al. (2004) report galaxy counts at 3.6 and 4.5 μm that converge very fast below fluxes of 100 μJy and corresponding to integrated background levels of 5.4 and 3.5 10^{-9} nWatt/m^2/sr, i.e., factors 4–5 less than the IRTS and COBE tentative direct estimates, implying either a truly diffuse background or zodiacal contamination in such measures.

2.8.1.3 The X-Ray Background

The x-ray background was historically the first cosmological background radiation discovered (Giacconi et al. 1962). The XRB is isotropic to a few % level on degree scales, and shows the dipole signal due to the motion of our rest-frame, proving its cosmic origin. The XRB spectrum and angular distribution have been accurately measured at 3 to 100 keV by a dedicated space experiment, the HEAO-1 (Boldt et al. 1987). At energies < 10 keV, measurements of the XRB with ASCA and XMM–Newton have produced somewhat discrepant results (by up to 30%), although these observatories are not optimized for background measurements.

The XRB sources remained unknown for more than 3 decades, although it was long suspected (Setti & Woltjer 1989; Franceschini et al. 1993; Comastri et al. 1995) that it was due to the integrated emission of populations of AGN.

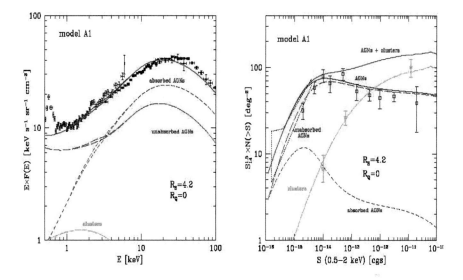

Figure 2.34. Left panel: the spectrum of the XRB compared with predictions of a standard population synthesis model (Gilli et al. 1999). XRB data below 6 keV are from Gendreau et al. (1995), while those above 3 keV are from the HEAO1 mission (Gruber 1992). The solid line is the overall fit; the other lines are the contributions of absorbed and unabsorbed AGNs. The radiation excess due to the iron line is also shown. Right panel: ROSAT integral counts at 0.5–2 keV, normalized to $S^{-1.5}$. The contributions from x-ray clusters, and of normal and obscured AGNs are indicated. The lines are the predictions of an XRB synthesis model by Gilli et al. (1999). (Taken from Gilli et al. 1999.)

Various models for the AGN evolution have been put forward in the meantime, attempting to combine the AGN statistics at various x-ray energies and the XRB spectral intensity at the same time. This has caused us to develop the so-called AGN population synthesis models, in which AGN classes with suitable distributions of the N_H column density consistent with the unification paradigm are called for to reproduce the whole dataset available (Madau et al. 1994; Comastri et al. 1995; Gilli, Salvati, & Hasinger 2001). Standard assumptions for these schemes were: the type-I and -II AGN fractions are ruled by the AGN unification scheme, hence type-II AGNs have the same luminosity function and the same evolution of type-Is; the spectral shapes and fractions of both types are independent of redshift. One of the nice features of this approach was to provide well-testable predictions, which are discussed in § 2.8.2.4.

After the operation of the Chandra and XMM–Newton observatories, the origin of the XRB has been firmly established. The extremely deep surveys by Chandra (Giacconi et al. 2001; Rosati et al. 2002; Brandt et al. 2001; Barger et al. 2002) have resolved into descrete sources most, if not all, the XRB emission at 1–10 keV. These sources appear to be a mix of type-I and -II AGNs, supporting the view that the bulk of the XRB flux peaking at 10–50 keV is due to a population of obscured AGN exceeding in number at these high x-ray energies the classical type-I objects.

2.8.2 Cosmological Evolutionary Patterns for AGNs

With the addition of the observational constraints from the background radiation to the direct statistical observables on distant active galaxy populations discussed in Chapter 3, we are now in a position to attempt a full characterization of their cosmological evolutionary patterns.

2.8.2.1 *Interpreting the Active Galaxy Evolution at Long-λ*

The 15 μm source counts shown in Figure 2.26 show evidence for strong evolution of the IR source population. The shape of these differential counts contains some indications about the properties of this evolving population: they show a roughly Euclidean slope from the brightest fluxes down to $S_{15} \sim 10$ mJy; a sudden upturn at $S_{15} < 3$ mJy, where the counts increase as $dN \propto S^{-3.1}dS$; and a convergence below $S_{15} \sim 0.2$ mJy (where $dN \propto S^{-2}dS$). This shape suggests that, for a fraction of the IR galaxies, the luminosities and spatial densities increase very fast with lookback time.

A best-fitting attempt (also involving data on the z-distributions, luminosity functions, and data at other IR and sub-mm wavelengths) was described in Franceschini et al. (2001). The model assumes the existence of two basic source populations with different physical and evolutionary properties: quiescent spirals (long dashed line in Figure 2.33) and a population of fast evolving sources (dotted line, including both starburst galaxies and type-II AGNs). The local fraction of the evolving starburst population is ~10% of the total. The contributions of quiescent galaxies and evolving starburst/AGNs are reported in the figure, while their total contribution to the CIRB appears as a thick line in Figure 2.33.

In this scenario, the active and the quiescent galaxies belong to the same population: each galaxy is expected to spend most of its lifetime in the quiescent state, but occasionally interactions or merger events with other galaxies trigger a short-lived (few to several 10^7 years) active starbursting phase. The inferred cosmological evolution for the latter may be interpreted as an increased chance to detect a galaxy during the active phase back in the past. This may follow from the higher probability of interactions during the past denser epochs (density evolution), and the larger gas masses available at higher z to form stars, which increases the rate of star formation and starburst luminosity (luminosity evolution), and the chance of triggering a nuclear AGN.

Our analysis in § 2.7.5 has found that only a fraction ($< 20\%$) of active galaxies show evidence for an energetically dominant AGN, and that for sources in which an AGN contribution is suspected, the bolometric emission is in any case dominated by star formation.

The evolution of the IR emissivity of galaxies from the present time to $z \sim 1$ is so strong that the combined set of constraints by the observed z-distributions and CIRB spectrum tend to impose it to turnover at $z > 1$ (Franceschini 2001): scenarios in which a relevant fraction of stellar formation occurs at very high-z (e.g., producing the bulk of stars in spheroids at $z > 3$) are not supported by these observations. We finally notice an interesting, though yet qualitative, match between this epoch of peak SF at $z \simeq 1$ to 2, as revealed by IR observations, and the epoch when the galaxy mass function shows first signs of a strong negative evolution (Zepf 1997; Franceschini et al. 1998; Franceschini & Rodighiero 2003; Dickinson et al. 2003).

2.8.2.2 *Matching the Background Energy Density and the Local Stellar Mass Density*

Relevant constraints on the high-redshift active galaxy populations can be inferred from the observed energy densities in the CIRB and optical backgrounds, under the assumption that they are mostly due to star-formation activity. Let us assume that a fraction f_* of the universal mass density in baryons Ω_b undergoes at redshift z_* a transformation with radiative efficiency ϵ. Note that for stellar processes, ϵ is determined by the IMF: $\epsilon = 0.001$ for a Salpeter IMF and a low-mass cutoff $M_{\mathrm{low}} = 0.1\ M_\odot$, while, for example, $\epsilon = 0.002$ for $M_{\mathrm{low}} = 2$. Note that the usually quoted value for such efficiency ($\epsilon = 0.007$) would apply for stellar generations including only high-mass stars ($M > 8 M_\odot$).

There is a straightforward relationship (e.g., Peebles 1993) between the bolometric intensity of a background radiation intensity and the energy production of its sources at $z = z_*$:

$$\nu I(\nu) = \left(\frac{c}{4\pi}\right)\left(\frac{\epsilon \rho c^2}{1 + z_*}\right), \qquad (2.3)$$

where ρ is the comoving baryon density and ϵ the total efficiency.

The local amount of baryons in stars, the total amount of baryons, and the Hubble constant, all ingredients of Equation (2.3), have been recently reviewed by Fukugita, Hogan & Peebles (1998), and Spergel et al. (2003) based on W-MAP CMB observations, and from the HST Key Program by Freedman et al. (2001). These results indicate the following set of values for the cosmological parameters: $H_0 = (70 \pm 5) h_{70}$ km/s/Mpc, $\Omega_b = 0.046 h_{70}^{-1}$

and $\Omega_* = 0.0035h_{70}^{-1}$ in units of the critical density, and a local fraction of baryons in stars of 8%.

Let us schematically assume that the optical/NIR background mostly originates by quiescent SF in spiral disks and by intermediate and low-mass stars. As observed in the solar neighborhood, a good approximation to the IMF in such moderately active environments is the Salpeter law with $M_{low} = 0.1$ ($\epsilon \sim 0.001$). Then the optical BKG intensity can be obtained by transforming at redshift $z_* \sim 1$ a fraction $f_* \simeq 5\%$ of baryons into (mostly low-mass) stars:

$$\nu I(\nu)|_{opt} \simeq 18 \ h_{70}^2 \left(\frac{\Omega_b}{0.046}\right) \left(\frac{f_*}{0.05}\right) \left(\frac{2.}{1+z_*}\right) \left(\frac{\epsilon}{0.001}\right) \ \mathrm{nW/m^2/sr},$$

(2.4)

which corresponds to roughly half the local density in low-mass stars being generated in this way and roughly the observed amount of heavy elements in stars.

On the other hand, given that luminous starbursts emit a negligible fraction of the energy in the optical-UV and most of it in the far-IR, we further assume that the CIRB (with its energy density of 35 nW/m²/sr, see § 2.8.2.1) originates from dusty star-forming galaxies at median $z_* \simeq 1$. This process has to explain the large energy content in the CIRB without exceeding the stellar remnants in local galaxies. Franceschini et al. (2001) argue that if during the starburst phase SF happens with a top-heavy IMF, this would ease the energy constraint from the CIRB intensity. Assuming $M_{low} \simeq 2 \ M_\odot$, we can obtain:

$$\nu I(\nu)|_{CIRB} \simeq 35 \ h_{70}^2 \left(\frac{\Omega_b}{0.046}\right) \left(\frac{f_*}{0.04}\right) \left(\frac{2.}{1+z_*}\right) \left(\frac{\epsilon}{0.002}\right) \mathrm{nW/m^2/sr},$$

(2.5)

This requires that a similar amount of baryons, $f_* \simeq 4\%$, as those transformed during the "secular" evolution, are processed with higher efficiency during the starbursting phases, producing a two times larger amount of energy and metals. Most of these metals have to be released by the galaxies into the diffuse cosmic media, as observed for example in the intracluster plasmas.

2.8.2.3 A Two-Phase Star-Formation Scheme: Origin of Galactic Disks and Spheroids

Our best-fit model for IR galaxy evolution implies that star formation in galaxies has proceeded along two phases: a quiescent one taking place during most of the Hubble time, slowly building stars with standard IMF from the regular flow of gas in rotational supported disks; and a transient actively starbursting phase, recurrently triggered by galaxy mergers and interactions. During the merger, *violent relaxation* redistributes old stars, producing de Vaucouleur profiles typical of galaxy spheroids. The energy constraints imposed by the CIRB suggest that during this phase young stars may be formed with a top-heavy IMF.

Because of the geometric (thin disk) configuration of the diffuse ISM and the modest incidence of dusty molecular clouds, the quiescent phase is only moderately affected by dust extinction, and naturally originates most of the optical/near-IR background.

The merger-triggered active starburst phase is characterized by a large-scale redistribution of the dusty ISM, with bar-modes and shocks compressing a large fraction of the gas into the inner galactic regions and triggering formation in molecular clouds. As a consequence, this phase is expected to be heavily extinguished and naturally originates the CIRB. Based on dynamical considerations, we expect that during this violent starburst phase the elliptical and S0 galaxies are formed in the most luminous IR starbursts at higher-z (corresponding to the SCUBA source population), whereas galactic bulges in later-type galaxies likely form in lower IR luminosity and lower-z objects (the ISO population).

Due to the local evidence that supermassive black holes are uniquely related with spheroidal components in galaxies, we expect that the past AGN phase is associated with the merger-triggered active starburst phase.

2.8.2.4 *Tracing the History of Gravitational Accretion in X-Rays*

New interesting facts have emerged from Chandra and XMM–Newton deep surveys, fairly radically changing our view of AGN evolution and showing some unexpected features of the sources of the XRB. These deep observations (some of them totaling 2 Msec of observing time with Chandra and up to 1 Msec with XMM) have essentially resolved the whole XRB at 2–10 keV (the exact resolved fractions depend on the somewhat uncertain normalization of the XRB spectrum).

A critical constraint on the XRB population synthesis models comes in particular from the redshift distributions of samples of several hundred faint hard x-ray sources, obtained through systematic follow-up efforts of the optical counterparts with large telescopes. The fraction of sources with spectroscopic redshifts already exceeds 60% (Hasinger 2003) and for the remaining sources techniques of photometric redshifts have been implemented (Zheng et al. 2004; Franceschini et al. 2004). These observed redshift distributions come out to be remarkably different from the predictions; with the majority of the sources falling below $z = 1$, whereas the expectation was for a large majority to be found at $z > 1$ (see Figure 2.35). The observed low-z peak is dominated by Seyfert galaxies with X-ray luminosities of 10^{42-44} erg/s, which therefore are bound to evolve differently from type-I QSOs.

Attempts to overcome these difficulties of the XRB population mod-

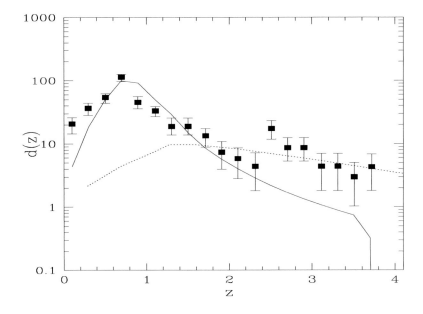

Figure 2.35. Redshift distributions from Hasinger (2003) of x-ray selected AGNs in deep Chandra and XMM surveys with flux limits between $S_{2-10} \sim 5 \ 10^{-16}$ and $\sim 1.5 \ 10^{-15}$ erg/s/cm^2, respectively. This is a representative sample of the x-ray population at the faint flux limits. The dotted line is a prediction based on the XRB *synthesis model* by Gilli et al. (2001) fitting the data only at $z > 1.5$. The continuous curve is the prediction by Franceschini, Braito & Fadda (2002) of the contribution of type-II AGNs from a model based on AGN evolution in the mid-IR. This provides a remarkable description of the data at $z < 2$, where the bulk of the XRB is produced.

els (other problems were to properly fit the counts showing evidence for steeper slopes at increasing x-ray energies, and the XRB spectral shape which appears harder than the population synthesis predictions, see Figure 2.34 left panel) were proposed by Franceschini, Braito, & Fadda (2002) and Gandhi & Fabian (2003). They have first relaxed the link between the type-I and type-II objects by assuming that the former evolve as found in optical and soft x-ray surveys, while the latter follow a different evolutionary history. They have then modeled the absorbed type-II AGN population starting from the evolutionary model of active galaxies needed to reproduce multiwavelength IR counts (see § 2.8.2.1) and exploiting the evidence (see Figure 2.29) that a roughly constant fraction $f_{AGN} \sim 15\%$ of this active galaxy population contains buried AGNs. By combining this information with the observed flux ratios between the 15 μm and x-ray fluxes, they were able to reproduce all the main x-ray observables: the counts as a function of energy, the z-distributions, and the XRB shape. Figure 2.35 shows how this model reproduces the observed z-distribution of faint x-ray AGNs.

Then combined deep x-ray and IR observations consistently find that the universe has experienced a violent phase of galaxy activity around $z \simeq 1$, which generates massive stars (whose flux is mostly re-radiated in the IR) and fuels nuclear BHs. Our analysis implies that roughly 10 to 20% of this activity has involved substantial AGN emission, mostly detectable as a faint hard x-ray flux. This fraction likely represents the duration of the AGN phase compared with that of the starburst during the activation process. While the IR re-radiated emission from AGNs can only contribute a minor fraction ($\sim 10\%$) of the CIRB, their hard x-ray emission is largely responsible for the XRB. Luminous type-I quasars turn out to be very biased tracers of this activity, their epoch of maximal emissivity being displaced to $z = 2$.

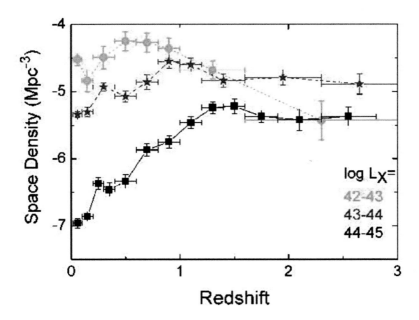

Figure 2.36. Preliminary evaluation of the comoving space densities of type-I x-ray AGNs from combined Chandra/ROSAT deep surveys as a function of redshift in 3 different luminosity classes: $\log(L_X)$ =42-43, 43-44 and 44-45. (From Hasinger 2003.)

This latter fact has been impressively proven by Hasinger (2003) based on preliminary evaluations of the AGN luminosity functions over very large redshift and luminosity intervals, obtained by combining the whole available dataset on x-ray AGNs (some of these results have been discussed in § 2.7.3, see Figure 2.23). To make the analysis as complete and homogeneous as possible, he has considered only type-I AGNs selected in the narrow energy interval 0.5–2 keV. Figure 2.36 shows the different evolutionary behavior for AGNs of different luminosities ($\log(L_X)$=42-43, 43-44 and 44-45). While the high-luminosity x-ray QSOs show the well known strong evolution with an increase by almost two orders of magnitude in comoving space density from $z = 0$ to 1.5, and keeping flat above $z \simeq 1.5$, the evolution of the lower luminosity classes is weaker and saturates at significantly lower redshifts. Seyferts of luminosity class $\log L$ =42-43 at $z = 0.5$ appear almost 100 times more numerous than QSOs at the same z, while the two AGN classes have comparable densities at $z = 2$. Beyond $z = 0.7$, there is a significant decline of the Seyfert space density. These, still preliminary, results reveal a dramatically different evolutionary picture for low-luminosity AGNs compared to the high-luminosity QSOs: in Hasinger's terms, "while the rare, high-luminosity objects can form and radiate very efficiently early in the universe, the bulk of the lower—luminosity AGNs have to wait much longer to grow." At first sight this evolutionary behavior may not fit completely into the simplest predictions of the hierarchical formation scenario.

Is there any evidence for such an *inverted* evolutionary pattern for the starburst population? Indications in this sense are discussed in § 2.8.3 below.

2.8.2.5   *Relating Fossil Remnants of the Past AGN Activity*

A very simple argument originally due to Soltan (1982) relates the total background light energy density ρ_γ, emitted by a population of sources from gravitational accretion of gas, with the density of the local remnants, in our case the density $\rho_\bullet$ of supermassive black holes in present-day galaxies. Further ingredients in this analysis are the efficiency η of the matter-radiant energy transformation and, if the energy is observed within a given photon waveband, the factor K correcting from bolometric to band emission. Then the energy observed within the band is:

$$E = \eta K M c^2, \qquad (2.6)$$

where $\eta = 0.06$ for Schwartchild and 0.40 for a maximally rotating Kerr black hole (e.g., Frank et al. 1992). The corresponding background radiation energy density is given by:

$$\rho_\gamma = \frac{\eta K \rho_\bullet c^2}{1 + z} \qquad (2.7)$$

where z is the mean redshift of the population. This result, coming from the same arguments justifying Equation (2.3), is remarkably general and independent of the assumed cosmology and requires only that the redshift distribution of the sources is known.

The local supermassive black hole density has been estimated by various authors using the tightest available correlation with black hole mass, the one between the mass and the velocity dispersion of the hosting bulge, the $M_\bullet - \sigma$ relation (Ferrarese & Merritt 2000; Gebhardt et al. 2000). The most recent update of such analyses reports a local black hole average mass density of (Ferrarese 2002, for $h_{70} = 1$):

$$\rho_\bullet \simeq 4.5 \pm 0.5 \ 10^5 \ M_\odot \mathrm{Mpc}^{-3}. \tag{2.8}$$

Searches of quasars in the optical have been very successful in characterizing their luminosity functions up to very high redshifts (§ 2.7.1). A recent analysis by Yu & Tremaine (2002) has compared their own (low) estimate of $\rho_\bullet \simeq 2.5 \pm 0.4 \ 10^5 \ M_\odot \mathrm{Mpc}^{-3}$ with the prediction of the QSO optical luminosity functions based on the 2dF redshift survey. They found that the QSO counts and their estimate of $\rho_\bullet$ are well in agreement with each other if a standard efficiency $\eta = 0.1$ is assumed. However, optical surveys are subject to the effects of dust extinction and do not allow detection of obscured AGNs. So the Yu & Tremaine account does not consider the obscured accretion.

X-ray studies, instead, provide tighter constraints on the history of gravitational accretion in the universe, being less sensitive to obscuration, particularly for high-z AGNs for which the rest-frame unabsorbed continuum at > 20 keV is redshifted into the observed 2–10 keV band. As previously discussed, the XRB itself and its spectral shape indicate very clearly that the majority of energy by AGNs is obscured: by considering a typical $\alpha_x = 1$ energy spectral index for the unobscured AGNs and comparing it with the XRB spectrum, Fabian & Iwasawa (1999) find that 80% or more of the XRB energy comes from obscured AGNs, while only 20% of it is from classical type-I QSOs (see also Franceschini et al. 1999).

We can now attempt to put together the information so far collected (§ 2.8.2.4 and this one) to see if there is a single self-consistent working scheme relating all the fossil records of the past AGN activity, the local BH mass density $\rho_\bullet$, the XRB spectral intensity, and the optical and x-ray AGN statistics. As discussed in Fabian (2003), from the Yu & Tremaine analysis [Equation (2.7)] we take that luminous type-I quasars, effectively found in optical surveys, with a redshift of peak activity $z \sim 2$ and accretion efficiency $\eta = 0.1$, explain the optical QSO counts and contribute $\sim 20\%$ of the XRB and a local mass density in supermassive black holes ($M > 10^8$ $M_\odot$) of $\sim 2 \ 10^5 \ M_\odot \ \mathrm{Mpc}^{-3}$. This assumes an energy density in the XRB of $6.7 \ 10^{-17}$ erg/cm^3 and a bolometric correction factor $K = 1/30$ typical of luminous AGNs (Elvis et al. 1994).

On the other hand, as discussed in Franceschini, Vercellone & Fabian (1998) and Fabian (2003), the bulk (80%) of the XRB is due to a population of lower-mass black holes ($M < 10^8$ M$_\odot$), undergoing obscured accretion at typically much lower redshift, $z \sim 0.7$. Assuming for them the same standard efficiency $\eta = 0.1$, a bolometric correction of $K = 1/20$ (typical of moderate-luminosity AGNs), and an energy density in the XRB of $6.7 \, 10^{-17}$ erg/cm^3, then from Equation (10) we can explain 80% of this energy with an additional contribution of $\simeq 2.5 \, 10^5$ M$_\odot$ Mpc^{-3} to the black hole mass density.

Altogether, the combined contributions of high-mass black holes descendant of the luminous QSO phase (mostly type-I QSO, efficiently sampled by optical surveys), and the lower-mass objects whose obscured ancestors have produced the bulk of the XRB, make up the local mass density, in nice consistency with the observations [Equation (2.8)].

As emphasized by Fabian (2003), in the local black hole mass budget there may be marginal room for only a minor fraction ($< 20\%$, or $< 10^5$ M$_\odot$ Mpc^{-3}) of black hole remnants from a completely obscured, Compton-thick AGN phase, not detectable in hard x-rays.

On the other hand, although dominating the XRB and despite that $\sim 80\%$ of their UV to x-ray spectrum is re-radiated into the far-IR, obscured AGNs contribute a likely small fraction of the energy of the CIRB background. The analyses of Fadda et al. (2002), Elbaz et al. (2002), and Franceschini et al. (2004), see § 2.7.5.1, have shown that the fraction of the CIRB due to AGN emission is $\simeq 15\%$ at 15 μm to 24 μm (the wavelength where the CIRB has been so far best resolved). Considering however that typical AGN IR spectra peak at shorter wavelengths compared to starbursts, the AGN fraction to the bolometric CIRB energy should be lower. Based on realistic spectral corrections, Elbaz et al. estimate a total AGN contribution to the CIRB of $\sim 4\%$. Indeed there cannot be room for a much larger contribution, unless the local black hole density $\rho_\bullet$ has been seriously underestimated.

To conclude, we have provided here and in the previous section an illustration of the power of combining data on (both local and high-redshift) source statistics with the upper limits set by the background radiation. The situation is particularly favorable if we consider that we can exploit independent constraints on both the star formation (from the CIRB) and the gravitational accretion (from the XRB) phenomena.

2.8.3 Coeval Evolution of Starbursts and AGNs

A pioneering attempt to compare the evolution with cosmic time of the comoving volume star-formation rate in galaxies with the evolution of the emissivity by AGNs has been discussed in Franceschini et al. (1999; see also Hasinger 1998), as a clue to understanding the relationship between black hole accretion and the formation of the surrounding structure. An interesting similarity has been found between the evolution rates for the total populations of galaxies and AGNs, which indicates that, on average, the history of BH accretion tracks that of stellar formation in the hosting galaxies. This result is illustrated in Figure 2.37. In the top figure the comoving volume emissivity based on a model of AGN evolution based on the ROSAT x-ray surveys selected from 0.5 to 2 keV (Miyaji et al. 1999, the upper lines) is compared with the comoving rate of star formation (SFR) in galaxies estimated from flux-limited spectroscopic surveys of galaxies (small filled circles; Lilly et al. 1995, Steidel et al. 1994). Both the SFR and AGN x-ray emissivity show a consistent increase by a factor ~ 10–20 going from $z = 0$ to $z \simeq 1$–1.5 and a flattening thereafter. Note that the curve of the comoving SFR is referred to the left axis and the AGN emissivity to the right one, while the relative normalization of the right and left axes is left free. More information is reported in the figure caption.

An attempt to disentangle the high-luminosity and high-mass contributions of galaxies and AGNs is also reported in Figure 2.37. The shaded region here corresponds to an estimate of the SFR history of stellar populations in high-mass spheroidal galaxies morphologically selected in the HDF-North. This evolutionary SFR density is roughly constant at $z \geq 1.5$, while showing a fast convergence at lower redshifts (correspondingly, 90% of stars in E/S0 galaxies appear to have been formed between $z \simeq 1$ and 3). This evolution pattern looks quite different from that of the general field population (small circles in the figure), which displays a much shallower dependence on cosmic time and a relatively sustained SFR to low redshifts. This is reminiscent of the results that we discussed in § 2.8.2.4 and Figure 2.36 about the differences in the evolution rates of high- and low-luminosity AGNs favoring higher evolution for the former.

The red filled squares in Figure 2.37 are, for comparison, the 0.5 to 2 keV comoving volume emissivities (in erg/s/Mpc^3, see right axis) of high-luminosity AGNs ($L_{0.5-2} \geq 10^{44.25}$ erg/s), consistent with that reported in Figure 2.36 for the high-luminosity AGNs (the novelty of that figure concerns the lower-luminosity population deeply traced with Chandra for the first time). Again, the two evolution histories of high-luminosity and high-mass galaxies and AGNs are interestingly similar.

In conclusion, we can summarize the main results of this and § 2.8.2.4 as follows. On one side, we have shown that the evolution of luminous quasars parallels that of the stellar populations in spheroidal galaxies, in keeping with the locally established association of supermassive BHs and galactic bulges. Furthermore, for both the AGN and galaxy populations there seems to be a systematic trend for the high-luminosity, high-mass systems to evolve on a faster cosmological timescale than the lower-mass lower-luminosity objects.

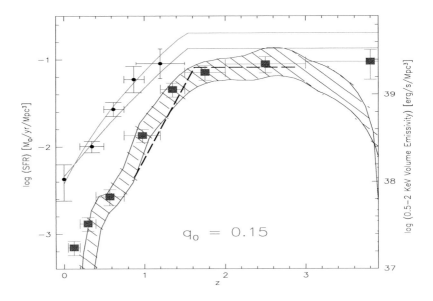

Figure 2.37. Comparison of the redshift evolution of the AGN emissivity per comoving volume (to be read on the right-hand axis) with the evolution of the star-formation rate in galaxies (left-hand axis). Small filled circles are the evolutionary SFR of field galaxies estimated from conversion of the rest-frame optical-UV flux to SFR. The shaded region describes the evolution of the SFR in field massive galaxies morphologically classified as ellipticals and S0s (Franceschini, et al. 1998). The filled squares are the 0.5 to 2 keV comoving volume emissivities (in erg/s/Mpc^3) of high-luminosity AGNs ($L_{0.5-2} \geq 10^{44.25}$ erg/s). The top two lines are the 0.5–2 keV volume emissivities of the total AGN population (normalized by the same factor), based on best-fit models of x-ray AGN evolution (Miyaji et al. 1999; Franceschini et al. 1999). The thick dashed line is the volume emissivity of optical quasars from Schmidt, Schneider & Gunn (1995). The scales of the two vertical axes are chosen in such a way that AGN volume emissivity and SFR density overplot each other. An open universe is assumed, with $q_0 = 0.15$.

2.8.4   Theoretical Issues

The first of our conclusions in the previous section, i.e., the tight parallelism of the evolution of host galaxies and nuclear supermassive black holes, are fairly naturally accounted for by current structure formation scenarios. Particularly, cold dark matter hierarchical models have been exploited to investigate their relationship. The two usually adopted modelistic approaches either rely on the Press–Schechter (1974) gravitationally-driven scheme, or to follow the semi-analytical approach implementing physical recipes for baryons into N-body simulations of the dark matter field. In such a framework, small-mass sub-galactic systems form first and then merge later into larger and larger aggregates. Proto-galaxies then experience multiple mergers during a protracted evolution covering a large fraction of the Hubble time from $z > 6$ to $z < 1$. Mergers between comparable-mass systems (major mergers) are expected to result in the formation of elliptical galaxies (see, e.g., Barnes & Hernquist 1992, 1996). This is probably the only way to achieve stellar systems with very high central concentrations as observed in early-type galaxies, an obviously favorable birth place for massive nuclear star clusters and a supermassive collapsed object.

In these hierarchical models, the black holes grow in mass during major mergers but are present in each galaxy at any redshifts. This process is then naturally *recurrent* and is able in principle to explain many of the observational evolutionary properties that we have discussed in previous sections (e.g., Haehnelt & Rees 1993; Rees & Silk 1998; Kauffmann & Haehnelt 2000; Cavaliere & Vittorini 2000; Wyithe & Loeb 2002; Volonteri, Haart, & Madau 2003). In particular, it naturally explains the *periodic and recursive* nature of the activity discussed in Sects. 2.7.5.1 and 2.7.5.2. The model also predicts that only a fraction of the supermassive black holes and of galaxies are active at any given time, in keeping with the two-phase evolution as in § 2.8.2.3.

However, there are important aspects of the scenario that require special attention and which may not easily fit. On one side, the enormous range in redshifts during which enhanced activity has taken place and the evidence for luminous quasars in seemingly massive, metal enriched hosts at $z > 6$, are difficult to explain without violating the constraints set on the local mass-density in black holes $\rho_{\bullet} \simeq 4.5\ 10^5\ M_{\odot}/\mathrm{Mpc}^3$ and the optical QSO luminosity functions at $z \sim 2 - 3$ (Volonteri et al. 2003; Bromley, Somerville & Fabian 2004). In particular, when the models are normalized to reproduce the QSO counts at $z \sim 6$, the local mass budget is grossly exceeded. Bromley et al. discuss some alternatives to overcome the problem. They indicate the need for a scaling ($f_m \propto v_{vir}^2$) of the gas accretion fraction f_m with the halo mass and virial velocity $v_{vir}^2 = GM/r$.

A probably even more serious difficulty may be to explain the observed effect of the high-luminosity, high-mass systems evolving on a faster cosmic timescale than the lower-mass lower-luminosity objects. This is rather clearly inconsistent with simple schemes for the formation of galaxies and AGNs based on purely gravitationally driven formalisms. For example, Haiman & Menou (2000) obtain from the Press-Schechter theory a much faster decay with cosmic time of the AGN accretion rate (hence of the AGN luminosity) for low-mass BHs in low-mass dark-matter halos, while that of massive objects is barely expected to decrease from $z = 3$ to $z = 0$. A similar problem likely arises when attempting to explain with a process of mostly gravitational clustering the origin of galaxies, and their spheroidal components in particular (see, e.g., the difficulty of explaining the evidence for E/S0 galaxies at high redshifts, e.g., Cimatti et al. 2004).

Progress is obtained by complementing the theories of structure formation based on the gravitational collapse of dark matter halos with some more detailed description of the dynamical processes in the baryonic component, particularly important in the high-density environments.

For example, feedback processes on the accreting baryonic gas, due to both young stars and the quasar itself (e.g., Silk & Rees 1998), may explain the strong scaling of the accreted fraction on the halo circular velocity v_{vir} and the somehow *anti-hierarchical* observed evolution. A solution in this sense has been recently proposed by Granato et al. (2003), although this model forms the whole observed local $\rho_\bullet$ at $z > 2$, in contradiction to our discussed constraints in § 2.8.2.5 (where at least half of $\rho_\bullet$ should be assembled at $z < 2$), and although this model may exceed the number of spheroidal galaxies at $z > 1.5$ in K-band selected samples (e.g., Somerville et al. 2004).

Another important, often neglected, effect is due to the progressive exhaustion of gas needed to fuel the AGN and the starburst, which is likely to rule the fast downward evolution to $z = 0$. Such exhaustion is due to freezing of gas, partly in low-mass stars, partly in a hot plasma, never able to cool back again as observed in galaxy clusters and groups. An interesting attempt at explaining in this sense the data in Figure 2.36 is reported in Menci et al. (2004).

As resulting from the Bromley et al. (2004) analysis, the dependence of evolution on the environment and the local environmental density, is in any case a likely key element. This, together with the effects of merging/interactions and the progressive exhaustion of the fuel, allow us to understand the accelerated evolution of luminous AGNs, and also the fast decay with time of the SF in massive spheroidal galaxies (Franceschini et al. 1999). In cosmic environments with higher-than-average density, the forming galaxies have experienced a high rate of interactions already at high redshifts ($z > 2$), this leading shortly to a population of spheroid-dominated galaxies (observable at low-z mostly in groups and clusters of galaxies) containing massive black holes whose accretion history is the same as that of the host galaxy. In lower density environments the interaction rate was slower, and proportionally lower was the mass locked in a bulge with respect to that making stars quiescently in a disk. Galaxies in these low-density environments keep forming stars and accreting matter on the black hole down to the present epoch and form the low-redshift AGN population.

These explanations seem to find remarkable confirmation by recent results of systematic analyses of galaxy populations in the local universe.

Thomas et al. (2004), in particular, have derived ages, total metallicities, and element ratios of 124 early-type galaxies in high and low density environments, and find evidence for an influence of the latter on the stellar population properties. They find that most star-formation activity in early-type galaxies is expected to have happened between redshifts $z \simeq 3$ and 5 in high density concentrations and between redshifts 1 and 2 in the field. They also infer that significant stellar mass growth in the early-type galaxy population below $z < 1$ must be restricted to less massive objects.

Based on a rich Sloan Digital Sky Survey (SDSS) low-z galaxy sample, Kauffmann et al. (2004) confirm the evidence that stellar formation and AGN activity has stopped at $z > 1$ in massive galaxies within high-density regions, while still going on in low density, and also confirm that both environmental (due to gas stripping) and intrinsic (due to gas exhaustion) effects rule the evolution of galaxies and AGNs, in addition to gravity.

2.8.5   Conclusions and Perspectives

An enormous amount of new results about the relationship of cosmological star formation and gravitational accretion in AGNs has been obtained in the last several years. These data concern not only source populations directly inspected back at the time of their formation, but also a variety of local remnants of past activity, including the multi-wavelength background radiation, the local stellar content of galaxies, and the eventually discovered population of nuclear supermassive black holes.

A general scheme for galaxy formation appears to get further support as new data accumulate, one explaining the structures on all the observed scales as a time-progressive aggregation, driven by gravity, and starting from tiny primordial fluctuations (the hierarchical clustering model).

An essential requisite of the latter is that the cosmological forming structure is dominated gravitationally by a non-baryonic cold dark matter field, whose predicted evolution seems to be reasonably consistent with the observations on large scales. Galaxies and AGNs are foreseen to progressively form by merging inside the dark matter halos, and this expectation seems also generically consistent with the data.

However, several important details of this picture require further inspection, particularly an apparently inverted evolutionary trend, such that higher-luminosity sources, containing the most massive SMBHs within the most massive spheroids, evolve on a faster cosmic timescale than the lower-luminosity, lower-mass objects. To reconcile this with the expectation of the hierarchical model, in which more massive systems grow up on longer timescales than lower-mass objects, it seems very likely that we need a much better understanding of the complex physical processes ruling the evolutionary history of baryons.

As usual, this progress will likely be driven during the next years by important instrumental and observational developments. A critical one for studying *in situ* and with unrivalled spatial resolution diffuse media is the ALMA interferometer. Our understanding of the relationship between star formation and black hole accretion will enormously benefit by the ALMA capability to investigate the highly extinguished active nuclei. The Herschel Space Observatory will also observe dusty and gaseous media at the peak of dust emission ($\lambda \sim 100~\mu$m), roughly at the same time.

Primeval objects and young stellar populations at very high redshifts will be studied with the James Webb Space Telescope, an observatory that will provide superior spatial resolution and photon collecting power than HST, at wavelengths 1 to 28 μm, where these sources would be observable.

A critical test of the obscured accretion in AGNs would require imaging capabilities at 10 to 100 keV, not possible so far with current grazing-incidence x-ray telescopes. At least Constellation-X, but other space telescopes as well, will provide this, and allow us to completely resolve the XRB at its peak emission energy around 40 keV.

Finally, a fundamental prediction of the hierarchical model for the formation of galaxies and AGNs, the coalescence of supermassive black holes in merging galaxies at any redshifts, will be testable by measuring the expected gravitational wave flow with laser space interferometry (LISA). All this new instrumentation will definitely move cosmology from the current exploratory phase to a fully mature science.

References

Alexander D M, Bauer F E, Brandt W N et al. 2003 *AJ* **125** 383

Alexander D M, Bauer F E, Chapman S C, Smail I et al. in ESO/USM/MPE Workshop on "Multiwavelength Mapping of Galaxy Formation and Evolution" eds. R. Bender and A. Renzini (astroph/0401129)

Andreani P, Franceschini A and Granato G 1999 *MNRAS* **306** 161

Antonucci R and Miller J 1985 *ApJ* **297** 621

Aussel H, Cesarsky C, Elbaz D and Starck J L 1999 *A&A* **342** 313

Barger A J, Cowie L L, Brandt W N et al. 2002 *AJ* **124** 1838

Barnes J E and Hernquist L E 1992 *ARAA* **30** 705

Bennett C L et al. 2003 *ApJ* **583** 1

Bernstein R A, Freedman W and Madore B F 2002 *ApJ* **571** 56

Bertoldi F, Menten K M, Kreysa E, Carilli C L and Owen F 2001 in *Highlights of Astronomy* Vol. 12 (San Francisco ASP)

Best P N, Longair M S and Roettgering H J 1997 *MNRAS* **292** 758

Blain A W, Smail I, Ivison Kneib J -P and Frayer D T 2002 *Physics Reports* **369** 111

Boldt E et al. 1987 *Phys. Rep.* **146** 215

Boyle B J, Shanks T, Croom S M et al. 2000 *MNRAS* **317** 1014

Braito V et al. 2004 *A&A* **420** 79

Brandt W N et al. 2001 *ApJ* **558** L5

Bromley J M, Somerville R S and Fabian A C 2004 *MNRAS* **350** 456

Burbidge G Burbidge 1967 in *Quasi Stellar Objects* New York Freeman

Cavaliere A and Padovani P 1989 *ApJ* **340** L5

Cavaliere A and Vittorini V 2000 *ApJ* **543** 599

Chapman S C, Blain A, Ivison R and Smail I 2003 *Nature* **422** 695

Chapman S C, Smail I, Windhorst R, Muxlow T and Ivison R J 2004 *ApJ* **611** 732

Cimatti A, Daddi E, Renzini A, Cassata P et al. 2004 *Nature* **430** 184

Comastri A et al. 1995 *A&A* **296** 1

Conselice C J, Chapman S C and Windhorst R A 2003 *ApJ* **596** L5

Dietrich F et al. 2003 *A&A* **398** 891

Dickinson M 2004 *AAS* 204 3313

Dickinson M, Papovich C, Ferguson H C and Budavari T 2003 *ApJ* **587** 25

Dole H, Gispert R, Lagache G and Puget J -L 2001 *A&A* **372** 364

Dunlop J S 2003 in *Coevolution of Black Holes and Galaxies* ed. L. C. Ho

Carnegie Observatories Astrophysics Series Vol. 1 (Cambridge: Cambridge Univ. Press)

Dunlop J S and Peacock J 1990 *MNRAS* **247** 19

Eales S et al. 2000 *AJ* **120** 2244

Elbaz D, Aussel H, Cesarsky C J, Desert F X, Fadda D, Franceschini A, Puget J L and Starck J L 1999 *A&A* **351** L37

Elbaz D, Cesarsky C J, Chanial P, Franceschini A et al. 2002 *A&A* **384** 848

Ellis R S 1997 *ARAA* **35** 389

Elvis M, Wilkes B J, McDowell et al. 1994 *ApJS* **95** 1

Fabian A.C 2003 in *Coevolution of Black Holes and Galaxies* Carnegie Observatories Astrophysics Series Vol. 1 ed. L. C. Ho (Cambridge: Cambridge Univ. Press) (astroph/0304122)

Fabian A C and Iwasawa K 1999 *MNRAS* **303** L34

Fadda D, Flores H, Hasinger G et al. 2002 *A&A* **383** 838

Fadda D, Lari C, Rodighiero G, Franceschini A, Elbaz D, Cesarsky C and Perez-Fournon I 2004 *A&A* **427** 23

Fan X et al. 2003 *AJ* **125** 1649

Fazio G et al. 2004 *ApJS* **154** 39

Ferrarese L 2002 in *Current high-energy emission around black holes* Proceedings of the 2nd KIAS Astrophysics Workshop Korea (astroph/0203047)

Ferrarese L and Merritt D 2000 *ApJ* **539** L9

Fixsen D J et al. 1998 *ApJ* **508** 123

Flores H, Hammer F, Elbaz D, Cesarsky C J, Liang Y C, Fadda D and Gruel N 2004

Flores H, Hammer F, Thuan T et al. 1999 *ApJ* **517** 148.

Franceschini A 2001 IAU Symposium 204 *The Extragalactic IR Background* M. Harwit and M. Hauser Eds 283

Franceschini A and Gratton R 1997 *MNRAS* **286** 235

Franceschini A and Rodighiero G 2003 in *The emergence of cosmic structure*: Thirteenth Astrophysics Conference. AIP Conference Proceedings Volume 666 pp. 215

Franceschini A, Aussel H, Cesarsky C, Elbaz D and Fadda D 2001 *A&A* **378** 1

Franceschini A, Bassani L, Cappi M et al. 2000 *A&A* **353** 910

Franceschini A, Berta S, Rigopoulou D, Aussel H et al. 2003 *A&A* **403** 501

Franceschini A, Braito V and Fadda D 2002 *MNRAS* **335** L51

Franceschini A, Hasinger G, Miyaji T and Malquori D 1999 *MNRAS* **310** L5

Franceschini A, Manners J, Polletta M et al. 2004 *AJ* in press

Franceschini A, Vercellone S and Fabian A C 1998 *MNRAS* **297** 817

Franceschini A et al. 1993 *MNRAS* **264** 35

Franceschini et al. 1998 *ApJ* **506** 600

Franceschini A et al. 2003 *MNRAS* **343** 1181

Frank J et al. 1992 *Accretion Power in Astrophysics* (Cambridge: Cambridge Univ. Press)

Frayer D T, Ivison R J, Scoville N Z et al. 1998 *ApJ* **506** L7

Frayer D T et al. 1999 *Astrophys. J.* **514** L13

Freedman et al. 2001 *ApJ* **553** 47

Freudling W, Corbin M R and Korista K T 2003 *ApJL* **587** L67

Fukugita M, Hogan C J and Peebles P J E 1998 *ApJ* **503** 518

Gandhi P and Fabian A C 2003 *MNRAS* **339** 1095

Gebhardt K et al. 2000 *ApJ* **539** L13

Gendreau K C, Mushotzky R F, Fabian A C et al. 1995 *PASJ* **47** L5

Genzel R and Cesarsky C. 2000 *ARAA* **38** 761

Genzel R, Tacconi L J, Rigopoulou D, Lutz D and Tecza M 2001 *ApJ* **563** 527

Genzel R et al. 1998 *ApJ* **498** 579

Giacconi R et al. 1962 *Phys. Rev. Lett.* **9** 439

Giacconi R et al. 2001 *ApJ* **551** 624

Gilli R, Risaliti G and Salvati M 1999 *A&A* **347** 424

Gilli R, Salvati M and Hasinger G 2001 *A&A* **366** 407

Gorjian V, Wright E L and Chary R R 2000 *ApJ* **536** 550

Granato G et al. 2004 *ApJ* **600** 580

Gruber D E 1992 in the Proceedings of The Xray background eds. X Barcons & A C Fabian (Cambridge: Cambridge Univ. Press) p. 44

Ivison R J, Smail I, Barger A J, Kneib J -P, Blain A W, Owen F N, Kerr T H and Cowie L L 2000 *MNRAS* **315** 209

Haiman Z and Menou K 2000 *ApJ* **531** 42

Hamann F and Ferland G 1999 *Annu. Rev. Astron. Astrophys.* 37 487

Hasinger G 1998 *Astronomische Nachrichten* **319** 37

Hasinger G 2003 in *The emergence of cosmic structure* Thirteenth Astrophysics Conference. AIP Conference Proceedings Volume 666 pp. 227 (astroph/0302574)

Hauser M G and Dwek E 2001 *ARAA* **39** 249

Hauser M G, Arendt R G, Kelsall T et al. 1998 *ApJ* **508** 25

Helou G, Soifer B T and Rowan-Robinson M 1985 *ApJ* **298** L7

Hughes D et al. 1998 *Nature* **394** 241

Kauffmann G and Haehnelt M 2000 *MNRAS* **311** 576

Kauffmann G, White S, Heckman T, Menard B, Brinchmann J, Charlot S, Tremonti C and Brinkmann J 2004 *MNRAS* in press (astroph/0402030)

Kormendy J and Sanders D 1992 *ApJL* **390** 53

Kurk J D, R'ottgering H J A, Pentericci L and Miley G K 2000 *A&A* **358** L1

La Franca F, Gruppioni C, Matute I et al. 2004 the *AJ* **127**

Lagache G, Abergel A, Boulanger F, Desert F X and Puget J L 1999 *A&A* **344** L322

Laurent O, Mirabel I F, Charmandaris V, Gallais P, Madden S C, Sauvage M, Vigroux L and Cesarsky C 2000 *A&A* **359** 887

Lawrence A 1991 *MNRAS* **252** 586

Liang Y C, Hammer F, Flores H, Elbaz D, Marcillac D and Cesarsky C J 2004 *A&A* **423** 867

Lilly S J, Le Fevre O, Hammer F and Crampton D 1996 *ApJ* **460** L1

Lilly S J et al. 1999 *ApJ* **518** 641

Lonsdale C, Polletta M, Surace J, Shupe D et al. 2004 ApJS **154** 54

Lutz D et al. 1996 *A&A* **315** L137

Lutz D et al. 1998 *ApJ* **505** L103

Madau P and Pozzetti L 2000 *MNRAS* **312** L9

Madau P, Ghisellini G and Fabian A C 1994 *MNRAS* **270** L17

Madau P et al. 1996 *MNRAS* **283** 1388

Maiolino R 2002 in *Active Galaxies* (Piano Triennale INAF Allegato 9 A. Franceschini et al. Eds.)

Maiolino R and Rieke G 1995 *ApJ* **454** 95

Matsumoto T 2001 *IAUS* **204** 87

Mattila K 1976 *A&A* **47** 77

Mattila K 2003 *ApJ* **591** 119

McCarthy P J, Miley G. K and de Ko S 1997 *ApJS* **112** 415

Menci N, Fiore F, Perola G C and Cavaliere A 2004 *ApJ* **606** 58

Miyaji T, Hasinger G and Schmidt M 2000 *A&A* **353** 25

Mushotzky R F and Loewenstein M 1997 *ApJL* **481** 63

Nandra K P et al. 2002 *ApJ* **576** 625

Norman C A and Scoville N Z 1988 *ApJ* **332** 124

Oliver S, Rowan-Robinson M, Alexander D M et al. 2000 *MNRAS* **316** 749

Omont A, Cox P, Bertoldi F et al. 2001 *A&A* **374** 371

Omont A, McMahon R G, Cox P et al. 1996 *A&A* **315** 1

Peebles P.J.E 1993 *Principles of Physical Cosmology*" (Princeton NJ: Princeton Univ. Press)

Petitjean P, Rauch M and Carswell R F 1994 *A&A* **291** 29

Poggianti B M and Wu H 2000 *ApJ* **529** 157

Poggianti B M, Bressan A and Franceschini A 2001 *ApJ* **550** 195

Puget G L, Lagache G, Clements D et al. 1999 *A&A* **345** 29

Puget J -L et al. 1996 *A&A* **308** L5

Reuland M, van Breugel W, R'ottgering H et al. 2003 *ApJ* **592** 755

Rigopoulou D, Spoon H W W and Genzel R 1999 *AJ* **118** 2625

Risaliti G et al. 2000 *A&A* **357** 13

Rodighiero G, Lari C, Franceschini A, Gregnanin Aand Fadda D 2003 *MNRAS* **343** 1155

Rosati P et al. 2002 *ApJ* **566** 667

Rowan-Robinson M et al. 1997 *MNRAS* **289** 482

Rush B, Malkan M A and Spinoglio L 1993 *ApJS* **89** 1

Sanders D and Mirabel I.F 1996 *ARAA* **34** 749

Sanders D et al. 1988 *ApJ* **325** 74

Schlegel D J, Finkbeiner D P and Davis M 1998 *ApJ* **500** 525

Schmidt M, Schneider D P and Gunn J E 1995 *AJ* **110** 68

Setti G and Woltjer L 1989 *A&A* **224** L21

Shaver P A, Hook I M, Jackson C A, Wall J V and Kellermann K I in *Highly Redshifted Radio Lines* ASP Conf. Series Vol. 156 Ed. by C. L. Carilli S. J. E. Radford K. M. Menten & G. I. Langston. ISBN 1-886733-76-7 p. 163

Smail I, Ivison R J and Blain A W 1997 *ApJL* **490** L5

Smith H E, Lonsdale C J and Lonsdale C.J 1998 *ApJ* **492** 137

Soifer B T, Neugebauer G, Matthews K et al. 2000 *AJ* **119** 509

Soltan A 1982 *MNRAS* **200** 115

Somerville R et al. 2004 *ApJ* **600** L135

Spergel D et al. 2003 *ApJS* **148** 175

Spinoglio L and Malkan M A 1989 *ApJ* **342** 83

Spinoglio L and Malkan M A 1992 *ApJ* **399** 504

Stanev T and Franceschini A 1998 *ApJ* **494** L159

Stecker F, De Jager O and Salamon M 1992 *ApJL* **390** L49

Steidel C, Dickinson M and Persson S.E 1994 *ApJ* **437** L75

Steidel C C, Hunt M, Shapley A, Adelberger K, Pettini M, Dickinson M and Giavalisco M 2002 *ApJ* **576** 653

Stoughton C et al. 2002 *AJ* **123** 485

Tacconi L, Genzel R, Lutz D et al. 2002 *ApJ* **580** 73

Thomas D, Maraston C, Bender R and Mendes de Oliveira C 2004 *ApJ* in press (astroph/0410209)

Toller G N 1983 *ApJL* **266** L79

Toomre A 1977 in *The evolution of galaxies and their stellar populations* Yale University Observatory 401

van der Werf P 1996 in *Cold Gas at High Redshift* M. Bremer et al. Eds Kluwer Academic Publishers 37

Veilleux S, Kim D -C and Sanders D B 1999 *ApJ* **522** 113

Veilleux S, Sanders D and Kim D -C 1997 *ApJ* **484** 92

Volonteri M, Haart F and Madau P 2003 *ApJ* **582** 559

Wolf C, Wisotzki L, Borch A et al. 2003 *A&A* **408** 499

Wright E L and Reese E D 2000 *ApJ* **545** 43

Wright G, James P, Joseph R and McLean I 1990 *Nature* 344 417

Wyithe Loeb 2002 *ApJ* **581** 886

Xu J et al. 2002 *ApJ* **580** 653

York D et al. 2000 *AJ* **120** 1579

Yu and Tremaine S 2002 *MNRAS* **335** 965

Zepf S E 1997 *Nature* **390** 377

Zheng W, Mikles V, Mainieri V et al. 2004 *ApJ* accepted (astroph/0406482)

Zirm A W, Dickinson M and Dey A 2003 *ApJ* **585** 90

Chapter 3

Galaxy Formation in the Hierarchical Universe

Rachel S. Somerville
Space Telescope Science Institute
Baltimore, Maryland

The modern hierarchical paradigm of structure formation arises from a theory in which most of the matter in the universe is composed of an as yet unidentified form of Cold Dark Matter (CDM), the seeds of structure are sowed by inflation, and these seeds grow into large-scale structures through gravitational instability. The details of inflation and the physics of the growth of perturbations in the early universe give rise to the characteristic shape of the power spectrum of matter density fluctuations in a CDM universe: more power on small scales than large means that small objects form first, and merge together to form larger and larger systems. The "bottom up" nature of this process gives rise to the term "hierarchical".

The general CDM picture is supported by a wide variety of observations, from the Cosmic Microwave Background (CMB) to hot gas in galaxy clusters, from the Lyman-α forest to galaxy clustering and gravitational lensing. These combined observations now not only confirm the general picture of CDM, but provide quite accurate constraints on the values of the main *cosmological parameters*. Because in this picture the evolution of the dark matter density field is governed just by the well-understood physics of gravity within an expanding universe, theorists can make accurate and robust predictions for how the dark matter structures form and evolve over time.

One of the great successes of the CDM theory is that it provides a natural and fairly successful framework within which to understand and attempt to model galaxy formation. Within the gravitationally collapsed structures that we refer to as "dark matter halos", conditions are such that we expect gas to radiatively cool on a fairly short timescale. It is then reasonable to expect that some of this (atomic) gas will be converted into dense molecular clouds and eventually proceed to form stars, however, the details of how these steps proceed are far from understood. As well, the energy and heavy elements produced especially by the most massive of these stars and supernovae complicate the system further by returning metals and energy to the interstellar medium and perhaps to the hot gas halo. Massive stars and supernovae may even drive large scale winds that spew material far beyond the potential well of the galaxy, into the intergalactic medium, polluting the material that will collapse to form future generations of galaxies.

While I think that most astrophysicists today would agree that this basic picture of galaxy formation is now on a fairly sure footing, and detailed calculations based upon it have yielded some non-trivial successes in predicting or interpreting various galaxy observations, most would also agree that many questions and problems remain. Why does such a small fraction of the total baryon budget end up in the form of cold gas and stars today? What sets the "characteristic" mass scale(s) for galaxies? Are the predicted properties of dark matter halos on very small scales ($\lesssim 1$ kpc) consistent with observations? How did disk galaxies apparently manage to survive the tumult of the young universe predicted by CDM, in which mergers were commonplace, with most of their angular momentum intact? Why is it that the most massive galaxies today seem to be the oldest, in apparent contradiction to the most basic principle of hierarchical formation?

Over the past five years or so, another paradigm shift has occurred because of our growing realization that most or all galaxies host SuperMassive Black Holes (SMBH)—and moreover, that the properties of these SMBH are intimately connected with those of their host galaxies. While "normal" galaxies and Active Galactic Nuclei (AGN) or QSOs used to be regarded as separate, possibly unrelated classes of objects, and were studied in general by two different communities, we now realize that most or all galaxies probably experienced a phase as a QSO or AGN. These realizations have led inescapably to the conclusion that we must understand the *co-evolution* of SMBH and galaxies in order to solve many of these puzzles.

In these notes, I will present a brief overview of the main physical processes that are believed to shape galaxy formation, including the formation of dark matter halos, gas cooling, star formation, photo-ionization and supernova feedback. Along the way, I will discuss some of the methods commonly used to model these processes, focusing mainly on so-called "semi-analytic" (as opposed to numerical, or N-body) techniques. In conclusion, I will discuss the successes of current hierarchical models of galaxy formation, some of the remaining problems, and how the co-evolution of SMBH and AGN might affect our picture of galaxy formation.

3.1 Formation and Evolution of Structure

Most students reading these lecture notes will be familiar with the basic story of the formation of structure in the CDM or hierarchical scenario. For those who are not, there are many excellent textbooks that present a basic introduction to these ideas. Because the material in this section is extensively covered in these textbooks, I will gloss it briefly and at a relatively superficial level. For those who have not studied this material, I particularly recommend Padmanabhan (1993), Peacock (1998), and Peebles (1993) as useful background to these notes. I shall give references to specific sections of these texts in the relevant locations in these notes.

3.1.1 Talking About Lumpiness

The basic starting point of this story is known as the Cosmological Principle: on large scales, the universe is *homogeneous* (same average density everywhere) and *isotropic* (the same in all directions). Luckily, the cosmological principle is clearly violated on *small* scales, or the universe would be profoundly boring (no galaxies, no stars, no planets, no people!). The interesting part of the story is in how the seeds of the inhomogeneities that give rise to all of these structures were formed, and how they have evolved over the past ~ 13.5 billion years.

Suppose we define the dimensionless *density perturbation field*

$$\delta(\mathbf{x}) \equiv \frac{\rho(\mathbf{x}) - \langle\rho\rangle}{\langle\rho\rangle} \qquad (3.1)$$

where $\rho(\mathbf{x})$ is the matter density field and $\langle\rho\rangle$ is the spatial average of $\rho(\mathbf{x})$. There are several statistics commonly used to describe matter density fluctuations as a function of spatial scale (see e.g., Peebles 1980, Chapter III). For example, the *correlation function* (sometimes called the two-point function) of the density field is defined as

$$\xi(\mathbf{r}) \equiv \langle\delta(\mathbf{x})\delta(\mathbf{x} + \mathbf{r})\rangle_V \qquad (3.2)$$

where the averaging is over a volume V. The *power spectrum* is the Fourier transform of the density correlation function, sometimes written $P(k) \equiv \langle|\delta_k|^2\rangle$ (see e.g., Peacock 1998, p. 497).

We now define a new measure of scale: the mass within a spherical "tophat" of radius r with mean density $\bar{\rho}$: $M = \frac{4\pi}{3}R^3\bar{\rho}$, where $\bar{\rho}$ is the mean background density of the universe. We can then define the *mass variance*:

$$\sigma^2(M) = \frac{\langle(M - \langle M\rangle)^2\rangle}{\langle M\rangle^2} = \frac{\langle\delta M^2\rangle}{\langle M\rangle^2}$$

$$= \frac{1}{2\pi^2} \int P(k) W^2(kR) k^2 dk$$

$$W(kR) = \frac{3[\sin(kR) - kR\cos(kR)]}{(kR)^3}$$

The function $W(kR)$ is known as the "window function". Various choices are possible for this function. Here, it is the Fourier transform of a *tophat in k space*, or sharp k-space filter. This window function has certain advantages, which I will discuss later. All three statistics $\xi(r)$, $P(k)$, and $\sigma(M)$ are *second order* in δ; they carry the same information and can be readily transformed from one to the other. Higher-order statistics such as the three-point and four-point correlation function (skewness and kurtosis) and the corresponding Fourier transforms (the bi-spectrum and tri-spectrum) may be defined in a similar manner (see e.g., Peebles 1980).

3.1.2   The Primordial and Linear Power Spectra

Now that we have the language to talk about lumpiness in the universe, where do these lumps come from and what are their properties? In the early 1980s, Starobinsky and Guth independently suggested that shortly after the Planck time (10^{-43} s after the Big Bang), the universe experienced a brief ($\sim 10^{-24}$ s) period of extremely rapid expansion known as *inflation*, during which the expansion factor grew by about a factor of 10^{50}. Inflation solves a number of classical problems such as *flatness* (why the geometry of the universe is so close to flat), *causality* (how parts of the sky that are too far apart for photons to travel between in a Hubble time know to be at the same temperature), and *monsters* (absence of observed singularities, such as monopoles, strings, and domain walls). It also provides a natural mechanism for producing the "seeds" of structure. These seeds are produced by quantum fluctuations which became macroscopic and "frozen in" when they expanded to a size larger than the horizon. The properties of the lumps left behind at the end of inflation (or *primordial power spectrum*) depend on the dynamics of the *inflaton field* (the scalar field that is invoked to drive the expansion). Suppose that the inflaton field is denoted ϕ and one location in space is perturbed by an amount $\delta\phi$ relative to another location. These two universes finish inflating at different times, and wind up with different energy densities. If the time difference is δt, then the induced density fluctuation is just $\delta = H\delta t$.

It is commonly argued that inflation "naturally" predicts that the primordial fluctuation spectrum is *scale invariant*; i.e., if we express the primordial power spectrum as $P_k \propto k^{n_s}$, then $n_s = 1$. This is known as the Harrison–Zeldovich–Peebles spectrum. The crux of the argument is that because de Sitter space is invariant under time translation, at a given time, the only length scale is the horizon size c/H. It then follows that fluctuations on this scale are expected to have the same amplitude at all times, and resulting in a Harrison–Zeldovich–Peebles spectrum at the end of inflation. The problem with this argument is that it applies only to fluctuations that are on the scale of the horizon when inflation ends. Fluctuations that are *outside* the horizon at the end of the inflation can re-enter it later, and continue to grow while they are outside the horizon. If we knew the behavior of the *inflaton potential*, commonly denoted $V(\phi)$, we could compute the expected primordial power spectrum including this effect. Unfortunately, we do *not* know the nature of $V(\phi)$, and one can easily come up with "well-motivated" models that produce power spectra with $n_s \neq 1$ and even with n_s a function of scale (k) itself (these are referred to as running spectral index (RSI) models). While observations such as the CMB and galaxy clustering can constrain the allowed parameter space for a given assumed class of inflationary models (or $V(\phi)$), this remains a fundamental uncertainty in cosmology.[1]

From the end of inflation to the era of recombination at $z \sim 1000$, a number of physical processes (gravity, pressure, free streaming) modify the primordial power spectrum. Gravity causes perturbations to grow, while pressure opposes gravity on small scales and *free streaming* damps out fluctuations on all scales up to the horizon until the particles go non-relativistic. This is why the existence of a light massive neutrino would wash out power on small scales, and is the origin of the terminology "hot", "warm", and "cold" dark matter (referring to whether the dark matter is relativistic at the epoch of matter/radiation equality). A qualitative understanding of these processes may be obtained using analytic arguments (see e.g., Peacock 1998, Chapter 15) but a precise computation of the processed power spectrum requires numerical techniques. The complex machinery that integrates the coupled set of differential equations for the distribution functions of the collisionless components (e.g., photons, neutrinos, and dark matter) simultaneously with those describing the density and pressure of the collisional baryon fluid is referred to as a "cosmological Boltzmann code." The CMBFAST code of Seljak & Zaldarriaga (1996) is an example of such a code, and it is publicly available. Alternatively, for applications that do not require great precision (such as semi-analytic modeling), one can use fitting functions such as those presented by Hu & Sugiyama (1996) to obtain the processed power spectrum. The relationship between the primordial power spectrum at the end of inflation and the processed power spectrum is called the *transfer function*.

Figure 3.1 shows the variance for four different cosmological models, computed using the Hu & Sugiyama (1996) analytic fits. The models are: SCDM, an Einstein–de Sitter model with a standard CDM power spectrum

---

[1] This discussion has of necessity been extremely schematic. I recommend Peacock (1998), Chapter 11, for a concise and clear summary of inflation theory and the origin of the primordial fluctuation spectrum.

and $H_0 = 50$ km s^{-1} Mpc^{-1}; τ CDM, an Einstein–de Sitter model with reduced power on small scales due to a decaying dark matter particle ($H_0 = 50$ km s^{-1} Mpc^{-1}), OCDM, an open model with $\Omega_m = 0.3$ and $H_0 = 70$ km s^{-1} Mpc^{-1}, and ΛCDM, a model with $\Omega_m = 0.3$ and a cosmological constant $\Omega_\Lambda = 0.7$ making the geometry flat ($H_0 = 70$ km s^{-1} Mpc^{-1}). All four models have a Harrison–Zeldovich–Peebles primordial power spectrum $n_s = 1$, and the *amplitude* of the power spectrum was chosen to produce the same number of galaxy clusters at the present day. A common way to specify the amplitude of the power spectrum is in terms of σ_8, the root variance at a scale of 8 h^{-1} Mpc.

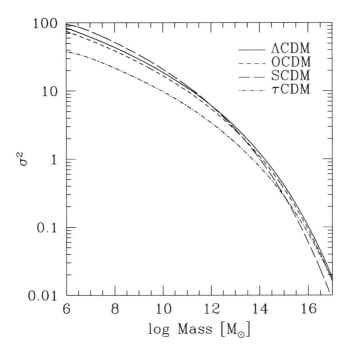

Figure 3.1. The mass variance function for four different cosmologies (see text).

Measurements of the temperature variations in the CMB radiation directly constrain the power spectrum at the last scattering surface (about 380,000 years after the Big Bang). The best constraints from CMB data currently come from the Wilkinson Microwave Anisotropy Probe (WMAP) experiment. A beautiful summary of the findings from the first year of WMAP data is given by Spergel et al. (2003). While these results have greatly improved the constraints on the cosmological parameters and the primordial power spectrum, there is a tendency to believe that the power spectrum is now *known*. It is important to realize, however, that WMAP constrains the power spectrum *only on relatively large scales*. The smallest scales probed by the CMB correspond to ~ 20–30 comoving Mpc. The power spectrum on the scale of *galaxies*, especially dwarf galaxies, is not directly constrained by WMAP nor any other existing or planned CMB experiment.

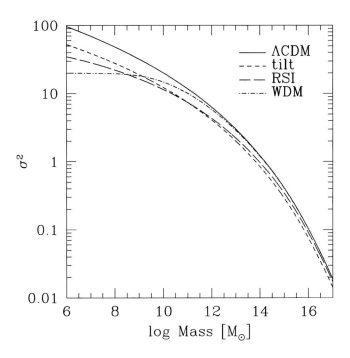

Figure 3.2. The mass variance function for four WMAP-normalized ΛCDM models with modified primordial power spectra (see text).

This is why it is important to combine the information from the CMB with other data such as gravitational lensing, galaxy clustering, and clustering of the Lyman-α forest. This has been done very elegantly by the WMAP team (Spergel et al. 2003). However, when combining data from different experiments, it is much more difficult to account accurately for possible systematic errors. In addition, extracting the linear power spectrum from the Lyman-α forest and galaxy clustering data is not nearly as clean a problem as obtaining it from the CMB observations. Non-linear evolution and the messy physics of gas and stars make this a complicated exercise. As we have discussed above, the main uncertainties in the predicted amplitude and shape of the linear power spectrum derive from our ignorance about the inflaton potential and the nature of the dark matter. Without stringent constraints on the small scale power spectrum, a fairly wide variety of inflationary variants are still allowed, as are certain kinds of Warm Dark Matter (WDM) models. Figure 3.2 shows the mass variance for several models, including a model with a "tilted" primordial power spectrum ($P_k \propto k^{n_s}$, with $n_s = 0.95$), a model with a running spectral index (n_s a function of k; this is the same RSI model discussed in Spergel et al. (2003)), and a WDM model. All of these models are normalized to agree with WMAP on large scales, but clearly would have very different implications for galaxy formation.

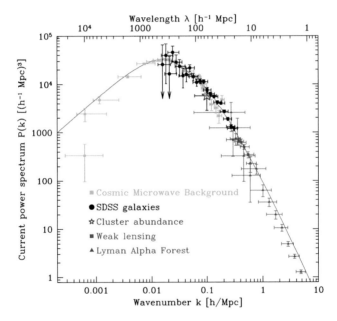

Figure 3.3. Constraints on the linear power spectrum, obtained from the CMB, galaxy clustering, cluster abundance, weak lensing, and the Lyman-α forest. The solid line shows the scale free ($n_s = 1$) CDM model favored by the first-year WMAP results ($\Omega_m = 0.28$, $h = 0.72$, $\Omega_b/\Omega_m = 0.16$; Spergel et al. 2003). Reproduced from Tegmark et al. (2003a).

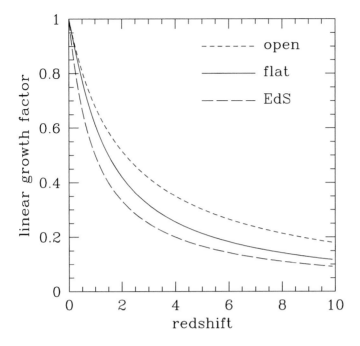

Figure 3.4. The linear growth function for three cosmologies (see text).

3.1.3 Linear Growth of Perturbations

The little lumps are amplified by gravity as time progresses: overdensities get more overdense and underdensities get more underdense. It is straightforward to write down the differential equations describing the evolution of an overdensity δ as a function of time. This is done in the standard texts and I will not reproduce these calculations here (e.g., Peacock 1998, p. 460). In the limit where the overdensity is small ($\delta << 1$), we can *linearize* the equations by keeping only terms that are first order in δ (again, see Peacock or other standard texts for details). This is known as the *linear growth regime*. The growing mode solution to the linearized equations is often written D_{lin}. This is very convenient, as it means that in the linear regime, the evolution of the power spectrum with time can be written simply $\delta(t) \propto D_{\mathrm{lin}}(t)$, or in terms of the variance, $\sigma(M,t) = D_{\mathrm{lin}}(t)\sigma(M)$. In an Einstein–de Sitter ($\Omega_m = 1$) universe, this function has the very simple form $D_{\mathrm{lin}} \propto (1+z)^{-1}$. In a universe with less than critical matter density ($\Omega < 1$), there is less matter and so overdensities grow more slowly with time. The behavior of D_{lin} with redshift for three cosmologies, Einstein–de

Sitter, $\Omega = 0.3$ flat, and $\Omega = 0.3$ open, is shown in Figure 3.4. This means that if we normalize two models, one with $\Omega = 1$ and one with $\Omega < 1$, so that their power spectrum on a given scale has a certain amplitude *at the present day*, then the $\Omega < 1$ model will have more structure at high redshift. This is why it is often said that "structure forms earlier in low Ω models." Of course, if we fixed the normalization at some early time (such as at the last scattering surface), then we would say that structure forms *later* in low Ω models.

3.1.4   Non-Linear Evolution

To capture the behavior of lumps in the non-linear regime (when $\delta > 1$), and higher-order clumping such as skewness, kurtosis, etc., numerical N-body experiments are extremely useful. A discussion of these experiments is given in Peacock (1998) and other texts. The growth of structure in different cosmological models can be seen graphically in the N-body simulations produced by the VIRGO collaboration (Jenkins et al. 1998), shown in Figure 3.5. The figure shows the projected matter density field at different epochs (dark colors represent underdense, and light colors overdense regions) in boxes of comoving length $239.5h^{-1}$Mpc on a side. The same four models (SCDM, τCDM, OCDM, and ΛCDM) that were discussed in §3.1.2 are shown. The four rows represent the four models (*not* in the order listed) and the three columns show different epochs $z = 3$, $z = 1$, and $z = 0$.

As an exercise, see if you can guess by looking at Figure 3.5 which row corresponds to which model. How would you go about this? First let's focus on the $z = 0$ pictures (the rightmost column). Notice that the largest structures appear in the same places because the random phases of the realization of the power spectrum were chosen to be the same in each of the simulations. Notice also that if you imagined counting the number of regions above a certain density threshold on scales of about $10\,h^{-1}$Mpc (or counted the really bright knots, which correspond to rich clusters), they would be about the same in all four rows. However, if you look at the number of *small* bright patches, on scales of less than about a Mpc, the four panels look quite different. In particular, the middle two columns clearly have more of this small scale lumpiness, corresponding to more power on small scales. The second row has the *most* small scale power. Which of the four models do we expect to have the most power on small scales (you might want to look back at Figure 3.1)?

Now let's look at the redshift evolution. Which row changes the most over the redshift interval from $z = 3$ to $z = 0$? Which changes the least? Again, we notice that we could divide the panels into two groups that look more similar: the middle two rows and the top and bottommost rows. The middle two rows seem to change a lot in appearance between $z = 3$ and

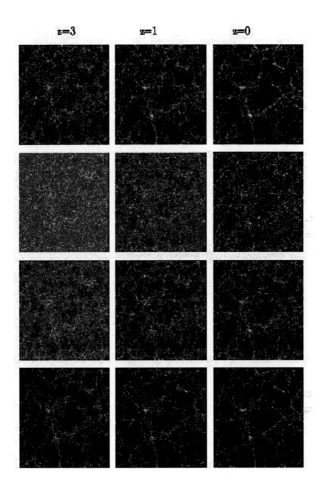

Figure 3.5. The projected mass distribution in slices of N-body simulation boxes. The four rows show four different cosmological models (see text), and the three columns show different epochs $z = 3$, $z = 1$, and $z = 0$. Image by Joerg Colberg from http://www.mpa-garching.mpg.de/Virgo/virgopics.html, based on the simulations presented in Jenkins et al. (1998).

$z = 0$, while the top and bottom rows change much less. One can see that in the top and bottom rows, the largest structures (voids, walls, filaments, and clusters), are largely in place even at $z = 3$, while in the middle two rows they are difficult to discern. Now looking at Figure 3.4 for help if necessary: which model do you expect to evolve the most from $z = 3$ to $z = 0$? Which the least? If the reproduction in the book is not good enough to make out the features that I am talking about here, go look at it at http://www.mpa-garching.mpg.de/Virgo/virgopics.html —which will also tell you whether you guessed right!

The evolution of the lumpiness on different scales is shown in a more quantitative manner in Figure 3.6, in terms of the correlation function $\xi(r)$ for the same four models discussed above. Note the growth of the amplitude of $\xi(r)$ with time on small scales. This change in the *shape* of the correlation function is the signature of non-linear evolution.

3.2   The Formation of Dark Matter Halos

3.2.1   The Spherical Tophat Collapse Model

Imagine spray painting a spherical region of the universe and following its evolution. We choose a region in which the mean density contained within the spray-painted sphere is larger than the average background density of the universe, and define the overdensity $\delta(r) \equiv \rho/\rho_b - 1$. Initially, the matter within this sphere will expand more or less with the Hubble expansion. As time goes on, however, the self-gravity of the local mass concentration will cause the expansion to become progressively slower. Eventually, the self-gravity will overcome the pressure of the expansion and actually cause the matter within our imaginary sphere to stop expanding and begin to collapse instead. The end result is a gravitationally bound object.

We can write down the equation of motion for this spherical "tophat", assuming that mass shells do not cross and random motions are small. These equations turn out to have a familiar form, as the spherical symmetry that we assumed means that the gravitational force depends only on the mass contained inside the sphere. The equations and their solutions are given in e.g., Padmanabhan (1993, pp. 273–285). They are the same as the equation of motion and solution for the scale factor in a Friedman–Robertson–Walker universe with $\Omega > 1$. Figure 8.1 on p. 282 of Padmanabhan shows the evolution of the density in such a region during the transition from the linear to the non-linear regime. We can use the equation of motion to solve for the time or redshift when the tophat turns around (stops expanding and begins to collapse), which occurs when $\dot{r} = 0$, and for its density when this occurs. The next important stage occurs when the equation of motion predicts that the tophat has collapsed to $r = 0$. Of course, before this can happen, several assumptions inherent

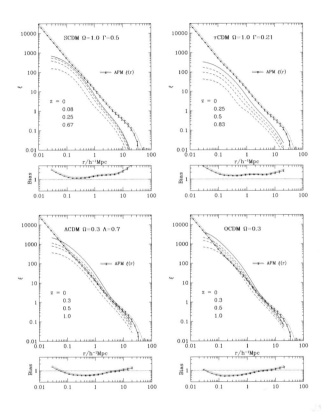

Figure 3.6. Evolution of the mass correlation function, $\xi(r)$, from the VIRGO simulations. The large square panels show the two-point correlation function in the four models at the redshifts given in the legend, with results at $z = 0$ plotted as a bold solid line. The observed galaxy correlation function from the APM galaxy survey (Baugh 1996) is shown as a solid line with error bars and as a dotted line. The former (solid line) corresponds to the assumption that clustering is fixed in comoving coordinates and the latter (dotted line) to the assumption that clustering evolves in proportion to the scale factor. The small panels below each $\xi(r)$ plot show the square root of the ratio of the observed galaxy to the theoretical mass correlation functions at $z = 0$. This ratio is the bias in the galaxy distribution that would be required for the particular model to match the observations. Reproduced from Jenkins et al. (1998).

in the equations have broken down, and the rapidly changing potential of the material leads to a process known as 'violent relaxation', which occurs on the dynamical timescale $t_{\mathrm{dyn}} \propto (G\rho)^{-1/2}$. The relaxed object is in virial equilibrium and has a characteristic size r_{vir}, density ρ_{vir}, and internal velocity V_{vir}. Once a system has virialized, it is stable and its density and size no longer change. Using the virial theorem, which says that the potential energy U and kinetic energy K are related by $|U| = 2K$, we can derive a relationship between the mass and virial radius or virial velocity for a tophat that has collapsed at a given redshift (Somerville & Primack 1999, appendix).

We define $\Delta_c(z)$, the critical overdensity for virialization as

$$\Delta_c(z) \equiv \frac{\rho_{\mathrm{vir}}\Omega(z)}{\Omega_0 \rho_c^0 (1+z)^3}, \qquad (3.3)$$

where ρ_c^0 is the critical density at the present day. Bryan & Norman (1998) give convenient fitting functions for $\Delta_c(z)$:

$$\Delta_c = 18\pi^2 + 82x - 39x^2 \qquad (3.4)$$

for a flat universe and

$$\Delta_c = 18\pi^2 + 60x - 32x^2 \qquad (3.5)$$

for an open universe, where $x \equiv \Omega(z) - 1$. These formulae are accurate to 1% in the range $0.1 \leq \Omega \leq 1$, which is more than adequate for most purposes. We can then write down the expression for r_{vir}:

$$r_{\mathrm{vir}} = \left[\frac{M}{4\pi} \frac{\Omega(z)}{\Delta_c(z)\Omega_0\rho_{c,0}} \right]^{1/3} \frac{1}{1+z}. \qquad (3.6)$$

In conjunction with the virial relation:[2]

$$V_c^2 = \frac{GM}{r_{\mathrm{vir}}} - \frac{\Omega_\Lambda}{3} H_0^2 r_{\mathrm{vir}}^2 \qquad (3.7)$$

we can calculate the circular velocity and viral radius for a halo with a given mass at any redshift z.

If $\Omega_m = 1$, the overdensity at which the spherical tophat model predicts virialization will occur is $\Delta_{\mathrm{vir}} \simeq 178 \simeq 200$. Thus the mass of a halo is often defined as the mass contained within a radius such that the mean overdensity with respect to critical density is 200 (M_{200}). However, the actual value Δ_{vir} predicted by the tophat model depends on cosmology, as described above. For example, for the concordance ΛCDM model, with

---

[2] The second term is small and can be neglected in all observationally plausible models, in particular the concordance ΛCDM model.

$\Omega_m = 0.3$ and $\Omega_\Lambda = 0.7$, $\Delta_{\mathrm{vir}} \simeq 100$ and M_{vir} or V_{vir} strictly refer to the mass or velocity internal in a region with this overdensity relative to the *background*. Both conventions are currently in use, so one must take care to use consistent quantities.

3.2.2 The Dark Matter Halo Mass Function: Press–Schechter

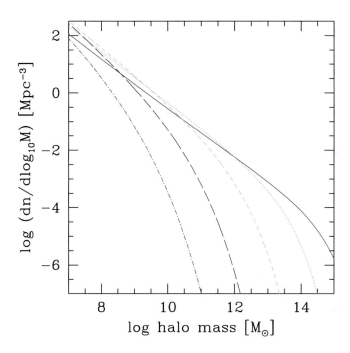

Figure 3.7. The dark matter halo mass function (the number of halos per logarithmic unit in mass and per comoving cubic Mpc) predicted by the standard Press–Schechter model in a concordance ΛCDM cosmology. Results are shown for redshifts $z = 0$ (solid), 1 (dotted), 3 (short dash), 6 (long dash), and 10 (dot-dash).

Press & Schechter (1974) developed a simple but surprisingly effective method to estimate the number density of dark matter halos as a function of their mass (the mass function). The model is based on a combination of linear growth theory, spherical collapse theory, and the properties of Gaussian random fields. Suppose that we have smoothed the initial density field on a scale R, as we discussed in §3.1.1. The fraction of the mass in the universe that is contained within spheres of radius R with mean density exceeding δ_c is

$$F(R,z) = \int_{\delta_c}^{\infty} d\delta \frac{1}{\sqrt{2\pi}D(z)\sigma(R)} \exp\left[-\frac{\delta^2}{2D(z)^2\sigma(R)^2}\right], \qquad (3.8)$$

where $D(z)$ is the linear growth function defined in §3.1.3 and $\sigma(R)$ is the mass variance on a scale R, defined in §3.1.1. The Press–Schechter ansatz is that this fraction should be identified with the fraction of the mass contained in collapsed clumps with mass *exceeding* $M = 4\pi\bar{\rho}_0 R^3/3$. There is a problem with this expression: as $M \to 0$, $\sigma \to \infty$, and $F \to \frac{1}{2}$, and so only half of the universe is in clumps of any mass. The ad hoc solution is to simply multiply by a fudge factor of 2. Later, in the excursion set derivation of the same formula, we will see where this factor comes from.

We can now obtain the mass function by differentiating this cumulative distribution function and dividing by the mass M. We then obtain

$$\frac{dn}{dM}(M,z)\,dM = \sqrt{\frac{2}{\pi}}\frac{\bar{\rho}_0}{M}\frac{\delta_c}{D(z)\sigma^2(M)}\frac{d\sigma}{dM}\exp\left[-\frac{\delta_c^2}{2D^2(z)\sigma^2(M)}\right]dM$$
$$(3.9)$$

which is the number of halos with mass between M and $M+dM$ that exist at a redshift z. We still need to specify the "critical density" parameter δ_c. Recall from the last section that in spherical collapse theory we calculated the overdensity for a perturbation that collapsed at time t. If we now extrapolate this to t_0 using linear theory we have $\delta_c(t) = \delta_c(t_0)/D_{\text{lin}}(t) \equiv \delta_{c0}/D_{\text{lin}}(t)$. In an $\Omega = 1$ universe, δ_{c0} has the value 1.69. It varies only slightly for different cosmologies; for an open universe with $\Omega_m = 0.2$, it is $\delta_{c0} = 1.64$ (Lacey & Cole 1993, LC93). Another approach is to consider δ_c to be a free parameter and adjust it to obtain good agreement with simulations. However, the best-fit values usually do not differ very much from the spherical collapse prediction (Gross et al. 1998).

Press–Schechter theory was placed on a more sturdy theoretical footing by an alternative derivation due to Bond et al. (1991). The "excursion set" derivation also leads to additional applications of the Press–Schechter approach, which I will discuss in the next section. For most choices of window function, the smoothed field $\bar{\delta}(R)$ contains correlations between different scales R, making analytic calculations impossible. However, if we use the *sharp k-space* filter, the correlations simplify and we can proceed analytically. The k-space tophat is defined as $W(k; k_c) = 1$ for $k \leq k_c$ and $W(k; k_c) = 0$ for $k > k_c$. In real space this function has the form

$$W(r) = \frac{(\sin k_c r - k_c r \cos k_c r)}{2\pi^2 r^3} \qquad (3.10)$$

which implies a mass $M = 6\pi^2 \rho_0 k_c^{-3}$. The smoothed field on a scale

$M \propto k_c^{-1}$ is now

$$\bar{\delta}(M) = \int_{k<k_c} \frac{d^3k}{(2\pi)^3} \delta_{\mathbf{k}}. \qquad (3.11)$$

The smoothed field $\bar{\delta}(M)$ is a Gaussian random variable with zero mean and variance $\sigma(M)$. Following LC93, we define $S(M) \equiv \sigma^2(M)$ to be the variance and $\omega \equiv \delta_c(t)$, and assume δ to refer to the smoothed field for the remainder of this section. If we smooth on a large enough scale, the variance will be zero. As we smooth on smaller and smaller scales, the change in $\delta(M)$ for an increase from k_c to $k_c + dk$ is just the integral over the $\delta_{\mathbf{k}}$ within the spherical shell in k-space. Because of the properties of the sharp k-space filter, the $\delta(k_c)$ for different scales are uncorrelated. The smoothed field $\delta(M)$ executes a random walk as the smoothing scale is changed (it may be helpful to look at Figure 1 in LC93). We can now make an ansatz similar to the original Press–Schechter one to arrive at the expression for the mass function. We associate the fraction of matter in collapsed objects of mass M at time t with the fraction of trajectories that make their *first upcrossing* through the threshold ω at the smoothing scale k_c (which can be translated to a mass $M(k_c)$ or variance $S(M)$). Trajectories can be divided into three categories (e.g., White 1994):

(1) $\delta > \omega$ for $k_c = K_c$
(2) $\delta < \omega$ for $k_c = K_c$ but $\delta > \omega$ for some $k_c < K_c$
(3) $\delta < \omega$ for all $k_c \leq K_c$

Category (1) corresponds to objects that have not yet collapsed. Category (2) includes objects that have collapsed on the scale $M(K_c)$ at an earlier time and have been incorporated into a larger object by the time t. Category (3) is what we want: objects that are collapsing on the scale M for the first time. An important realization is that the fractions in category (1) and category (2) are equal: each trajectory in (1) can be reflected around its first upcrossing of ω to obtain a trajectory that will fall into category (2) and vice versa. The distribution of δ is a Gaussian with variance S, so the fraction of trajectories in category (1) is just

$$f_1 = \frac{1}{2\pi} \int_\omega^\infty d\delta \exp\left[-\frac{\delta^2}{2S}\right] \qquad (3.12)$$

from which the fraction of mass in halos with mass less than M, with the factor of 2 in place, immediately follows.

$$F(<M) = 1 - 2 \times \frac{1}{2\pi} \int_\omega^\infty d\delta \exp\left[\frac{-\delta^2}{2S}\right]. \qquad (3.13)$$

The differential mass function (in terms of the mass fraction in halos of a given mass) follows as before (here in the notation of LC93, which we will use in the next section):

$$f(S,\omega)dS = \frac{\omega}{\sqrt{2\pi}S^{3/2}} \exp\left[-\frac{\omega^2}{2S}\right]dS \qquad (3.14)$$

Examples of the halo mass function predicted by the Press–Schechter model at various redshifts are shown in Figure 3.7. This model is computed for the same parameters as the ΛCDM model discussed in § 3.1.2 ($\Omega_m = 0.3$, $\Omega_\Lambda = 0.7$, $H_0 = 70$ km s^{-1} Mpc^{-1}, $\sigma_8 = 0.9$) which are consistent with the concordance model. Note that on the mass scales of $\lesssim 10^{12}$ $M_\odot$, the number density of halos at a given mass first *increases*, and then decreases with increasing redshift. We can see easily from this plot that hierarchical models predict that there should be more small mass objects, and fewer large mass objects, as we go back in time.

3.2.3 The Conditional Halo Mass Function: Extended Press–Schechter

One of the beautiful things about the excursion set derivation of the mass function described above is that one can immediately write down the *conditional* mass function, the fraction of the mass in halos with mass M_1 at z_1 that have merged to form halos with mass M_0 at z_0 ($M_1 < M_0$, $z_1 > z_0$). It is exactly the same problem as before, but with the source of the trajectories moved from the origin to the point (S_0, ω_0). This amounts to making the substitution $S \to S_1 - S_0$ and $\omega \to \omega_1 - \omega_0$ in Equation (3.14):

$$f(S_1,\omega_1 \mid S_0,\omega_0)dS_1 = \frac{(\omega_1 - \omega_0)}{\sqrt{2\pi}(S_1 - S_0)^{3/2}} \exp\left[-\frac{(\omega_1 - \omega_0)^2}{2(S_1 - S_0)}\right]dS_1. \quad (3.15)$$

This expression can be manipulated to derive estimates for halo formation times and survival times, the rate of mergers with objects of a given mass scale, and other quantities (LC93). The probability that a halo of mass M_0 at redshift z_0 had a progenitor in the mass range $(M_1, M_1 + dM_1)$ is given by (LC93):

$$\frac{dP}{dM_1}(M_1, z_1 \mid M_0, z_0)dM_1 = \frac{M_0}{M_1} f(S_1,\omega_1 \mid S_0,\omega_0)dS_1, \qquad (3.16)$$

where the factor M_0/M_1 converts the counting from mass-weighted to number-weighted.

3.2.4 Merger Trees

The Press–Schechter (PS) and Extended Press–Schechter (EPS) models outlined above make it very straightforward to predict the mean properties of an ensemble of halos as a function of mass and redshift. In order to model

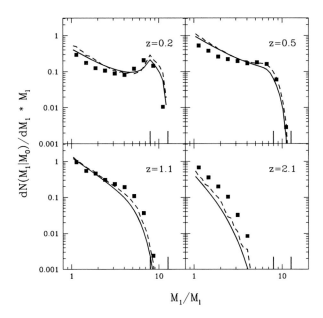

Figure 3.8. The conditional mass function of halos predicted by the Extended Press–Schechter model (solid lines), found in N-body simulations (filled squares), and in the merger trees (dashed lines). The Extended Press–Schechter prediction and the merger trees have been averaged over the bin in parent mass used for the simulations, indicated by the vertical lines on the bottom of the box. The panels show redshifts 0.2, 0.5, 1.1, and 2.1 as indicated on the figure. The mean parent mass is $M_0 = 10M_l$, where M_l is the minimum mass of the merger tree. Reproduced from Somerville et al. (2001a).

galaxy formation in detail, however, we would like to trace the merger histories of *individual objects*. This entails the construction of a "merger tree", i.e. the masses of ensembles of progenitor halos and the redshifts at which they merge to form larger halos. Several methods have been proposed that utilize the EPS formalism to "plant" merger trees (Kauffmann & White 1993; Cole 1991; Somerville & Kolatt 1999; Cole et al. 2000). Unfortunately, no method yet proposed is both practical and truly elegant or rigorous, because of difficulties that I now describe. This discussion is based on Somerville & Kolatt (1999), to which interested students should refer for more details.

Consider a halo with a given mass M_0 at redshift z_0. The EPS model provides the *average distribution* of progenitor masses M_1 at a higher redshift z_1, $dP/dM(M_1, z_1 \mid M_0, z_0)$. It might appear that it would be a simple matter to construct a merger tree by simply choosing the masses of the progenitors of our parent halo at some earlier redshift z_1 from the expression for $dP/dM(M_1, z_1 \mid M_0, z_0)$ given by Equation (3.16) above, then repeating this process starting from each progenitor in turn for the next step back in time. Two difficulties immediately arise in implementing this approach. First, from inspection of Equation (3.16), the number of halos clearly diverges as the mass goes to zero. However, note that the mass contained in small halos (Equation 3.15) does not diverge as $M \rightarrow 0$. In order to pick masses from the number weighted probability function numerically, it is necessary to introduce a cutoff mass, or effective mass resolution, M_l. For the purposes of galaxy formation modeling, this is not inherently a problem, since one should be able to set M_l to a small enough value that any halo with $M < M_l$ will not be able to form galaxies because any gas within such small halos will be photoionized or unable to cool efficiently. However, we do still need to treat the mass below and above the cutoff in a self-consistent way.

The second, and more serious, problem is that the progenitor masses must simultaneously be drawn from the distribution $dP/dM(M)$ *and* add up to the mass of the parent, M_0. The root of the problem is that dP/dM is just the *average* number of halos that one can make out of the mass $M_0 f(M) \, dM$. What we really want is the *joint* probability function for the set of progenitors $\{M_1, ..., M_n\}$, i.e. $dP_n/dM(\{M_1, ..., M_n\}, z_1 \mid M_0, z_0)$. An obvious problem with the use of the single halo probability rather than the joint probability is that there is no guarantee that we will not at some stage pick a progenitor that does not "fit" in the halo: i.e. $M > M_0 - \sum_i M_i$ where M_i are the masses of all the previously picked progenitors. In addition, since the expression $dP/dM(M)$ only gives the probability that there was *a* progenitor of mass M at an earlier time, we do not know *a priori* how many progenitors were present at the redshift z_1. In fact, one can demonstrate that for certain choices of the power spectrum (including CDM-like spectra), *it is impossible to simultaneously exactly satisfy the conditional mass function constraint and mass conservation*. All proposed methods of constructing merger trees therefore involve some black magic, and indeed violate one or both of these constraints at some level.

3.2.5   Agreement Between Semi-Analytic and N-Body Methods

Before we ask the question, "How well do these analytic models agree with the results of N-body simulations," it is important to keep in mind that identifying and characterizing dark matter halos in N-body simulations is not entirely straightforward. Clearly, we want to identify the gravitationally bound objects that have separated from the Hubble flow and collapsed.

In practice, this is complicated because these collapsed objects are not always in virial equilibrium, they are in general not perfectly spherical or even axisymmetric, and they have substructure (gravitationally bound objects within other bound, virialized objects). Operationally, there are two or three basic methods for finding dark matter halos, and numerous variations on these have been proposed and used. The *Spherical Overdensity* method searches for regions within spheres such that the enclosed overdensity exceeds some specified value (generally 200 or $\Delta_{\rm vir}$; see § 3.2.1). A drawback to this method is that halos are assumed to be spherical, while this is not always a good approximation. Another commonly used method is *Friends of Friends* (FOF), in which particles are linked together if they lie within a given linking length of one another. While this method has the advantage that it does not impose any specific geometry, it has the disadvantage that physically disjoint objects can be linked together by a narrow bridge of particles. While differences between masses assigned to individual halos by different methods can be quite large, statistical quantities such as multiplicity functions agree reasonably well (Jenkins et al. 2001).

When we consider how crude the assumptions made in the Press–Schechter model really are, it is a wonder that it gives sensible results at all. Surprisingly, though, the halo multiplicity function extracted from N-body simulations not only has the same functional form predicted by the Press–Schechter model, this simple model predicts roughly the correct amplitude and characteristic mass scale M_* (where M_* is defined as the mass of a halo that is just collapsing at redshift z, so that $\sigma(M_*) = \delta_c/D(z)$). However, in detail, the mass function predicted by the Press–Schechter model does deviate significantly from the results from N-body simulations. Press–Schechter predicts about a factor of two too many halos at masses less than M_*, and systematically underestimates the number of halos with $M > M_*$. For very rare, massive halos, this error can be as large as five orders of magnitude.

Recently, several variants on the Press–Schechter model have been proposed, which produce improved agreement with N-body simulations. Of these, the easiest to use are probably the functions given by Sheth & Tormen (1999) or Jenkins et al. (2001). Although Sheth, Mo, & Tormen (2001) propose a physical basis for their model in terms of ellipsoidal rather than spherical collapse, it is probably best to consider these convenient fitting functions, rather than physical models. In particular, it is somewhat dangerous to use them for different cosmologies, or outside the redshift or mass range in which they were derived.

The conditional mass function obtained from the extended Press–Schechter model suffers similar discrepancies—it overestimates the number of small progenitors, and underestimates the number of large progenitors (Somerville et al. 2001a); see Figure 3.8. The Sheth–Tormen function may

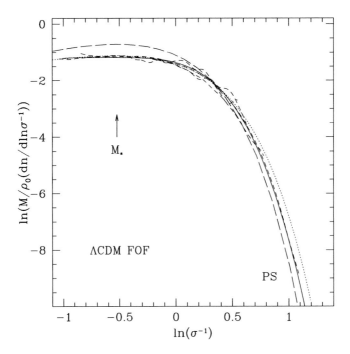

Figure 3.9. A comparison of the halo mass function at $z = 0$ predicted by analytic models with the results of N-body simulations in the ΛCDM cosmology. Halos were found using the FOF algorithm. The short dashed lines show results from the individual ΛCDM simulations used in this chapter and the solid curve is a fit to the combined results of the simulations. The long dashed line shows the Press–Schechter prediction and the dotted line, the Sheth–Tormen prediction. The arrow marks the characteristic mass scale, M_*. Note that natural logarithms in this plot have been used in this plot. Reproduced from Jenkins et al. (2001).

be extended to obtain an improved conditional mass function in the same manner as the original Press–Schechter model, however, because the modified conditional mass function violates the Markov property (a step from t_1 to t_2, then t_2 to t_3 does not give the same result as going from t_1 to t_3), it cannot be used to construct a merger tree. Thus, all existing methods for constructing semi-analytic merger trees are based on the original extended Press–Schechter model, and thus are inaccurate at this level. This results in a systematic error in predicted quantities such as halo formation times: the semi-analytic trees predict halo formation times that are systematically later than merger trees extracted from N-body simulations (Wechsler et al. 2002).

3.3 Spatial Clustering and Bias

3.3.1 Bias

Generally, the term *bias* represents the possibility that galaxies do not exactly trace the dark matter. The concept of bias was first discussed in the early days of the CDM model, before the results from the COBE satellite told us how to normalize the overall amplitude of the matter power spectrum, and at a time when at least theorists were strongly predisposed to believe that we live in a universe with critical density $\Omega_m = 1$. The concept of biased galaxy formation was discussed by Kaiser (1984), in the context of the observed strong clustering of Abell clusters. Davis et al. (1985), using some of the first cosmological N-body simulations of the evolution of clustering in a CDM universe, identified the epoch when the slope of the correlation function equals the value observed for present-day galaxies ($\gamma = -1.8$) as redshift zero (recall that in CDM models the clustering strength of dark matter becomes greater on all scales as time progresses, and also that the shape of the correlation function changes, indicating that clustering evolves at different rates on different spatial scales; see Figure 3.6). For $\Omega = 1$, the amplitude of the dark matter correlation function ($r_0 = 1.3h^{-1}$ Mpc) was then much smaller than the observed value ($r_0 = 5h^{-1}$ Mpc). This problem could be solved by assuming that galaxies cluster more strongly than the dark matter, $\xi_g = b^2\xi_{DM}$, where a bias value $b > 1$ corresponds to positive bias. A value of $b < 1$ is known as anti-bias. Curiously enough, in modern COBE/WMAP normalized ΛCDM, the dark matter correlation function length is nearly identical to that of fairly luminous optically selected galaxies ($r_0 = 5h^{-1}$ Mpc), so that $b \sim 1$ on large scales for luminous, optically selected galaxies (see Figure 3.6). Luminous galaxies are *anti-biased* on small scales in this model.

The implicit assumptions that bias is local (depends only on the local density), deterministic, linear, and independent of scale (at least on scales large enough to be in the linear regime) have remained common to the present day, although there is direct observational evidence that at least some of these assumptions cannot be correct. The mismatch in the shape of the DM correlation function and that of galaxies implies that biasing *must* be scale dependent. Galaxy clustering (and therefore bias) is found to be a function of luminosity and morphological type: for example, luminous, red, or early-type galaxies are more clustered than less luminous, blue, or late-type galaxies (e.g., Norberg et al. 2002). Moreover, in addition to having different amplitudes, the correlation functions for these populations have different slopes Zehavi et al. (2002), indicating that bias is *scale dependent in a type-dependent way*. Measurements of the clustering of Lyman break galaxies at $z \sim 3$ have shown that bias must be a strong function of cosmic epoch, in the sense that galaxies were more biased in the past (e.g., Adelberger et al. 1998).

A cautionary note here is that, as pointed out in Dekel & Lahav (1999), differing and non-equivalent definitions of bias are used in the literature. For example, the bias may be defined as the ratio of the galaxy to mass overdensity δ: $b_\delta \equiv \delta_g/\delta_m$, or as the square root of the ratio of the galaxy and matter correlation function ($b_\xi \equiv \sqrt{(\xi_g/\xi_m)}$), or as the ratio of the root variances ($b_\sigma \equiv \sigma_g/\sigma_m$).

3.3.2   Analytic Models of Bias

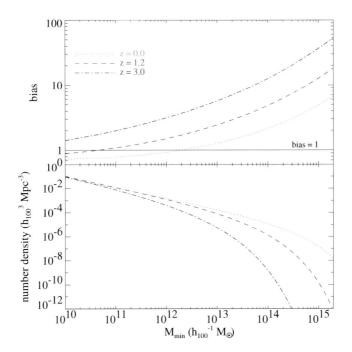

Figure 3.10. The average bias and comoving number density as a function of minimum host halo mass predicted by the Sheth–Tormen model and the Mo & White clustering model, for the simple case of one galaxy per dark matter halo above the minimum mass. Reproduced from Moustakas & Somerville (2002).

Why *should* we expect galaxies to be biased with respect to the dark matter? Our modern understanding of bias is based on the same high peaks model originally proposed by Kaiser. The basic idea is that in CDM models, there are waves upon waves upon waves. Based on the idea that we developed in §§ 3.2.1 and 3.2.2, that gravitational collapse will occur

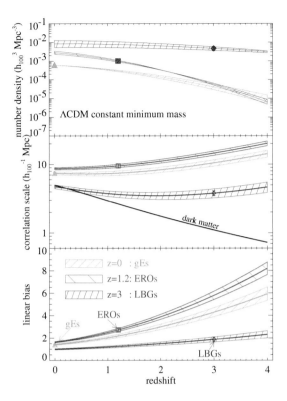

Figure 3.11. The redshift evolution of the comoving number density (top panel); the comoving correlation scale r_0 (middle panel); and the linear bias (lower panel), for three different populations: Lyman break galaxies, observed at $z = 3$, Extremely Red Objects at $z \sim 1$, and $z = 0$ giant ellipticals. Points show the observed properties of the objects at their redshift. A halo occupation model has been used to extrapolate each population forwards and/or backwards in time, assuming that the parameters of the HOD do not change with time. Reproduced from Moustakas & Somerville (2002).

when a region exceeds a given threshold overdensity δ_c, it makes sense that a small peak of a given mass will collapse earlier if it is riding on the back of a large swell (see Figure 17.4 on p. 580 of Peacock). Cole & Kaiser (1989) proposed a model known as the peak-background split, based on the same ingredients as the Press–Schechter model, to describe linear bias. This model was later extended by Mo & White (1996), Sheth & Tormen (1999), and Sheth et al. (2001). On large scales, the Mo & White model for halo bias as a function of halo mass and redshift gives

$$b(M, z) = 1 + \frac{\nu^2(M, z) - 1}{\delta_c} \qquad (3.17)$$

where $\nu(M, z) \equiv \delta_c / [D(z)\sigma(M)]$.

This model qualitatively reproduces the observed behavior of the biasing of dark matter halos in N-body simulations. We can immediately see that in this model, we expect halos that are more massive than M_* are positively biased ($b > 1$), halos with $M = M_*$ ($\nu = 1$) are unbiased ($b = 1$), and halos with $M < M_*$ are anti-biased ($b < 1$). It also follows that halos of a given mass are rarer and more biased in the past.

A first-order approach to understanding galaxy bias as a function of type and redshift is to simply associate a galaxy population with a given magnitude limit with dark matter halos above a minimum mass. An example of the expected bias in such a model is shown in Figure 3.10. This simple view would lead us to conclude that red, early-type galaxies dwell in more massive halos than blue, late type galaxies, and that more luminous galaxies live in heavier halos than fainter ones. Another conclusion is that high redshift galaxies which are as clustered as present-day galaxies (for example, Lyman break galaxies at $z \sim 3$ have about the same correlation length as bright galaxies today) live in less massive halos than today's galaxies (Wechsler et al. 2001; Bullock, Wechsler, & Somerville 2002; Moustakas & Somerville 2002).

Such an approach does help to understand the observed trends with galaxy type and with redshift, but a one-galaxy per halo model clearly cannot be correct in detail. Especially if we are interested in understanding galaxy clustering on scales less than a few Mpc, or smaller than the scale of typical dark matter halos, we must consider the possibility that dark matter halos are inhabited by more than one galaxy. The Halo Model is a term covering a general class of models that attempts to draw an empirical connection between dark matter halos and galaxies by parameterizing the number of galaxies per halo as a function of halo mass. More generally, this function (the number of galaxies of a given type and luminosity occupying a dark matter halo of a given mass) is referred to as the *Halo Occupation Distribution* (HOD). The global number density and clustering properties of galaxies may be predicted using an assumed HOD plus the known clustering behavior of dark matter halos (which can be obtained from N-body simulations or analytic models). This approach can be extended further by assuming a *conditional luminosity function*, the luminosity function of galaxies within a halo of a given mass (e.g., van den Bosch, Yang, & Mo 2003). A large literature has sprung up on this subject, which I cannot hope to even summarize here. I refer interested students to the recent review by Cooray & Sheth (2002).

3.4 Dark Matter Halos:
Internal Properties and Correlations

Over the past five to ten years, the vast improvement in resolution and dynamic range accessible to N-body techniques with modern supercomputers has led to a characterization of the *internal structure* of dark matter (DM) halos that form in the CDM paradigm. In this section I summarize some important results from this work.

3.4.1 Halo Profiles

Navarro, Frenk, & White (1995, 1996) presented the important result that the (spherically averaged) density profiles of dark matter halos in CDM simulations, when scaled properly, obey a universal form independent of halo mass and cosmology. The functional form that they proposed (now known as the Navarro–Frenk–White or NFW profile) is:

$$\rho(r) = \frac{\rho_s}{(r/r_s)\left(1 + r/r_s\right)^2}, \qquad (3.18)$$

The NFW function is described by two parameters, here r_s and ρ_s, and predicts that density scales as $\rho \propto r^{-1}$ at small radius, $\rho \propto r^{-2}$ at intermediate radius, and $\rho \propto r^{-3}$ at large radius. The parameter r_s represents the radius at which the log slope of ρ is –2. An alternative way of expressing this parameter is in terms of the concentration $c \equiv r_{\rm vir}/r_s$. The log slope in the very inner part of halos has been a subject of much controversy. The latest results suggest that the slope in the inner one percent of the virial radius probably lies between the NFW slope of -1 and the Moore profile which predicts an inner slope of -1.5 (Moore et al. 1999). The concentration parameter c may easily be generalized to variants such as the Moore profile.

Although the NFW function has two parameters that are in principle independent, NFW found a strong correlation between the two parameters. This is usually expressed as a correlation between virial mass and concentration, in the sense that more massive halos are less concentrated. NFW suggested that this may be because halo concentrations reflect the background density of the universe at the time when the halo formed. Larger mass halos form later, when the universe is less dense, and therefore are expected to be less concentrated. NFW presented a simple toy model based on this idea that described the correlation they observed in their simulations at redshift zero.

The studies of NFW and Moore et al. were based on a technique in which individual dark matter halos were extracted from a large, low resolution simulation and re-simulated at high resolution. Bullock et al. (2001b, B01) studied halo profiles for a much larger, statistically complete

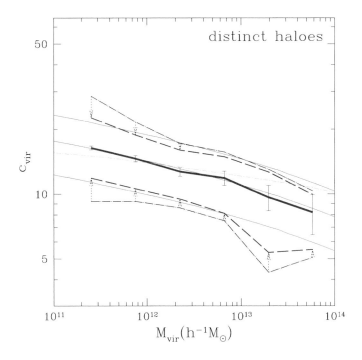

Figure 3.12. Concentration versus mass for halos at $z = 0$. The thick solid curve is the median at a given $M_{\rm vir}$. The error bars represent Poisson errors of the mean due to the sampling of a finite number of halos per mass bin. The outer dot-dashed curves encompass 68% of the $c_{\rm vir}$ values as measured in the simulations. The inner dashed curves represent only the true, intrinsic scatter in $c_{\rm vir}$, after eliminating both the Poisson scatter and the scatter due to errors in the individual profile fits due, for example, to the finite number of particles per halo. The central and outer thin solid curves are the predictions for the median and 68% values from the toy model of Bullock et al. (2001b). The thin dot-dashed line shows the prediction of the toy model of NFW. Reproduced from Bullock et al. (2001b).

sample of halos from a simulation with high spatial and mass resolution throughout the box. Although they found the same correlation between halo mass and concentration as NFW, they found that there is a significant scatter in halo concentration at fixed mass. B01 also found that the mean halo concentration at fixed mass is a fairly strong function of redshift, $c \propto (1+z)^{-1}$. This result was not well described by the toy model of NFW. B01 presented a different toy model, based on a similar idea (that halo profiles reflect the epoch of formation), that described the correlation as a function

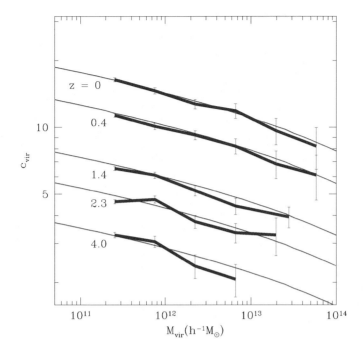

Figure 3.13. Median c_{vir} values as a function of M_{vir} for distinct halos at various redshifts. The error bars are the Poisson errors due to the finite number of halos in each mass bin. The thin solid lines show the toy model predictions. Reproduced from Bullock et al. (2001b).

of halo mass and redshift for the ΛCDM cosmology they considered. Eke, Navarro, & Steinmetz (2001) tested this model for a broader variety of cosmologies, including some WDM cases, concluded that it did not work well for all the cases they tested, and proposed a new toy model. The net impression, though, is that none of these toy models has really captured the basic physics that sets the mass-concentration relationship for dark matter halos. The question also remains as to what determines the size of the scatter in this relation, and whether there is a second parameter.

3.4.2 Angular Momentum

Another important intrinsic property of dark matter halos is their internal *angular momentum*. A convenient way to express this quantity is in terms of the dimensionless *spin parameter* $\lambda \equiv J|E|^{1/2}G^{-1}M^{-5/2}$, where J is the angular momentum, M is the mass, and E is the energy of the dark matter halo (Peebles 1969). An alternative definition of the spin parameter is $\lambda' \equiv J/(\sqrt{2}MVR)$, where M, V, and R are the halo virial mass, velocity, and radius, respectively. The advantage of this definition is that it is numerically very similar to the standard λ parameter for NFW halos, but it is easier to measure in simulations (Bullock et al. 2001a).

It is well established that the distribution function of spins for DM halos formed in CDM simulations has the form of a log-normal (Barnes & Efstathiou 1987), and is invariant with cosmic epoch, halo mass, or environment, and cosmology or power spectrum (Lemson & Kauffmann 1999). I shall return to the origin of the characteristic distribution of average halo spin in §3.4.3.

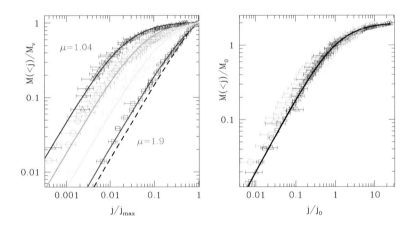

Figure 3.14. Mass distribution of specific angular momentum in four halos spanning a range of μ values from 1.04 to 1.9. Symbols and errors correspond to the ranked j measurements in cells, while the curves are the functional fits, $M(< j) = M_{\mathrm{vir}}\mu j/(j_0 + j)$. (a) All profiles are normalized to coincide at M_{vir}, where $j = j_{\mathrm{max}}$. The value of μ measures the relative extent of the power-law regime until it bends over. Shown for comparison is the distribution for a uniform sphere in solid-body rotation (dashed line). (b) All profiles are normalized to coincide at j_0 and on top of the universal profile (curve). The value of μ now correlates with the uppermost point, j_{max}/j_0, along the universal curve. Reproduced from Bullock et al. (2001a).

In the meantime, I turn to the distribution of angular momentum as a function of radius *within* dark matter halos. Bullock et al. (2001a) found that, in analogy with the universal dark matter density profile, when scaled appropriately, there is also a *universal angular momentum profile* for dark matter halos. Like the NFW profile, the angular momentum profile is characterized by two parameters:

$$M(< j) = M_{\rm v} \frac{\mu j}{j_0 + j}, \quad \mu > 1. \tag{3.19}$$

The profile has an implicit maximum specific angular momentum $j_{\rm max} = j_0/(\mu - 1)$. It is roughly a power law for $j \lesssim j_0$, and flattens out for $j \gtrsim j_0$. As can be seen by its relation to $j_{\rm max}$, the quantity μ (> 1) acts as a shape parameter: for $\mu \gg 1$, $M(< j)$ is a pure power law, while $\mu \to 1$ means that only half the mass falls within the power-law regime and the bend is pronounced. The pair of parameters μ and j_0 fully define the angular momentum distribution of the halo. The global spin parameter is related to μ and j_0 via

$$j_0 \, b(\mu) = \sqrt{2} V_{\rm v} R_{\rm v} \, \lambda', \tag{3.20}$$

where

$$b(\mu) \equiv \int_0^1 \frac{m}{\mu - m} dm = -\mu \ln(1 - \mu^{-1}) - 1. \tag{3.21}$$

The angular momentum profiles are shown in Figure 3.14.

3.4.3 Correlation Between Halo Properties and Formation History

These studies of statistical ensembles of halos have revealed several important properties of dark matter halos forming in a CDM universe. To summarize the relations discussed above:

(i) The density profiles of DM halos can be characterized by a universal functional form with two parameters.
(ii) DM halos exhibit correlations between their structural parameters (e.g., mass and concentration), but with a significant level of scatter.
(iii) The halo mass–concentration relationship evolves with redshift ($c \propto (1 + z)^{-1}$ at fixed mass).
(iv) The average spins of DM halos can be characterized by a universal distribution with a log-normal form.
(v) The angular momentum profiles of DM halos can be characterized by a universal functional form with two parameters.

Our understanding of the physical origin of all of these "observed" properties of halos in N-body simulations is still rather incomplete. The toy models describing the concentration–mass relation, discussed above, are based on the idea that the halo density reflects the density of the universe at the time when the halo formed. They qualitatively reproduce the observed trends of concentration with mass and redshift, but do not explain the scatter in concentration at fixed mass. Furthermore, these trends have been derived for statistical ensembles of halos—they do not tell us how the density profile of *individual halos* builds up over time.

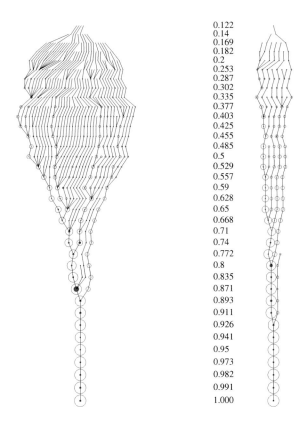

0.122
0.14
0.169
0.182
0.2
0.253
0.287
0.302
0.335
0.377
0.403
0.425
0.455
0.485
0.5
0.529
0.557
0.59
0.628
0.65
0.668
0.71
0.74
0.772
0.8
0.835
0.871
0.893
0.911
0.926
0.941
0.95
0.973
0.982
0.991
1.000

Figure 3.15. Structural merger trees for two halos. This diagram illustrates the merging history of a cluster-mass halo (left; $M_{\rm vir} = 2.8 \times 10^{14} h^{-1} {\rm M}_\odot$ and $c_{\rm vir} = 5.9$) and a galaxy-mass halo (right; $M_{\rm vir} = 2.9 \times 10^{12} h^{-1} {\rm M}_\odot$ and $c_{\rm vir} = 12.5$) at $a = 1$. The radii of the outer and inner (filled) circles are proportional to the virial and inner NFW radii, $r_{\rm vir}$ and r_s, respectively, scaled such that the two halos have equal sizes at $a = 1$. Lines connect halos with their progenitor halos. All progenitors with $M > 2.2 \times 10^{11} h^{-1} {\rm M}_\odot$ are shown for the cluster-mass halo; all progenitors with $M > 2.2 \times 10^{10} h^{-1} {\rm M}_\odot$ are shown for the galaxy-mass halo. The scale factor a is listed in the center of the plot. The width of the diagram is arbitrary. Reproduced from Wechsler et al. (2002).

3.4.3.1   The Origin of Halo Density Profiles

To study these questions, one must build a *structural merger tree*, a merger tree in which the structural properties of the halos (such as the density profile, or spin) are traced along with the mass. This was first done by Wechsler et al. (2002, W02). Each dark matter halo, in addition to having its mass, concentration, and spin parameter measured, is traced through time as it builds up through mergers and accretion. This allows us to study the relationship between halo *structural properties* and *mass accretion/merger history*. W02 found that *at fixed mass*, dark matter halos that form earlier have higher concentrations than those that form later. This neatly explains both the origin of the mean mass-concentration relation—more massive halos form later on average, than less massive halos—and the scatter at fixed mass.

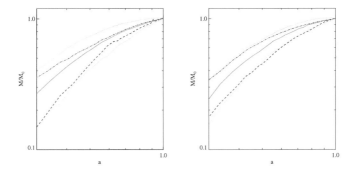

Figure 3.16. Average mass accretion histories, normalized at $a = (1+z)^{-1} = 1$. **Left**: binned in 3 bins by final halo mass: $M_0 = 1-4\times10^{12}h^{-1}M_\odot$ (dot-dashed), $M_0 = .4 - 3 \times 10^{13}h^{-1}M_\odot$ (solid), and $M_0 > 3 \times 10^{13}h^{-1}M_\odot$ (dashed). The three (green) curves connect the averages of $M(a)/M_0$ at each output time. The pair of dotted lines shows the 68% spread about the middle case (the spread is comparable for the other bins). We see that massive halos tend to form later than lower-mass halos, whose mass accretion rate peaks at an earlier time. **Right**: binned in 3 bins by formation epoch a_c. Dot-dashed lines correspond to early formers (typically low-mass halos), dashed lines to late formers (typically higher-mass halos). The averages and spread are displayed in analogy to the left panel. Reproduced from Wechsler et al. (2002).

A key realization made by W02 was that none of the definitions of "halo formation time" used in the previous toy models were appropriate. Tracing the mass of the largest progenitor halo over time (Mass Accretion History; MAH) for many halos, they found that the majority of halo MAHs can be described reasonably well over most of their lifetime by the simple functional form:

$$M(z) = M_o e^{-\alpha z}. \tag{3.22}$$

The parameter α can be related to a characteristic epoch of formation a_c, defined as the scale factor ($a = (1 + z)^{-1}$) when the logarithmic slope of the mass accretion rate, $d \log M/d \log a$, falls below some specified value, S (the value of S is arbitrary; W02 choose $S = 2$). The functional form defined in Equation 3.22 implies $a_c = \alpha/S$. This means that halos have a period of rapid growth, followed by a leveling off, and the formation time marks the transition between the two. Remarkably, the halo concentration is then simply given by $c = c_1/a_c$, where c_1 is a constant. This provides a means of relating the *halo mass accretion history* and the concentration. When ensembles of MAH trajectories are produced using EPS, this relation reproduces the concentration–mass relation, its scatter at fixed mass, and its evolution with redshift (W02).

3.4.3.2   *The Origin of Halo Spin*

The internal angular momentum of dark matter halos in a CDM universe is generally attributed to tidal torques from the large-scale density field, experienced before turnaround, while the proto-halo perturbations are in the linear regime (Peebles 1969; Doroshkevich 1970; White 1984). This picture, combined with the invariance of the *distribution* of spins with cosmic epoch, may give the impression that spin is a property that remains invariant over the lifetime of a halo. However, by studying the angular momentum histories of individual halos, we find that this is not at all the case. It appears that halos acquire angular momentum in *mergers*, via transfer of orbital angular momentum to internal spin. Because minor mergers (i.e., events in which one member of the merging pair is much smaller in mass than the other) are frequent and more or less isotropic, the angular momentum from these events mostly cancels out. However, major mergers (in which the lumps have comparable masses) are rare and therefore tend to result in a significant net gain of angular momentum (depending, of course, on the impact parameter of the merger). This is illustrated, based on the same structural merger trees discussed above, in Figure 3.17, and has obvious significance for our understanding of the structure of galactic disks (discussed in more detail below).

Based on these ideas, Vitvitska et al. (2002) and Maller, Dekel, & Somerville (2002) investigated a simple recipe implemented within an EPS merger tree, in which halos gain internal spin angular momentum from the orbital angular momentum in each merger they experience. Both studies found that such a scheme reproduces the log-normal distribution obtained in the N-body simulations (see Figure 3.18). Maller et al. (2002) further showed that this approach could reproduce the known insensitivity of halo spin to mass, redshift, and cosmology, as well as the correlation with merger

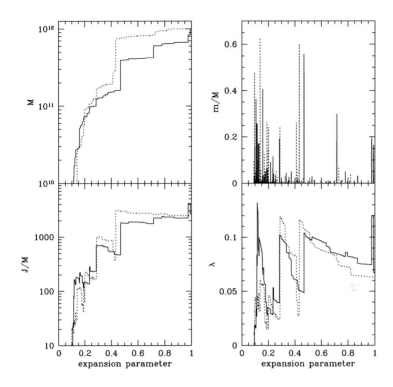

Figure 3.17. The build-up of mass (top) and specific angular momentum (bottom) for two different halos (solid and dotted) as a function of expansion factor $a = (1 + z)^{-1}$. *Right panels:* Mass ratios of merging satellites (top) and spin parameter (bottom) as a function of scale factor. Reproduced from Vitvitska et al. (2002).

history noted above. Even more remarkably, Bullock et al. (2001a) and Maller & Dekel (2002) found that by tracing the incoming angular momentum in merging satellites in an EPS merger tree, they could reproduce the *angular momentum profile* found in simulations (see §3.4.2).

3.5 Galaxy Formation Within the CDM Paradigm

3.5.1 Consumer's Guide to Galaxy Formation Models

There are currently two main kinds of tools that are used to study galaxy formation within the hierarchical paradigm that we have been discussing.

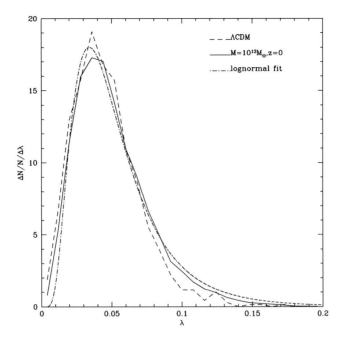

Figure 3.18. Distribution of halo spin parameters at redshift $z = 0$. The solid line represents the distribution of λ derived using the merger model. The distribution of λ from an N-body simulation is shown by the dashed curve. The dot-dashed curve is a log-normal fit to the merger model. Reproduced from Vitvitska et al. (2002).

Numerical N-body techniques represent the dark matter, gas, and stellar density field with particles or grid cells, and solve the physical equations describing the evolution of these components. These equations may be cast in either Lagrangian (e.g., Smoothed Particle Hydrodynamics, SPH) or Eulerian form. A further distinction can be made with respect to the initial conditions; cosmological simulations start with the primordial power spectrum at some high redshift and allow galaxies to form as they will, according to the laws of gravity, hydrodynamics, thermodynamics, etc. In order to study specific problems, one can bypass some of the uncertainties associated with galaxy formation by setting up individual galaxies according to a desired set of initial conditions, then evolving them. For example, this approach has been used very effectively to study galaxy collisions (e.g., Barnes & Hernquist 1996). These techniques have the advantage that one obtains detailed spatial information about where galaxies are relative to one another; and (subject to the spatial/force resolution of the simulation) about the internal structure of the stars, gas, and dark matter within the galaxies. The main disadvantages are the limitations imposed by the finite availability of computational resources, which limit the number of particles or grid cells that can be simulated, and hence the volume that can be simulated and the mass, spatial, and force resolution. The other difficulty is that the physics governing star formation and feedback occurs on scales well below those that can be simulated directly in a simulation of even a single galaxy. Therefore this physics must be treated by sub-grid "recipes", which are quite uncertain, and inevitably contain several free parameters. Exploring this parameter space with sufficiently high resolution is difficult with current computing capabilities.

The other main class of tools is referred to as "semi-analytic" models, but this term has become a catch-all for a wide variety of models, some of which have very different ingredients. Here are some examples of different kinds of semi-analytic models.

Press–Schechter Merger Tree models use merger trees built using the Press–Schechter-based approach discussed above. They generally include cooling, star formation, feedback, and stellar population modeling. Some include simple chemical evolution modeling; some do not. Examples include Kauffmann, White, & Guiderdoni (1993), Cole et al. (1994), Cole et al. (2000), and Somerville & Primack (1999, SP99).

Hybrid Merger Tree models use merger trees extracted from N-body simulations. The dynamics (merging) of sub-halos within virialized halos are often treated semi-analytically. These models generally include cooling, star formation, feedback, and stellar population modeling. Some include simple chemical evolution modeling, some do not. Examples include Kauffmann et al. (1999), Springel et al. (2001), and De Lucia, Kauffmann, & White (2004).

Disk models neglect galaxies' formation histories (Mo, Mao, & White 1998; van den Bosch 2000) or treat halo formation using a smooth "mass accretion history" (Avila-Reese, Firmani, & Hernández 1998; van den Bosch 2002). Merging and non-central galaxies are neglected. Some of these models include cooling, star formation, and feedback (e.g., van den Bosch 2002); some do not (Mo et al. 1998).

3.5.2 Gastrophysics

In this section, I summarize the standard ingredients for modeling gas cooling, star formation, and feedback in merger-tree-based semi-analytic models (the first two categories mentioned above). More details may be found in e.g., Somerville & Primack (1999), Kauffmann et al. (1999), Cole et al. (2000), Somerville, Primack, & Faber (2001b), and Springel et al. (2001). There are of course many, many other papers on these techniques, but I will not attempt to provide a full list here.

3.5.2.1 Gas Cooling

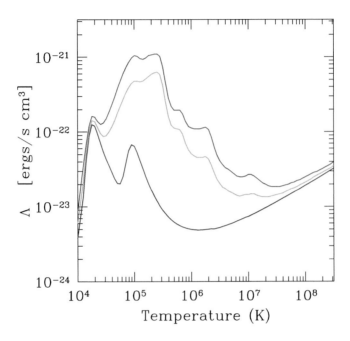

Figure 3.19. Metallicity dependent radiative cooling function from Sutherland & Dopita (1993). Shown from top to bottom are solar, 0.3 solar, and primordial metallicity.

Semi-analytic models almost universally adopt a variant of the cooling model introduced by White & Frenk (1991). This model supposes that each newly formed halo, at the top level of the tree, contains pristine hot gas that has been shock heated to the virial temperature of the halo (T) and that the gas traces the dark matter. The rate of specific energy loss due to radiative cooling is given by the cooling function $\Lambda(T)$ times the electron density n_e (see e.g., Binney & Tremaine 1987, p. 580). The critical density that will enable the gas to cool within a timescale $\tau_{\rm cool}$ is:

$$\rho_{\rm cool} = \frac{3}{2} \frac{\mu m_p}{\chi_e^2} \frac{k_B T}{\tau_{\rm cool} \Lambda(T)} \tag{3.23}$$

where μm_p is the mean molecular weight of the gas and $\chi_e \equiv n_e/n_{\rm tot}$ is

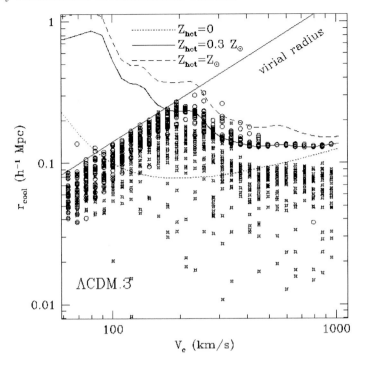

Figure 3.20. Cooling radius of halos as a function of circular velocity. The straight diagonal line shows the virial radius, which the cooling radius may not exceed. The curved lines show the cooling radius predicted by the literal static halo cooling model (see text), assuming that the hot gas has primordial, 0.3 solar, or solar metallicity. Open circles show the application of the static halo model within the merger trees, and crosses show the dynamic halo model (see text), assuming a fixed metallicity of 0.3 solar. Earlier conversion of gas from the hot to cold phase and reheating of hot gas by halo mergers results in a lower cooling efficiency for large halos in the dynamic halo model. Reproduced from Somerville & Primack (1999).

the number of electrons per particle. Assuming that the gas is fully ionized and has a helium fraction by mass $Y = 0.25$,

$$\rho_{cool} = 3.52 \times 10^7 \frac{k_B T}{\tau_{cool} \Lambda_{23}(T)}, \qquad (3.24)$$

where $k_B T$ is in degrees Kelvin, τ_{cool} is in Gyr, and $\Lambda_{23}(T) \equiv \Lambda(T)/(10^{-23}$ ergs s^{-1} cm^3). The virial temperature is approximated as $k_B T = 71.8\sigma_{vir}^2$, where σ_{vir} is the virial velocity dispersion of the halo. Assuming a form for the gas density profile $\rho_g(r)$, one can now invert this expression to obtain the "cooling radius", defined as the radius within which the gas has had time to cool during the timescale τ_{cool}. Assuming $\rho(r) \propto r^{-2}$ (a singular isothermal sphere), this gives

$$r_{\text{cool}} = \left(\frac{\rho_0}{\rho_{\text{cool}}} \right)^{1/2} , \qquad (3.25)$$

where $\rho_0 = f_{\text{hot}} V_c^2 / (4\pi G)$, f_{hot} is the hot gas fraction in the cooling front, and $V_c = \sqrt{2}\sigma_{\text{vir}}$ is the circular velocity of the halo.

The cooling function $\Lambda(T)$ is also strongly metallicity dependent; naturally, metal-enriched gas cools much more efficiently. A standard reference for the metallicity-dependent cooling function (shown in Figure 3.19) is Sutherland & Dopita (1993). Thus the cooling rate of the gas depends on its prior enrichment history. This cooling rate may be computed by dividing the time interval between halo mergers (branchings) into small time-steps. For a time-step Δt, the cooling radius increases by an amount Δr and the mass of gas that cools in this time-step is $dm_{\text{cool}} = 4\pi r_{\text{cool}}^2 \rho_g(r_{\text{cool}})\Delta r$. For small halos and at early times, the cooling radius computed in this way may exceed the virial radius, and two additional limitations are generally applied: only gas within the current virial radius is allowed to cool, and cold gas falls onto the disk on a free fall time (or at the gas sound speed).

If we imagine a "static" dark matter halo, forming at time $t = 0$ with circular velocity V_c and a corresponding virial temperature T, and ignore the chemical enrichment of the hot gas, this model is quite straightforward to implement. The halo simply grows isothermally, gradually cooling at the rate given by the cooling function $\Lambda(T)$ as described above. The predicted cooling radius as a function of halo circular velocity is shown by the curved lines in Figure 3.20. Here, the cooling time τ_{cool} is the age of the universe at the current redshift, and the gas fraction in the cooling front f_{hot} is always equal to the universal baryon fraction f_{bar}.

However, in the hierarchical framework of the merger trees, most halos are built up from merging halos that have experienced cooling, star formation, and feedback in previous time-steps. This modifies the gas fraction f_{hot} in the cooling front. Gas cools at different temperatures, depending on the progenitor virial temperature. The cooling radius in hierarchically forming dynamic halos is shown by the open circles in Figure 3.20. They show some scatter, but generally follow the expectations of the simple static halo model.

Violent mergers between halos may shock heat the hot gas. In the models of SP99, all hot gas is assumed to be reheated to the virial temperature of the new halo following a "major" halo merger ($M_1/M_0 > 0.5$, where M_1 is the mass of the largest progenitor and M_0 is the mass of the merged halo). This tends to suppress cooling in the largest mass halos (which are more likely to have experienced a recent major merger), as seen in Figure 3.20.

3.5.2.2 Disk Sizes

As discussed in § 3.4.2, CDM halos have spin angular momentum acquired from tidal torques or mergers. It is likely that the specific angular momentum of the hot shocked gas within these halos is similar to that of the dark matter component (van den Bosch et al. 2002). The cooling gas collapses until it is supported either by this angular momentum or by pressure. In the former case, assuming that the specific angular momentum is conserved and that the gas collapses into a disk with an exponential profile, one can use the formalism introduced by Fall & Efstathiou (1980) and refined by e.g., Blumenthal et al. (1986), Flores et al. (1993), and Mo et al. (1998) to compute the exponential scale radius of the disk. Assuming that the dark matter halo profile is a singular isothermal sphere, and neglecting the modification of the inner profile of the dark matter due to the self-gravity of the infalling baryons, the disk size is just $r_s = 1/\sqrt{2}\lambda r_i$, where r_i is the radius before collapse.

In the more realistic case of a NFW profile and including the self-gravity of the baryons using the *adiabatic invariant* formalism, the scale radius is given by

$$r_s = 1/\sqrt{2}\lambda r_i f(c, \lambda, m_d), \qquad (3.26)$$

where c is the concentration of the halo and m_d is the fraction of baryons within the virial radius that are in the galactic disk. The function $f(c, \lambda, m_d)$ is given in Mo et al. (1998) and is an increasing function of c and m_d and a decreasing function of λ (i.e., the more concentrated the halo or the heavier the disk the smaller the disk size, but gas with more angular momentum contracts less). There is a large literature on detailed modeling of disk sizes, though most *merger tree* based semi-analytic models still incorporate a rather rudimentary treatment of galaxy sizes (the GALFORM model described in Cole et al. 2000 is currently the one exception of which I know).

3.5.2.3 Photoionization Squelching

The accretion of gas by low mass halos is suppressed or "squelched" in the presence of a photoionizing background. The critical mass, below which the gas is simply too hot to be contained by the potential well, corresponds to the Jeans mass (Rees 1986). However, because dark matter halos in CDM models have complex and varied formation histories, a sort of time-averaged Jeans Mass is needed to determine the gas content of any given halo at any given time. The recent analysis of Gnedin (2000, G00), based on hydrodynamic simulations, suggests that the filtering mass, which corresponds to the length scale over which baryonic perturbations are smoothed in linear theory, provides a good description of this time-averaged, effective Jeans mass. G00 provides a fitting function for the mass of gas within halos

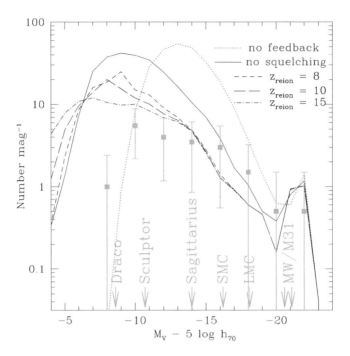

Figure 3.21. The V-band luminosity function of the Local Group. Symbols with error bars show the observed local group luminosity function, with Poisson errors. The dotted line shows the results of the models with no feedback of any kind, and the solid line shows the results of the models with supernova feedback but no photo-ionization squelching. Short dashed, long dashed, and dot-dashed lines show the model results with squelching, for three different assumed epochs of reionization as indicated on the figure panel. The magnitudes of several familiar Local Group members are indicated. Reproduced from Somerville (2002).

after reionization as a function of halo mass:

$$M_g = \frac{f_b M_{\text{vir}}}{[1 + 0.26 M_C/M_{\text{vir}}]^3}, \qquad (3.27)$$

where M_C is a characteristic mass representing the mass of objects that on average retain 50% of their gas. Somerville (2002) approximated M_C as the mass corresponding to a halo with virial velocity $V_c = 50\,\text{km s}^{-1}$ collapsing at redshift z. This provides a good description of the redshift dependence of M_C found in the simulations of G00: M_C increases from about $10^8\,\text{M}_\odot$ at $z = 8$ to several $10^{10}\,\text{M}_\odot$ at $z = 0$. Then, in agreement with the results shown in Figure 2 of G00 and with previous results (e.g., Thoul & Weinberg

1996; Quinn, Katz, & Efstathiou 1996), halos with virial velocity $V_c \sim 50$ km/s are able to capture about half of the gas that they would accrete in the absence of a background, while halos with $V_c < 30\,\mathrm{km\,s^{-1}}$ are able to accrete very little gas, and halos with $V_c > 75\,\mathrm{km\,s^{-1}}$ are nearly unaffected.

Figure 3.21 shows the impact of squelching on the luminosity function of satellite galaxies in the Local Group. The earlier the universe is reionized, the fewer halos manage to collapse and cool gas before the ionizing field (and the effect of squelching) is switched on, and the fewer small-mass, low-luminosity galaxies are formed. Squelching greatly improves the agreement of the semi-analytic model predictions for the LG luminosity function at $M_V \gtrsim -14$ (about the magnitude of Sagittarius), while having almost no effect on galaxies brighter than the Large Magellanic Cloud ($M_V \lesssim -18$). Similar conclusions were reached by Kauffmann et al. (1993) and Benson et al. (2002a).

3.5.2.4 Star Formation

A standard approach for modeling star formation in galaxy formation models is the Schmidt Law: $\dot{\rho}_* \propto \rho_g{}^N$, where $\dot{\rho}_*$ is the star formation rate density, ρ_g is the cold gas density, and typically $N \simeq 1.5$. Similarly, observations of nearby star-forming galaxies suggest that $\dot{\Sigma}_* \propto \Sigma_g{}^N$, where $\dot{\Sigma}_*$ and Σ_g are the 2-D star formation and gas surface densities (Kennicutt 1983, 1998).

In semi-analytic models, the total star formation rate in a galaxy is often modeled by an expression of the form:

$$\dot{m}_* = \frac{m_{\mathrm{cold}}}{\tau_*}, \tag{3.28}$$

where m_{cold} is the total mass of cold gas in the disk and our ignorance is piled in the efficiency factor τ_*. This factor is often assumed to be proportional to the dynamical time of the disk $t_{\mathrm{dyn}} = r_{\mathrm{disk}}/V_c$, which in the case of a self-gravitating disk reduces to the Schmidt Law. Some modelers have also added a dependence on redshift, e.g., $\tau_* \propto t_{\mathrm{dyn}}(1+z)^\alpha_*$ (Kauffmann & Haehnelt 2000) or galaxy circular velocity (Cole et al. 2000; De Lucia et al. 2004):

$$\tau_* \propto t_{\mathrm{dyn}} \left(\frac{V_c}{V_0}\right)^{\alpha_*}. \tag{3.29}$$

It is worth noting some subtleties of the modeling, and some implications of making different choices for these scalings. First, while really t_{dyn} should represent the dynamical time *of the galactic disk*, as I mentioned earlier, most published merger-tree-based SAMs do not model the size or internal velocity of disks in detail and so in practice, most of the models use a fixed fraction of the *DM halo dynamical time* (e.g., Kauffmann et al. 1993; Cole et al. 1994; Kauffmann et al. 1998; Somerville & Primack 1999;

Somerville et al. 2001b). The updated GALFORM model described in Cole et al. (2000) is an exception.

The dynamical time $t_{\rm dyn} \equiv r_{\rm vir}/V_{\rm vir}$ of dark matter halos at a given redshift is independent of the halo circular velocity; recall that the spherical collapse model predicts that the virial radius scales like $r_{\rm vir} \propto V_c$ (see § 3.2.1). However, the virial radius of a halo with a given circular velocity increases with time (e.g., $r_{\rm vir} \propto (1 + z)^{-3/2}$ in an Einstein–de Sitter universe), and so the scaling with $t_{\rm dyn}$ builds in a higher *star formation efficiency* (more star formation per unit mass of cold gas) at earlier times. When the additional dependence on circular velocity is included, α_* is generally chosen to have a value $\alpha_* \sim$ 1.5–2, so that star formation is *less efficient in smaller halos*. Since the mass and circular velocity of typical objects is smaller at high redshift, this has the opposite effect of making star formation (on average) *less* efficient at high redshift.

The above recipes are assumed to apply to quiescent, isolated disks. Motivated by observed starburst galaxies and N-body simulations (Mihos & Hernquist 1994, 1995, 1996), most SAMs also include a more efficient bursting star formation mode, triggered by galaxy–galaxy mergers. It is also standard to divide mergers into two categories: major and minor, where the distinction is drawn based on the mass ratio of the merging pair $\mu = m_{\rm small}/m_{\rm big}$. In major mergers, $\mu > f_{\rm bulge} \simeq 0.3$, the usual assumption is that most or all of the cold gas is consumed in a very efficient starburst. As well, any disk component is destroyed and all pre-existing stars are left in a spheroidal remnant. The treatment of minor mergers varies more from model to model. Some modelers assume that there is no starburst, so that the stars and cold gas of the two galaxies are simply combined (e.g., Kauffmann et al. 1999; Cole et al. 2000). Others assume that the efficiency of the starburst is a function of the mass ratio μ (Somerville et al. 2001b), or of the specific angular momentum transferred in the merger (Menci et al. 2004). Although minor mergers then produce less efficient starbursts than major mergers, because they are much more common, minor bursts can still have a significant effect on the overall star formation rate, especially at high redshift (Somerville et al. 2001b; Menci et al. 2004).

3.5.2.5 *Supernova Feedback*

Like star formation, supernova feedback is treated with a set of ad hoc but somewhat physically motivated recipes. Observationally, it is known that the rate at which cold gas is heated and ejected is proportional to the star formation rate (Martin 1999). It also seems that the winds driven by SNae can eject the gas from the halos of small galaxies ($V_c \lesssim$ 100–150 km/s) but not from larger objects (Martin 1999).

The ansatz often used to model feedback is that the mass of reheated gas is equal to the total energy injected by the SNae divided by the specific internal energy of the gas (e.g., Kauffmann et al. 1993). The internal energy of the gas is given by $U = 3/2NkT = 3/4N\mu m_H V_c^2$, and the specific internal energy is then just $3/4V_c^2$. The energy injected by the SNae is $\propto \epsilon\eta_{SN}E_{SN}\dot{m}_*$, where η_{SN} is the number of SNae per solar mass of stars, and E_{SN} is the average energy of the ejecta from one SN. The mass of reheated gas is then:

$$\Delta M_{rh} = \epsilon\frac{4}{3}\frac{\eta_{SN}E_{SN}}{V_c^2}\Delta M_*, \tag{3.30}$$

where ϵ is the fraction of the SN energy injected into the gas. The rate of gas reheating in SAMs is often written:

$$\dot{m}_{rh} = \epsilon_{SN}^0(V_c/V_0)^{-\alpha_{rh}}\dot{m}_*, \tag{3.31}$$

where ϵ_{SN}^0 and α_{rh} are free parameters and $\dot{m}_*$ is the star formation rate. Based on the energy ansatz above, which implies $\alpha_{rh} = 2$, the scaling factor $\epsilon_{SN}^0 V_0$ is given by:

$$\epsilon_{SN}^0 V_0^2 = \frac{4}{3}\eta_{SN}E_{SN}, \tag{3.32}$$

where η_{SN} depends on the stellar initial mass function (IMF) ($\eta_{SN} = 7.4 \times 10^{-3}$ for a Salpeter IMF), and $E_{SN} \simeq 10^{51}$ ergs.

Another seemingly subtle but important point is the fate of the reheated gas. If it is deposited in the hot halo, then the gas tends to quickly recool and the feedback is much less effective. If it is ejected from the halo entirely, then future star formation is delayed at least until gas becomes incorporated in a larger structure at some later stage in the merger tree.

3.5.2.6 Chemical Evolution

Some models include a simple treatment of chemical evolution, assuming a constant "effective yield", or mean mass of metals produced per mass of stars, and using the instantaneous recycling approximation. Newly produced metals are deposited in the cold gas. Subsequently, the metals may be ejected from the disk and mixed with the hot halo gas, or ejected from the halo, typically following the fate of the reheated gas as discussed above. The metallicity of each batch of new stars is set equal to the metallicity of the cold gas at the time when the stars are formed. Note that because enriched gas may be ejected from the halo, and primordial gas is constantly being accreted by the halo, this approach is not equivalent to a classical "closed box" model of chemical evolution, although it is similar in spirit. As the hot gas becomes enriched with metals, cooling becomes more efficient (see § 3.5.2.1).

3.5.2.7 Galaxy Merging

In this picture, merging of galaxies within a common dark matter halo is important in several ways: mergers trigger starbursts, and build spheroids. Even in the hybrid SAMs, in which one could in principle use the N-body information to track mergers, often the resolution is not good enough to do this and one must use a semi-analytic recipe to model the merging of sub-halos within virialized host halos (e.g., Kauffmann et al. 1999).

The standard recipe is based on the idea that after two dark matter halos merge, the galaxies orbit within them and lose energy via dynamical friction. The Chandrasekhar approximation provides the differential equation for the distance of the satellite from the center of the halo (r_{fric}) as a function of time:

$$r_{\text{fric}} \frac{dr_{\text{fric}}}{dt} = -0.428 f(\epsilon) \frac{Gm_{\text{sat}}}{V_c} \ln \Lambda \qquad (3.33)$$

(e.g., Binney & Tremaine 1987, p. 424). In this expression, m_{sat} is the combined mass of the satellite's gas, stars, and dark matter halo, and V_c is the circular velocity of the parent halo. Not to be confused with the cosmological constant, here $\ln \Lambda$ is the Coulomb logarithm, which can be approximated as $\ln \Lambda \approx \ln(1 + m_h^2/m_{\text{sat}}^2)$, where m_h is the mass of the parent halo. The circularity parameter ϵ is defined as the ratio of the angular momentum of the satellite to that of a circular orbit with the same energy: $\epsilon = J/J_c(E)$. (Lacey & Cole 1993) showed that the approximation $f(\epsilon) = \epsilon^{0.78}$ is a good approximation for $\epsilon > 0.02$.

As the satellite falls in, its dark matter halo is tidally stripped by the background potential of the parent halo. While this effect is ignored by many of the models, it may significantly slow down merging especially for small mass objects (Taffoni et al. 2003; Colpi, Mayer, & Governato 1999). Recently, more detailed modeling of satellite merging, including both tidal stripping and tidal heating, has been presented by several groups (Benson et al. 2002b; Taylor & Babul 2004; Zentner & Bullock 2003).

3.5.2.8 Stellar Population Synthesis

Stellar population synthesis models (e.g., Bruzual & Charlot 1993; Fioc & Rocca-Volmerange 1997; Leitherer et al. 1999; Devriendt, Guiderdoni, & Sadat 1999) provide the Spectral Energy Distribution (SED) of a stellar population of a single age and metallicity. These models must assume a stellar IMF. The model stars are then evolved according to theoretical evolutionary tracks for stars of a given mass. By keeping track of how many stars of a given age and metallicity are created according to our star-formation recipe, one can create synthesized spectra for the composite population in each galaxy at a given observation time t_0:

$$L_\lambda(t_0) = \int_0^{t_0} S_\lambda(t_0 - t, Z_*(t)) \, \dot{m}_*(t) dt. \tag{3.34}$$

Here $S_\lambda(t, Z)$ is the "single burst" SED for stars of age t and metallicity Z, and $\dot{m}_*(t)$ is the star formation rate at time t. The SED L_λ can then be convolved with the filter response functions appropriate to a particular set of observations.

3.5.2.9 Dust Absorption

Absorption of galactic light by dust in the interstellar medium causes galaxies to appear fainter and redder in the ultraviolet to visible part of the spectrum. The absorbed light is re-radiated in the infra-red and longer wavelengths. The treatment of dust absorption and emission in semi-analytic models in the literature ranges from non-existent (e.g., Kauffmann et al. 1993; Cole et al. 1994; Baugh et al. 1998), to very simple (e.g., Kauffmann et al. 1999; Somerville & Primack 1999; Somerville et al. 2001b), to quite sophisticated (e.g., Granato et al. 2000).

An example of a simple approach used by K99 and SP99 is based on the empirical results of Wang & Heckman (1996). These authors give an expression for the B-band, face-on extinction optical depth of a galaxy as a function of its blue luminosity:

$$\tau_B = \tau_{B,*} \left(\frac{L_{B,i}}{L_{B,*}} \right)^\beta, \tag{3.35}$$

where $L_{B,i}$ is the intrinsic (unextinguished) blue luminosity, and we use $\tau_{B,*} = 0.8$, $L_{B,*} = 6 \times 10^9 L_\odot$, and $\beta = 0.5$. The B-band optical depth is converted to other wavebands using a standard galactic extinction curve (e.g., Cardelli, Clayton, & Mathis 1989). The extinction in magnitudes is then related to the inclination of the galaxy using a standard "slab" model (a thin disk with stars and dust uniformly mixed together):

$$A_\lambda = -2.5 \log \left(\frac{1 - e^{-\tau_\lambda{}'' \theta}}{\tau_\lambda{}'' \theta} \right) \tag{3.36}$$

where θ is the angle of inclination to the line of sight (e.g., Guiderdoni & Rocca-Volmerange 1987). A random inclination is assigned to each model galaxy.

3.5.2.10 Free Parameters

In nearly all of the above recipes, we introduced one or more free parameters. Each model differs slightly in the parameters introduced, but I give here an example list of typical parameters:

- τ_*^0: the star formation timescale scaling factor
- ϵ_{SN}^0 : supernova reheating efficiency
- y : chemical evolution yield (mass of metals produced per unit mass of stars)
- α_*: exponent in optional V_c dependence in star formation law
- α_{rh}: exponent used in supernova reheating power-law
- f_{reheat}: gas reheated after a halo merger event if the largest progenitor of the current halo is less than a fraction f_{reheat} of its mass.
- f_{mrg} : fudge factor in dynamical friction merging recipe
- f_{spheroid} : the mass ratio that divides major mergers from minor mergers; determines whether a spheroid is formed

While the full parameter list may seem dauntingly long, most of the parameters are either fairly well constrained or the results do not depend on them sensitively. The most important and most uncertain parameters are what are called here ϵ_{SN}^0 and τ_*^0, the efficiency of supernova feedback and of star formation. However, the difficulty is that not only do we not know the values of these parameters, we do not even really know the *functional form* of the star formation and feedback, or even on which variables they should depend. Differences in *parameterization*, rather than in the parameter values, are generally responsible for the sometimes substantial differences in results obtained by different modeling groups.

The usual approach is to use a subset of properties of nearby galaxies to fix the values of the free parameters. Here also, the choice of which observations one uses to normalize the models can be important. For example, one approach (used by e.g., Kauffmann et al. (1999) and Somerville & Primack (1999)) is to normalize to the properties of a representative reference galaxy, chosen to have the circular velocity of the Milky Way ($\sim 220\,\mathrm{km\,s^{-1}}$). The parameters are then tuned so that on average, this halo forms central galaxies with the luminosity, cold gas mass, and metallicity of the Milky Way (for typical values used see e.g., Somerville & Primack 1999; De Lucia et al. 2004). An alternate approach (used by e.g., Cole et al. 1994; Cole et al. 2000) is to normalize to match the observed *distribution* of local galaxy properties, e.g., the luminosity function.

3.5.3 Basic Predictions of CDM-Based Galaxy Formation Models

When we put together all the ingredients summarized above we can predict a large number of observable properties of galaxies, for example:

- stellar mass (bulge and disk)
- cold gas mass
- age distribution of stars and mean stellar age
- metallicity distribution of stars and mean stellar metallicity

- metallicity of cold gas
- star formation rate
- integrated galaxy spectrum (bulge and disk)
- integrated galaxy luminosity (bulge and disk) in any waveband

We also obtain predictions for the properties of the hosting dark matter halos, such as:

- number of galaxies in the halo
- mass and temperature of hot gas
- metallicity of hot gas

In the hybrid models, we also have information on the position and velocity of each dark matter halo, and so with some assumptions about how galaxies are distributed within the host halos, we can compute clustering statistics like correlation functions.

The large number of predicted observables is what makes it possible to make meaningful predictions despite the large number of free parameters and the uncertainty in the physical recipes in the models. As well, one may obtain leverage by confronting the models, normalized at zero redshift, with observations at high redshift.

3.5.3.1 Characteristic Scales of Galaxy Formation

A fundamental observation is the characteristic shape of the global distribution of galaxy luminosities and stellar masses: this distribution is a power law on small scales, with a rather sharp "knee" and an exponential cutoff at large scales. Apparently, there is a characteristic *upper mass scale* for galaxies. Mainly from observations of very faint galaxies in the Local Group, it also seems that there is a *lower mass scale* for galaxies. One can immediately see that the *mass function of dark matter halos* predicted by the CDM model does not help us to understand either of these features. The mass function continues to rise indefinitely at low masses, and the exponential cutoff at large masses occurs at a mass value much larger than the corresponding cutoff for galaxies. The answer, then, must come from the way that baryons collapse, cool, and form stars within these halos.

The physics of gas cooling should introduce both a lower and upper mass cutoff. From the cooling curve for atomic hydrogen shown in Figure 3.19, we see that cooling by atoms becomes inefficient below temperatures of $\sim 10^4$ K. Also, from Figure 3.19, we see that the cooling time for halos larger than about $\sim 200\,\mathrm{km\,s^{-1}}$ is longer than the Hubble time at $z = 0$, and large mass halos are also more likely to have had a recent major merger, which may heat the gas. Cooling thus introduces both a lower mass and an upper mass scale for galaxies, which are within an order of magnitude of the observed scales. However, there are still several

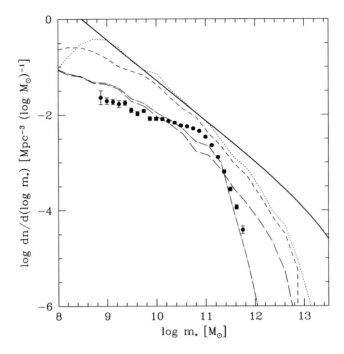

Figure 3.22. The stellar mass function of galaxies. Symbols with error bars show the estimated observed stellar mass function from the Sloan Digital Sky Survey (Bell et al. 2003b). The bold solid line shows the mass function of dark matter halos predicted by the Sheth-Tormen model, with the masses scaled by the universal baryon fraction. The dotted/dashed lines show the predictions from SAMs, broken down to show the impact of various ingredients. Dotted: cooling and star formation only; short-dashed: cooling, star formation, and squelching; long dashed: cooling, star formation, squelching, and SN feedback; long-dash: cooling, star formation, squelching, and SN feedback, with modified merging recipe (see text).

problems. First, a large fraction of the gas is able to cool by $z = 0$, in contradiction to the observation that only about 8^{+4}_{-5} percent of the baryons are in cold gas or stars today (Bell et al. 2003a). This is often referred to as the "overcooling problem", and it occurs both within numerical hydrodynamic simulations and semi-analytic models when no braking mechanism on cooling is implemented. Second, the shape of the baryonic mass function is wrong. It is too steep on the small mass end, and the large-mass cutoff is not sharp enough—there is an excess of galaxies with baryonic masses much larger than any observed galaxies.

Photo-ionization squelching pushes the minimum mass or luminosity for galaxies to larger values by suppressing gas collapse in halos with mass smaller than the filtering mass scale discussed in § 3.5.2.3. Figure 3.21 shows how this modifies the predicted luminosity function for satellite galaxies in the Local Group (Somerville 2002). Squelching also helps reduce the total fraction of baryons that cool and form stars in absence of other feedback. It is clear, however, that while squelching helps with a number of the problems discussed above, it does not solve them on its own.

Supernova feedback is often invoked as a way to cure these problems. Supernovae reheat cold gas and may even drive it out of the dark matter halo, where it will not be able to cool again for quite some time. As discussed above, however, the efficiency of SN feedback at reheating and expelling gas on galaxy-wide scales is poorly understood, and in both semi-analytic models or numerical simulations we must resort to fairly ad hoc approaches to modeling it. There seems to be a general consensus that in order for feedback to be effective, thermal heating is not enough — the gas tends to recool again too rapidly. What is needed is strong *mechanical* feedback, a *wind* that physically ejects gas from the potential well of the halo. Observational evidence for such powerful winds exists, both in nearby and distant vigorously star forming galaxies (Heckman et al. 2000; Martin 1999; Pettini et al. 2001). However, direct simulation of these galactic winds has not been very successful. Such powerful winds have been put in "by hand" in numerical simulations by Springel & Hernquist (2003a), and are a common feature in semi-analytic models. Benson et al. (2003) investigate several possible solutions to the luminosity/mass function problem, including a standard treatment of SN feedback, conduction, and winds. They conclude that the energy available from stellar/SN feedback may not be enough to drive the powerful winds needed to expel gas from large mass halos, and that another energy source such as AGN may be the key to the problem.

In Figure 3.22, following the spirit of Benson et al. (2003), we show an illustration of how the various physical processes included in a particular implementation of a semi-analytic model shape the stellar mass function at the present day. For this exercise, we assume a "standard" ΛCDM cosmology ($\Omega_m = 0.3$, $\Omega_\Lambda = 0.7$, $H_0 = 70$ km s^{-1} Mpc^{-1}, $\sigma_8 = 0.9$), and compute four model runs in which we introduce the gastrophysical ingredients described in §3.5.2 one by one, as follows: run 1 – cooling and star formation; run 2 – cooling, star formation, and photo-ionization squelching; run 3 – cooling, star formation, squelching, and SN feedback. The star formation is modeled using a recipe like Equation 3.28 with $\tau_* \propto t_{\rm dyn}$ and SN feedback is modeled as in Equation 3.31, assuming that the reheated gas is ejected from the halo if $V_c > V_{\rm crit,ej} = 200\,{\rm km\,s}^{-1}$. The free parameters are set to $\tau_*^0 = 50$ Gyr and $\epsilon_{SN}^0 = 0.1$. For this comparison, we have set the metallicity of the hot gas equal to 0.3 $Z_\odot$ for all four runs to decouple the effect of chemical enrichment. The resolution of the merger trees is taken to be $V_c = 30 \, \mathrm{km \, s^{-1}}$ (i.e., halos with circular velocity smaller than this do not have their histories followed).

First we show the mass function of dark matter halos predicted by the Sheth–Tormen model, with the masses scaled by the universal baryon fraction (i.e., the simplest prediction if each halo formed one galaxy with a mass $f_{\mathrm{bar}} M_{\mathrm{halo}}$). The estimated observed stellar mass function from the Sloan Digital Sky Survey (Bell et al. 2003b) is shown for comparison. One can immediately see the mismatch both in shape and amplitude. This implies that in large mass halos $\gtrsim 10^{12}$, at least half of the gas must either be kept hot or ejected from the halo. The second case (run 1) shows how this function changes if we include *cooling and star formation only* (i.e., no feedbacks of any kind). Cooling does introduce a "knee" in the mass function, but at $\sim 10^{13} M_\odot$, about an order of magnitude larger than the mass of the knee in the observed stellar mass function. Also, all the gas cools in halos with mass $\lesssim 10^{12} M_\odot$, as expected from Figure 3.20, so all the problems with the low mass end, and the overall over-cooling problem, remain (note that the turn-over in this curve at low masses $\lesssim 10^9 M_\odot$ is an artifact of the finite mass resolution used for the merger trees). The next case (run 2) shows the effect of introducing photo-ionization squelching, assuming that the universe was reionized at $z = 15$. This helps a bit on the small-mass end, but has a minor affect overall. After including SN feedback (run 3), the shape and normalization of the stellar mass function is much improved, but the small-mass slope is still a little bit too steep, there is a deficit around the knee, and there is a remaining small excess of massive galaxies.

This remaining shape-mismatch is a common feature in the luminosity function predictions of many semi-analytic models (e.g., Kauffmann et al. 1999; Somerville & Primack 1999). Springel et al. (2001) presented one of the few examples of a hybrid semi-analytic model in which the merging of sub-halos within a virialized cluster-sized halo was followed explicitly in a very high resolution N-body simulation. They showed that when they implemented the usual semi-analytic recipe for dynamical friction merging within their merger trees, they found the usual shape mismatch in the luminosity function (too few galaxies at L_*, and too many bright galaxies). However, when they followed the merging of the sub-halos in the N-body code, they found that this problem was mostly solved. The problem seems to be that the dynamical friction model predicts that massive galaxies in clusters merge more efficiently than they do in N-body simulations: thus galaxies from around L_* end up merging into the "monster" galaxies that create the bright "tail". In support of this explanation for the problem, if we assume that the dark halos of satellite galaxies are stripped off when they enter a large halo ($V_c > 350 \, \mathrm{km \, s^{-1}}$), thus greatly increasing the time it takes them to merge, we find a significant improvement in the stellar mass function shape (the final case shown in Figure 3.22).

Figure 3.23 shows the resulting fractions of baryons in cold and hot gas and stars as a function of halo mass at $z = 0$ in the model with cooling, star formation, squelching, and SN feedback. Ten merger-tree realizations at each mass are shown. We can see that the recipes we included have resulted in an overall fraction of baryons in halos that is significantly reduced from the universal value (even in cluster mass halos), and is a decreasing function of mass (small-mass halos hang on to a smaller fraction of their baryons). The fraction of baryons in stars ranges from less than 10 percent in small-mass halos to about 15 percent in cluster-mass halos. The fraction of baryons in cold gas ranges from 10 percent in small-mass halos to one percent in clusters, and is a strong function of mass (small-mass halos have a larger fraction of baryons in cold gas). The fraction of baryons in hot gas picks up rather steeply at around $10^{12}\,\mathrm{M_\odot}$, and rises to about 20–40 percent in cluster-mass halos. This is consistent with the lack of observed hot x-ray halos surrounding small mass isolated galaxies.

3.5.3.2 *The Global History of Star Formation*

Having tuned the rather ad hoc recipes and parameters of our models to fit local data such as luminosity/stellar mass functions, it is interesting to see what the models predict for the time dependence of basic observables. It is also instructive to examine the roles of structure formation, cooling, star formation, squelching, and feedback in shaping the history of galaxies. We make use of the same set of runs discussed in the previous section, in which we introduced the gastrophysical recipes one by one, but we now examine the evolution of global quantities over time.

Figure 3.24 shows the global star formation rate (for all galaxies) as a function of redshift in the three SAM runs (runs 1–3). Comparing run 1 and run 2, we see the effect of photo-ionization squelching on the total star formation rate, which is non-negligible especially at high redshift ($5 \lesssim z \lesssim 15$). The small bump at $z \sim 15$ occurs because gas is efficiently ejected from small halos, but cannot recollapse because of the squelching. Adding SN feedback mainly shifts the whole curve down by a large factor, but changes the shape very little. It is also interesting to compare the predictions of the semi-analytic models with results from a series of hydrodynamic N-body simulations (Springel & Hernquist 2003b, SH), which include the same basic gastrophysical ingredients (cooling, star formation, squelching, and strong SN-driven winds). The SH simulations predict about the same star formation behavior at low redshift ($z \lesssim 3$), but while the SAM star formation history (SFH) has a broad peak around ($z \lesssim 3$) and has started to decline by $z \sim 5$, the SFH in the numerical simulations peaks at a higher redshift $z \sim 5$–6. It is perhaps most interesting that in all the models, the SFH exhibits a characteristic rise and fall. The location of the

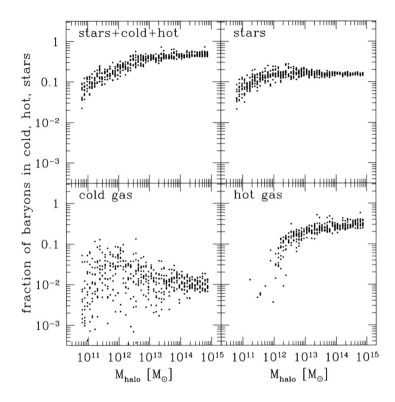

Figure 3.23. Fraction of baryons in cold gas, hot gas, and stars as a function of halo mass, in a semi-analytic model with cooling, star formation, photo-ionization squelching, and SN feedback.

peak in time/redshift depends (apparently, fairly weakly) on the details of the gastrophysical recipes included in the models. To understand the origin of this rise and decline of star formation, let us examine a few more of the relevant quantities in the models.

Figure 3.25 shows the mass density of cold gas in all four runs along with the mass in halos above the mass resolution of the SAM runs ($V_c = 30\,\mathrm{km\,s^{-1}}$) times the universal baryon fraction. Comparing this with the run with cooling only, we see that nearly all the available gas is cold at all redshifts in the absence of feedback. Around $z \sim 5$, the cold gas density begins to turn over in runs 1–3. The rate of this turn-over depends somewhat on the rate at which new gas collapses and cools (and hence on squelching) and how rapidly gas is reheated by SNae, but we can see from comparing runs 1, 2, and 3 that the dominant effect seems to be simple gas

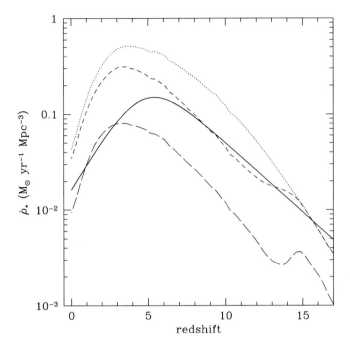

Figure 3.24. The total star formation rate density as a function of redshift, predicted by semi-analytic models with various gastrophysics included. Dotted line: run 1 (cooling and star formation only); short-dashed line: run 2 (cooling, star formation, and squelching), long-dashed line: run 3 (cooling, star formation, squelching, and SN feedback). The solid line shows the results from numerical N-body simulations with hydrodynamics, cooling, star formation, photoionization squelching, and SN feedback (Springel & Hernquist 2003a).

consumption by stars.[3] Figure 3.26 shows the mass in hot gas for the four runs. There is an interesting kink in the hot gas mass around $z \sim 2$ in all the runs, which is when cluster-mass halos with long cooling times start to form in significant numbers in this cosmology. Figure 3.27 shows the mass in each component (cold, hot, stars, all baryons) for run 3.

From this comparison, it seems that the rise in the SFH from high redshift to $z \sim 5$ is driven by the growth of structure in dark matter dominated halos. The fall-off in the SFH from $z \sim 5$ to $z = 0$ seems to trace the cold gas density, which is depleted mainly because it is consumed by

---

[3] Note that these models do not track and recycle the mass lost from stars and supernovae. This could contribute a significant budget of new cold gas at late times.

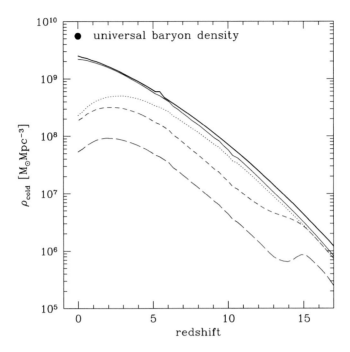

Figure 3.25. Global mass density of cold gas predicted by semi-analytic models with various gastrophysics included. Light solid line: cooling only (run 0); dotted and dashed lines as in Figure 3.24. The thick solid line is the mass in halos above the mass resolution of the simulation ($V_c = 30\,\mathrm{km\,s^{-1}}$) times the universal baryon fraction. The universal density of baryons is shown by the dot.

star formation. Hernquist & Springel (2003) formulated a similar argument based on their N-body hydrodynamic simulations, and presented a general *analytic* functional form describing the star formation history of the universe.

There is one feature of the models we have been considering that could be overlooked, however, that has an important impact on these results. Recall that we assumed that the star formation efficiency in each galaxy scales like the dynamical time t_{dyn}, where we define the star formation efficiency as the star formation rate per unit mass of cold gas ($e_* \equiv \dot{m}_*/m_{\mathrm{cold}}$). As we have seen, the density of halos reflects the redshift at which they formed, and so we expect t_{dyn} at fixed mass to be *shorter* at earlier times. This results in a rapidly increasing star formation efficiency as one looks further back in time. We can think of another reason that the star formation efficiency may have been higher at early times—we know that the efficiency of

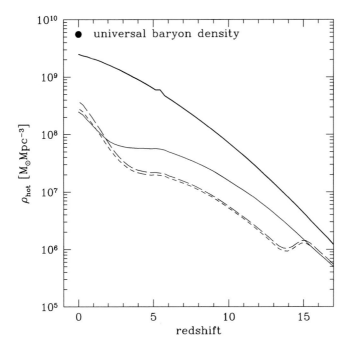

Figure 3.26. Global mass density of hot gas predicted by semi-analytic models with various gastrophysics included. Line types are as in Figure 3.25. The universal density of baryons is shown by the dot.

star formation in galaxy mergers is much higher than in quiescent galaxies, and the merger rate was also higher in the past. Somerville et al. (2001b) studied the role of the "accelerating" star formation efficiency in quiescent galaxies versus the increasing fraction of star formation in "collisional starburst" galaxies by separating out these two effects. In their Constant Efficiency Quiescent (CEQ) model, the star formation efficiency e_* was assumed to be constant in all galaxies and at all times. In the Accelerated Quiescent (AQ) model, $e_* \propto 1/t_{dyn}$, and mergers were assumed to have no impact on this efficiency. In the Collisional StarBurst (CSB) model, both major and minor mergers were assumed to trigger bursts with increased e_* as described in §3.5.2.4 of these notes, and the star formation in *quiescent* galaxies was assumed to be constant, as in the CEQ model. The star formation history in these three models (reproduced from Somerville et al. 2001b) is shown in Figure 3.28, and the cold gas mass density (this time expressed in units of the critical density, Ω_{cold}) is shown in Figure 3.29. First, it is interesting to note that the CEQ model SFH peaks at a *much*

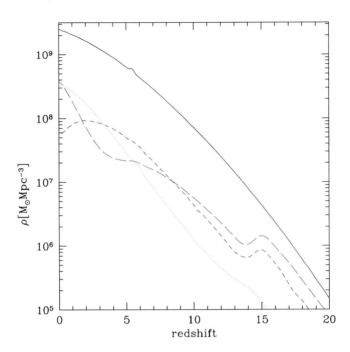

Figure 3.27. Predicted mass in stars (dotted line), cold gas (short dashed line), hot gas (long dashed line) and mass in halos above the mass resolution of the simulation ($V_c = 30\,\mathrm{km\,s^{-1}}$) times the universal baryon fraction (solid line).

lower redshift than the others ($z \sim 1$), and is clearly inconsistent with the observations of star forming galaxies at high redshift, implying that the increase in the star formation rate at $z \gtrsim 1$ that we saw previously is not primarily due to the increased availability of cold gas—it is because we are getting much more "bang for our buck" due to the increased star formation efficiency at early times. Remarkably, the SFH in the AQ and CSB models is nearly identical in shape. This is surprising when we consider that the physics responsible for the increasing e_* is completely different in the two models.

Examining Figure 3.29, we see the importance of complementary constraints on different phases of matter in the universe in breaking these degeneracies. The mass in cold gas can be constrained from observations of Damped Lyman-α systems at high redshift and from H$_\mathrm{I}$ surveys at low redshift. Clearly, a successful model must be able to produce enough stars to satisfy direct observations of star forming galaxies, while still leaving enough "left-over" cold gas to satisfy the kinds of observations shown in

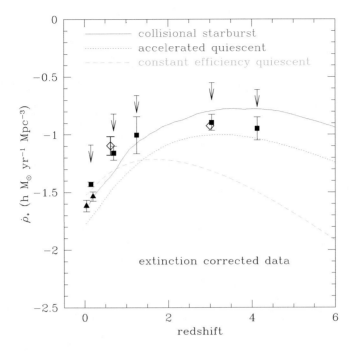

Figure 3.28. The global star formation history of the universe. Observational estimates are shown by symbols (see Somerville et al. 2001b, for references). The three curves show semi-analytic models in which the recipe for star formation has been varied (see text). The location of the characteristic peak in the star formation history in time depends on the details of how star formation efficiency in galaxies changes with time. Reproduced from Somerville et al. (2001b).

Figure 3.29. Although the CSB model had a slightly higher global star formation rate than the AQ model, it has more gas left over, indicating that the *overall* efficiency of star formation is higher. The predicted mass of cold gas is quite sensitive to the details of the modeling of supernova feedback, however.

3.5.3.3   Galaxy Colors

Galaxy colors reflect a degenerate combination of the ages and metallicities of their stars as well as the effects of dust extinction. They are therefore something of a blunt tool, but nonetheless they provide important constraints on galaxy formation models. Moreover, deep multi-color data for fairly wide fields are increasingly available from surveys such as the Great

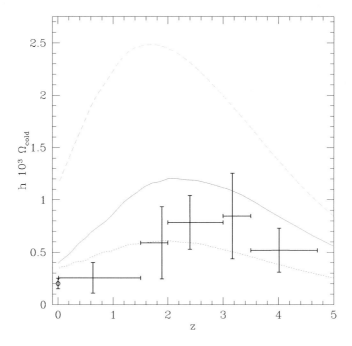

Figure 3.29. The evolution of the mass density in units of the critical density (Ω) in cold gas as a function of redshift. Data points show observational estimates of the matter density in neutral hydrogen from local H_I surveys (at $z \sim 0$) and from Damped Lyman-α systems at high redshift. Curves show the *total* mass in cold gas in the same three model variants shown in Figure 3.28. The mass budget in cold gas is an important complementary constraint on the star formation efficiency in models: a successful model must leave enough cold gas left over to account for the observed cold gas in the universe.

Observatories Origins Deep Survey (GOODS).[4] It is a familiar fact that in the local universe, the most massive galaxies have the reddest colors (the Color–Magnitude (CM) relation). This is in apparent contradiction to the expectations in a hierarchical model, in which the smallest objects form first and large mass objects form late. Indeed, hierarchical models have traditionally had difficulty reproducing the observed sense of the CM relation, often producing an inverted relation in which luminous galaxies are bluer than low-luminosity galaxies, or at best, a flat relation (Kauffmann & Charlot 1998). Reproducing the strong observed relation between

[4] http://www.stsci.edu/science/goods/

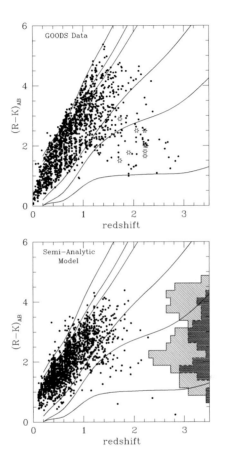

Figure 3.30. Observed frame color $(R - K)_{AB}$ versus redshift for the GOODS South field (left) and the semi-analytic mock catalog (right). Tracks are shown for single age, stellar metallicity populations with ages of 13.5, 5.8, 3.2, 1, 0.5, and 0.1 Gyr (unreddened), from top to bottom. Large star symbols show spectroscopically confirmed objects. The color distribution of galaxies with $K < 22$ in the redshift range $1.7 < z < 2.5$ is shown in the right panel as histograms (diagonal hatched: GOODS; horizontal fill: mock catalog). Reproduced from Somerville et al. (2004).

galaxy mass or luminosity and *metallicity* helps to reproduce the observed CM-relation slope, at least in galaxy clusters, (De Lucia et al. 2004), but particularly for the field, the observed distribution of galaxy colors as a function of luminosity has been extremely difficult to reproduce in hierarchical models.

The problem seems to be even worse for galaxies at intermediate redshift $z \sim$ 1–2. Near IR selected surveys have shown that there exists a population of fairly luminous galaxies at $z \sim$ 1–2 with very red colors, often called Extremely Red Objects (EROs). There has been much debate about whether these red colors indicate very old stellar populations or large quantities of dust, but at least in about half of the galaxies, the morphologies indicate that they are elliptical-type and presumably old galaxies (e.g., Moustakas et al. 2004). Semi-analytic models—the same models that did so well at reproducing the global observed star formation history—underproduce these red galaxies by about a factor of five to ten, depending on the precise color and magnitude cut (Firth et al. 2002; Somerville et al. 2004). A comparison of the observed R-K colors from GOODS and a semi-analytic mock catalog (Somerville et al. 2004), which illustrates this problem, is shown in Figure 3.30.

3.6 Concluding Thoughts

In these notes, I have tried to sketch out the basic foundation of our current understanding of the forces that shape galaxies in a CDM universe. We have seen that at early times, the rate of formation of dark matter halos large enough to cool gas and to survive supernova feedback is the limiting factor in the growth of galaxies. At later times, the efficiency of star formation and the limited availability of fuel for new stars comes into play. If we accept this framework, it is inevitable that the growth of SMBH in galactic centers is regulated by the same forces, at some level. In this picture, galactic spheroids are formed in violent major mergers, and the tight connection between BH mass and spheroid mass or velocity dispersion observed in nearby galaxies strongly suggests that the growth of SMBH occurs in these same events (Kauffmann & Haehnelt 2000). At the same time, we have argued that a large fraction of the baryons in collapsed halos must be prevented from cooling or expelled from the halo in order to prevent the overcooling problem, and in order to achieve the characteristic shape of the galaxy luminosity function and stellar mass function. It is still unclear whether stars and supernovae generate sufficient energy to heat or eject such a large fraction of the baryons (e.g., Benson et al. 2003).

Although the overall star formation and stellar mass assembly history sketched out by observations appear to be reproduced reasonably well by the hierarchical models, the mismatch of the predicted *colors* indicated that star formation is not occurring in the right pieces in the models. We seem to need a process that can shut off star formation in the most massive, spheroidal type galaxies at all redshifts (at least $z \lesssim 2$). The presumed presence of a massive black hole in these galaxies, and the associated AGN activity, seems very likely to be connected to this process.

The enormous energies produced by AGN must certainly have a dramatic impact on their host galaxies. The main difficulty at this stage is in understanding how the energy produced by the monster buried deep within the center of a galaxy couples to the interstellar gas on kpc scales, and the intergalactic/intracluster gas on Mpc scales. Here, observations provide some clues: we witness the powerful outflows from AGN and quasars, which may be driven by radiation pressure against dust in the cold gas component (Murray, Quataert, & Thompson 2004), and we also see the impact of giant radio jets on hot gas in clusters (e.g., McNamara et al. 2000; Fabian et al. 2003). However, a great deal more work is necessary in order to model these processes in a detailed manner within the context of the hierarchical universe that I have described here. It is now clear that there is a symbiotic relationship between galaxies and their black holes, and understanding their co-evolution will provide fertile material for many future Ph.D. theses.

References

Adelberger K L, Steidel C C, Giavalisco M, Dickinson M, Pettini M and Kellogg, M 1998 *ApJ* **505** 18

Avila-Reese V, Firmani C and Hernández X 1998 *ApJ* **505** 37

Barnes J and Efstathiou G 1987 *ApJ* **319** 575

Barnes J and Hernquist L 1996 *ApJ* **471** 115

Baugh C M 1996 *MNRAS* **280** 267

Baugh C M, Cole S, Frenk C S and Lacey C G 1998 *ApJ* **498** 504

Bell E F, McIntosh D H, Katz N and Weinberg M D 2003a *ApJL* **585** L117

Bell E F, McIntosh D H, Katz N and Weinberg M D 2003b *ApJS* **149** 289

Benson A J, Bower R G, Frenk C S, Lacey C G, Baugh C M and Cole S 2003 *ApJ* **599** 38

Benson A J, Frenk C S, Lacey C G, Baugh C M and Cole S 2002a *MNRAS* **333** 177

Benson A J, Lacey C G, Baugh C M, Cole S and Frenk C S 2002b *MNRAS* **333** 156

Binney J and Tremaine S 1987 *Galactic Dynamics* (Princeton, NJ: Princeton Univ Press)

Blumenthal G, Faber S M, Flores R and Primack J R 1986 *ApJ* **301** 27

Bond J R, Cole S, Efstathiou G and Kaiser N 1991 *ApJ* **379** 440

Bruzual A G and Charlot S 1993 *ApJ* **405** 538

Bryan G L and Norman M L 1998 *ApJ* **495** 80

Bullock J S, Dekel A, Kolatt T S, Kravtsov A V, Klypin A A, Porciani C and Primack J R 2001a *ApJ* **555** 240

Bullock J S, Kolatt T S, Sigad Y, Somerville R S, Kravtsov A V, Klypin A A, Primack J R and Dekel A 2001b *MNRAS* **321** 559

Bullock J S, Wechsler R H and Somerville R S 2002 *MNRAS* **329** 246

Cardelli J A, Clayton G C and Mathis J S 1989 *ApJ* **345** 245

Cole S 1991 *ApJ* **367** 45

Cole S and Kaiser N 1989 *MNRAS* **237** 1127

Cole S, Aragón-Salamanca A, Frenk C S, Navarro J F and Zepf S E 1994 *MNRAS* **271** 781

Cole S, Lacey C G, Baugh C M and Frenk C S 2000 *MNRAS* **319** 168

Colpi M, Mayer L and Governato F 1999 *ApJ* **525** 720

Cooray A and Sheth R 2002 *Phys. Rep.* **372** 1

Davis M, Efstathiou G, Frenk C S and White S D M 1985 *ApJ* **292** 371

De Lucia G, Kauffmann G and White S D M 2004 *MNRAS* **349** 1101

Dekel A and Lahav O 1999 *ApJ* **520** 24

Devriendt J E G, Guiderdoni B and Sadat R 1999 å**350** 381

Doroshkevich A G 1970 *Astrofizika* **6** 581

Eke V R, Navarro J F and Steinmetz M 2001 *ApJ* **554** 114

Fabian A C, Sanders J S, Allen S W, Crawford C S, Iwasawa K, Johnstone R M, Schmidt R W and Taylor G B 2003 *MNRAS* **344** L43

Fall S M and Efstathiou G 1980 *MNRAS* **193** 189

Fioc M and Rocca-Volmerange B 1997 å**326** 950

Firth A E, Somerville R S, McMahon R G, Lahav O, Ellis R S, Sabbey C N, McCarthy P J, Chen H-W, et al 2002 *MNRAS* **332** 617

Flores R, Primack J R, Blumenthal G and Faber S M 1993 *ApJ* **412** 443

Gnedin N Y 2000 *ApJ* **542** 535

Granato G L, Lacey C G, Silva L, Bressan A, Baugh C M, Cole S and Frenk C S 2000 *ApJ* **542** 710

Gross M A K, Somerville R S, Primack J R, Holtzman J and Klypin A 1998 *MNRAS* **301** 81

Guiderdoni B and Rocca-Volmerange B 1987 å**186** 1

Heckman T M, Lehnert M D, Strickland D K and Armus L 2000 *ApJS* **129** 493

Hernquist L and Springel V 2003 *MNRAS* **341** 1253

Hu W and Sugiyama N 1996 *ApJ* **471** 542

Jenkins A, Frenk C S, Pearce F R, Thomas P A, Colberg J M, White S D M, Couchman H M P, Peacock J A, et al 1998 *ApJ* **499** 20

Jenkins A, Frenk C S, White S D M, Colberg J M, Cole S, Evrard A E, Couchman H M P and Yoshida N 2001 *MNRAS* **321** 372

Kaiser N 1984 *ApJL* **284** L9

Kauffmann G and Charlot S 1998 *MNRAS* **294** 705

Kauffmann G and Haehnelt M 2000 *MNRAS* **311** 576

Kauffmann G and White S D M 1993 *MNRAS* **261** 921

Kauffmann G, Colberg J, Diaferio A and White S D M 1998 *MNRAS* **303** 188

Kauffmann G, Colberg J M, Diaferio A and White S D M 1999 *MNRAS* **303** 188

Kauffmann G, White S and Guiderdoni B 1993 *MNRAS* **264** 201

Kennicutt R C 1983 *ApJ* **272** 54

Kennicutt R C 1998 *ApJ* **498** 181

Lacey C and Cole S 1993 *MNRAS* **262** 627

Leitherer C, Schaerer D, Goldader J D, Delgado R M G, Robert C, Kune D F, de Mello D F, Devost D, et al 1999 *ApJS* **123** 3

Lemson G and Kauffmann G 1999 *MNRAS* **302** 111

Maller A H and Dekel A 2002 *MNRAS* **335** 487

Maller A H, Dekel A and Somerville R 2002 *MNRAS* **329** 423

Martin C L 1999 *ApJ* **513** 156

McNamara B R, Wise M, Nulsen P E J, David L P, Sarazin C L, Bautz M, Markevitch M, Vikhlinin A, et al 2000 *ApJL* **534** L135

Menci N, Cavaliere A, Fontana A, Giallongo E, Poli F and Vittorini V 2004 *ApJ* **604** 12

Mihos J C and Hernquist L 1994 *ApJL* **425** 13

Mihos J C and Hernquist L 1995 *ApJ* **448** 41

Mihos J C and Hernquist L 1996 *ApJ* **464** 641

Mo H J and White S D M 1996 *MNRAS* **282** 347

Mo H J, Mao S and White S D M 1998 *MNRAS* **295** 319

Moore B, Quinn T, Governato F, Stadel J and Lake G 1999 *MNRAS* **310** 1147

Moustakas L A and Somerville R S 2002 *ApJ* **577** 1

Moustakas L A, Casertano S, Conselice C J, Dickinson M E, Eisenhardt P, Ferguson H C, Giavalisco M, Grogin N A, et al 2004 *ApJL* **600** L131

Murray N, Quataert E and Thompson T in press *ApJ* astro-ph/0406070

Navarro J, Frenk C and White S 1995 *MNRAS* **275** 56

Navarro J F, Frenk C S and White S D M 1996 *ApJ* **462** 563

Norberg P, Baugh C M, Hawkins E, Maddox S, Madgwick D, Lahav O, Cole S, Frenk C S, et al 2002 *MNRAS* **332** 827

Padmanabhan T 1993 *Structure Formation in the Universe* (New York: Cambridge Univ Press)

Peacock J A 1998 *Cosmological Physics* (New York: Cambridge Univ Press)

Peebles P J E 1969 *ApJ* **155** 393

Peebles P J E 1980 *The Large-Scale Structure of the Universe* (Princeton, NJ: Princeton Univ Press)

Peebles P J E 1993 *Principles of Physical Cosmology* (Princeton, NJ: Princeton Univ Press)

Pettini M, Shapley A E, Steidel C C, Cuby J, Dickinson M, Moorwood A F M, Adelberger K L and Giavalisco M 2001 *ApJ* **554** 981

Press W H and Schechter P L 1974 *ApJ* **187** 425

Quinn T, Katz N and Efstathiou G 1996 *MNRAS* **278** L49

Rees M J 1986 *MNRAS* **218** 25

Seljak U and Zaldarriaga M 1996 *ApJ* **469** 437

Sheth R K and Tormen G 1999 *MNRAS* **308** 119

Sheth R K, Mo H J and Tormen G 2001 *MNRAS* **323** 1

Somerville R S 2002 *ApJL* **572** L23

Somerville R S and Kolatt T S 1999 *MNRAS* **305** 1

Somerville, R S and Primack, J R 1999 *MNRAS* **310** 1087

Somerville R S, Lemson G, Sigad Y, Dekel A, Kauffmann G and White S D M 2001a *MNRAS* **320** 289

Somerville R S, Moustakas L A, Mobasher B, Gardner J P, Cimatti A, Conselice C, Daddi E, Dahlen T, et al 2004 *ApJL* **600** L135

Somerville R S, Primack J R and Faber S M 2001b, *MNRAS* **320** 504

Spergel D N, Verde L, Peiris H V, Komatsu E, Nolta M R, Bennett C L, Halpern M, Hinshaw G, et al 2003 *ApJS* **148** 175

Springel V and Hernquist L 2003a *MNRAS* **339** 289

Springel V and Hernquist 2003b *MNRAS* **339** 312

Springel V, White S D M, Tormen G and Kauffmann G 2001 *MNRAS* **328** 726

Sutherland R S and Dopita M A 1993 *ApJS* **88** 253

Taffoni G, Mayer L, Colpi M and Governato F 2003 *MNRAS* **341** 434

Taylor J E and Babul A 2004 *MNRAS* **348** 811

Tegmark M, Blanton M, Strauss M, Hoyle F, Schlegel D, Scoccimarro R, Vogeley M, Weinberg D, et al 2003a, ArXiv Astrophysics e-prints

Tegmark M, Strauss M, Blanton M, Abazajian K, Dodelson S, Sandvik H, Wang X, Weinberg D, et al 2003b, ArXiv Astrophysics e-prints

Thoul A A and Weinberg D H 1996 *ApJ* **465** 608

van den Bosch F C 2000 *ApJ* **530** 177

van den Bosch F C 2002 *MNRAS* **332** 456

van den Bosch F C, Abel T, Croft R A C, Hernquist L and White S D M 2002 *ApJ* **576** 21

van den Bosch F C, Yang X and Mo H J 2003 *MNRAS* **340** 771

Vitvitska M, Klypin A A, Kravtsov A V, Wechsler R H, Primack J R and Bullock J S 2002 *ApJ* **581** 799

Wang B and Heckman T 1996 *ApJ* **457** 645

Wechsler R H, Bullock J S, Primack J R, Kravtsov A V and Dekel A 2002 *ApJ* **568** 52

Wechsler R H, Somerville R S, Bullock J S, Kolatt T S, Primack J R, Blumenthal G R and Dekel A 2001 *ApJ* **554** 85

White S D M 1984 *ApJ* **286** 38

White S D M 1994 in *Les Houches Lectures*, MPA preprint 831 astro-ph/9410043

White S D M and Frenk, C S 1991 *ApJ* **379** 52

Zehavi I, Blanton M R, Frieman J A, Weinberg D H, Mo H J, Strauss M A, Anderson S F, Annis J, et al 2002 *ApJ* **571** 172

Zentner A R and Bullock J S 2003 *ApJ* **598** 49

Chapter 4

Feedback in Cosmic Structures

Alfonso Cavaliere & Andrea Lapi
Department of Physics
University of Rome Tor Vergata
Rome, Italy

We review and discuss the processes that govern the amount and the thermal state of the hot plasma pervading clusters and groups of galaxies: the gravitational heating driven by the merging histories of the Dark Matter (DM), the radiative cooling of the baryons, and the energy fed back to them by SNe and by AGNs or quasars. We show that the x-ray emissions, the entropy levels, and the entropy profiles observed throughout the range from clusters to groups require the AGNs to contribute substantially to preheat the plasma and hinder its gathering in the poor clusters. The observations also require the quasars to blow some of the plasma out of groups and galaxies; this process can be caught in the act by resolved observations of enhanced Sunyaev–Zel'dovich signals.

4.1   Introduction

The formation of cosmic structures, i.e., galaxies, and their groups and clusters, takes place by the interplay of three nested building blocks: the overall space–time framework provided by the Hubble expansion; the collapse into bound "halos" of the DM, the *collisionless* inconspicuous component of the cosmic fluid; the response of the other component, the shining *collisional* baryons that cool and condense into stars then exploding as SNe, or accrete onto supermassive black holes powering the outbursts of AGNs and quasars.

The highest baryonic fractions are contained in rich clusters, which also constitute the largest and youngest DM aggregates in the universe. In fact, according to the hierarchical paradigm DM masses $M \sim 10^{15}$ M$_\odot$

virialize at redshifts $z \lesssim 0.5$ within sizes $R \sim$ Mpc; after the virial theorem they set gravitational potential wells of depth $GM/5R \approx \sigma^2$, betrayed by the velocity dispersion $\sigma \approx 10^3$ km s^{-1} of the galaxies zipping through the cluster (see Figure 4.1). But the overall stellar mass contributed by all such 10^3 galaxies constitutes only a small fraction $m_\star \sim m/5$ (in round numbers) of the cluster's baryonic content, in turn amounting to $m \sim M/5$ mostly constituted by *diffuse* baryons.

The latter fill out the DM potential wells, being in thermal equilibrium with the associated electrons and in virial equilibrium with the DM at energies $kT \approx kT_v \equiv \mu m_p \sigma^2 \approx 5$ keV ($\mu \approx 0.6$ is the mean molecular weight). These hot electrons and baryons are easily observed in x-ray rays, beginning with the first x-ray satellite *Uhuru* (Cavaliere, Gursky & Tucker 1971; see also Sarazin 1988). This is because they emit by thermal Bremsstrahlung huge integrated powers

$$L_X \approx 2 \times 10^{44} \left(\frac{n}{10^{-3}\,\mathrm{cm}^{-3}} \right)^2 \left(\frac{R}{\mathrm{Mpc}} \right)^3 \left(\frac{kT}{\mathrm{keV}} \right)^{1/2} \mathcal{F} \ \mathrm{erg\,s}^{-1}\,, \quad (4.1)$$

where $\mathcal{F} \equiv \int_0^1 dx\, x^2\, (n/n_2)^2\, (T/T_2)^{1/2}$ is a shape factor computed in terms of $x \equiv r/R$ and of the quantities n_2, T_2 at the virial radius R; hence number densities $n \sim 10^{-3}$ cm^{-3} are inferred, but then these diffuse baryons with high T and low n satisfying $kT/e^2\, n^{1/3} \sim 10^{12}$ constitute the best proton–electron *plasma* in the universe ever, the intracluster plasma or ICP (see Figure 4.2).

This is an apparently *simple* medium on the following accounts: microscopically, it is constituted by pointlike particles in thermal equilibrium; at the macroscopic end, the overall ICP masses in clusters make up baryonic fractions $m \approx 0.15\,M$ close to the cosmic values (White et al. 1993, Allen & Fabian 1994); finally, the chemical composition of the ICP is reasonably constant, and close to $1/3$ of the cosmic value (see Matteucci 2003). However, *complexity* arises at various levels: microscopic energy equipartition of electrons and ions may be broken on scales $\lesssim R/10$ (Ettori & Fabian 1998); macroscopically, events and processes of merging, inflow, and preheating affect the ICP from the *outside*, while the plasma in galaxy groups (intragroup plasma or IGP) is also affected from the *inside* by radiative cooling and by energy feedback. This is why the smaller and cooler groups with mass $M \sim 10^{13}$ M$_\odot$, sizes $R \sim 1/3$ Mpc, and temperatures $kT_v \sim 1$ keV do *not* constitute simple scaled down versions of the rich clusters.

4.2   Basics

The buildup of all cosmic structures with their potential wells is understood in terms of the *hierarchical* cosmogony (see Figure 4.3).

Figure 4.1. An optical view of the central region of the Coma cluster (A1656).

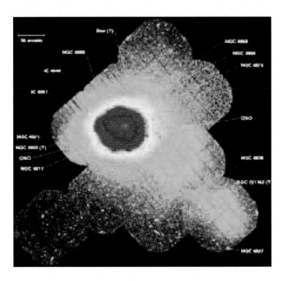

Figure 4.2. The Coma cluster at large with a merging group, as mapped in X rays by *XMM-Newton*.

The space–time framework is provided by the concordance cosmology. With density parameters $\Omega_M \approx 0.3$ for the DM and $\Omega_\Lambda \approx 0.7$ for the

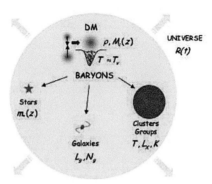

Figure 4.3. The formation of cosmic structures takes place by the interplay of three nested building blocks: baryons processed within DM halos that collapse from the expanding universe.

dark energy component (likely in the form of a cosmological constant; see Bennett et al. 2003), this envisages the framework to expand at the rate $H \equiv \dot{R}(t)/R(t) = H_0 \left[\Omega_M (1 + z)^3 + \Omega_\Lambda\right]^{1/2}$. The current value is $H_0 \approx 72$ km s^{-1} Mpc^{-1}; the expansion decelerates at early cosmic times, then for $z \lesssim 1$ it accelerates.

Within this framework, structure buildup occurs as growth of DM halos (and deepening of the associated wells) driven by the gravitational instability from minute initial perturbations; this is slowed down by the cosmic expansion, but eventually the local gravity prevails to enforce collapse. The standard sequence runs as follows: a slightly overdense region expands more slowly than its surroundings, progressively detaches from the Hubble flow and turns around; then it collapses and eventually virializes to form a halo (see Padmanabhan 2003).

4.2.1   Closely Scale-Invariant DM

These dynamical events governed by *weak* gravity are bound to yield nearly *self-similar* structures and evolution of the DM, once the basic quantities at the virialization are appropriately rescaled. The scaling laws (see Peebles 1993) contain the spectral index $\nu \sim -1.2 \div -1.5$ of the initial density perturbations at cluster-group scales; backward in time, these imply the earlier halos at virialization to be denser with mass density $\rho(z) \approx 10^2 \, \rho_M \propto (1 + z)^3$, and lighter with characteristic masses $M_c(z) \propto (1 + z)^{-6/(\nu+3)}$;

relatedly, such halos are also smaller with radii $R \propto (M/\rho)^{1/3} \propto M_c^{(5+\nu)/6}$, and shallower with well depths $v^2 \propto G M/R \propto M_c^{(1-\nu)/6}$.

These relations are precise in the critical cosmology with $\Omega_\Lambda = 0$ and $\Omega_M = 1$ but require modifications for $z \lesssim 1$ in the concordance cosmology. For example, the characteristic mass approaches the behavior $M_c(z) \propto (1+z)^{-3/(\nu+3)}$ for $z \lesssim 0.6$ (see Lapi 2004); this is affected by the linear growth function of the perturbations, being slowed down by the cosmic expansion at the full rate $H(z)$. In other words, the formation of large structures is being frozen out by the currently accelerated cosmic expansion.

Considerable scatter and dispersion around these basic quantities arise anyway from the stochastic (closely Gaussian) nature of the initial perturbation field. For example, the full halo mass distribution is described by the Press & Schechter (1974) function, updated by Sheth & Tormen (2002).

Several extensive and resolved N-body simulations (reviewed by Norman 2004) confirm the above picture, and add two lines of detailed information. First, the halo growth actually takes place through repeated and stochastic merging events with other clumps of comparable or smaller size, down to nearly smooth inflow (see Figure 4.4); the variance in such events gives rise to a "tree" of merging histories.

Second, the profiles of the halo gravitational potentials follow the closely self-similar NFW (Navarro, Frenk & White 1997) form

$$\Phi(x) \approx -3\,\sigma^2\,g(c)\,\frac{\ln(1+c\,x)}{x} \;, \tag{4.2}$$

where $x \equiv r/R$. In fact, the concentration parameter $c \approx 5\,(M/10^{15}\,\mathrm{M}_\odot)^{-0.13}$ slowly increases (Bullock et al. 2001) and the factor $g(c) = [\ln(1+c) - c/(1+c)]^{-1} \approx 1.04 \div 1.4$ weakly rises from clusters to groups. This causes the smaller, earlier halos to be somewhat more concentrated, a non-self-similar if minor feature concurring with the slight decrease with mass of the spectral index $\nu \approx -1.2 \div -2$ that enters $M_c(z)$.

4.2.2 Plasma in Hydrostatic Equilibrium

The diffuse baryons pervade the potential wells of clusters and groups, in overall virial equilibrium with the DM as with the member galaxies. This implies the time $R/(10\,kT/3\,m_p)^{1/2}$ for the sound to cross the bulk of these systems to be comparable to, or somewhat shorter than the dynamical time R/σ. So within R the gravitational forces of the DM can be *balanced* in detail by the gradient of the microscopic pressure $p = n\,kT/\mu$ of the ICP

$$\frac{1}{m_p\,n}\frac{dp}{dr} = -\frac{d\Phi}{dr} \;. \tag{4.3}$$

The solution of this differential equation requires one *boundary* condition, for example the value $n(R) = n_2$; it also requires an equation of *state*

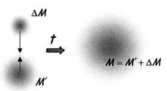

Figure 4.4. To illustrate the growth of DM halos by merging.

relating $T(r)$ to $n(r)$. The apparently simplest state is the isothermal one, with $T(r) \approx$ const; if so, from R toward the center the density increases as

$$n(r) = n_2 \, e^{\beta \, \Delta\phi(r)} \tag{4.4}$$

(see Figure 4.5). Here $\Delta\phi(r) = [\Phi(R)-\Phi(r)]/\sigma^2$ is the normalized potential difference provided by the DM, while

$$\beta \equiv \frac{\mu m_p \, \sigma^2}{kT} = \frac{T_v}{T} \tag{4.5}$$

is the ratio between the plasma and the DM scale heights (Cavaliere & Fusco-Femiano 1976) in the potential well.

Assuming, to begin with, that the DM is also distributed with constant σ, from the Jeans equation $\rho(r) \propto e^{\Delta\phi}$ is found to hold, and the standard β-model $n(r) \propto \rho^\beta(r)$ obtains. Simple expressions result on using the analytic King (1972) distribution for the DM, namely, $\rho \propto [1 + (r/r_c)^2]^{-3/2}$ with a core radius $r_c \approx R/12$; this yields the explicit surface brightness profile $\ell_X(r) \propto [1+(r/r_c)^2]^{-3\beta+1/2}$. Following Jones & Forman (1984), this handy model provides the standard fits to the resolved surface brightness profiles observed in clusters outside the cores.

With the NFW potential modifications are required. Here the DM density reads $\rho \propto [x\,(1+c\,x)^2]^{-1}$, and $\sigma(r)$ has a maximum around $r_s = R/c$; then the isothermal plasma equilibrium follows (Makino, Sasaki & Suto 1998)

$$\frac{n(r)}{n_2} = \left[\frac{(1+cx)^{1/x}}{1+c} \right]^{3\beta\,g(c)} . \tag{4.6}$$

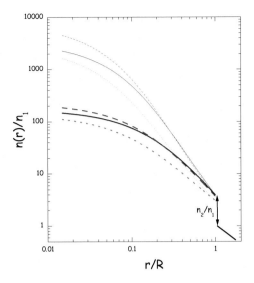

Figure 4.5. Density profiles for isothermal (*thin* lines) and polytropic ($\Gamma = 1.2$, *thick* lines) ICP in the NFW potential. The *solid* lines are for a rich cluster with $kT_v \approx 8$ keV; the *dashed* lines are for a poor cluster with $kT_v = 2$ keV; and the *dotted* lines are for a small group with $kT_v = 0.5$ keV. Note the density jump n_2/n_1 at the shock near R, decreased in the latter case by SN preheating at the level $k\Delta T = 1/4$ keV per particle.

Note that for $r \to 0$ the DM density $\rho(r)$ diverges to a formal cusp; however, an effective limit is set by the simulation resolution. On the other hand, the plasma density $n(r)$ is always less singular than the DM's, and has a finite angled shape if NFW is taken literally; however, such details are blurred by the uncertain physics of the central "cool cores" (see § 4.3.3). Perhaps more relevantly, with NFW the plasma $n(r)$ has lower central values and would extrapolate to higher wings than with King, when both are normalized at $r = R$.

Resolved x-ray observations indicate some deviations of the ICP from isothermality; specifically, $T(r)$ declines by about a factor $1/2$ beyond $R/2$ (e.g., Ettori & Fabian 1999; De Grandi & Molendi 2002). The trend may be approximated with a polytropic assumption $p(r) \propto n^\Gamma(r)$, in terms of the macroscopic index Γ that may range from 1 (isothermal) to 5/3 (adiabatic); in fact, the observed temperature profiles suggest intermediate values $\Gamma \approx 1.1 \div 1.2$. The corresponding equilibrium had been given by

Cavaliere & Fusco-Femiano (1978) in the form

$$\frac{n(r)}{n_2} = \left[\frac{T(r)}{T_2}\right]^{1/(\Gamma-1)} = \left[1 + \frac{\Gamma-1}{\Gamma}\beta\,\Delta\phi(r)\right]^{1/(\Gamma-1)}, \qquad (4.7)$$

where $\beta = T_v/T_2$ is the value at the boundary R. As shown in Figure 4.5, lower central densities (and higher central temperatures) obtain compared with the isothermal case; a polytropic plasma cloud would extend out to a finite, if large, radius. In § 4.3.3 we shall relate the index Γ with the entropy profiles.

Keep in mind for forthcoming use that around the boundary $r \approx R$ the slope of the density profile $n(r)$ given by Equation (4.3) may be approximated as $d\ln n/d\ln r \approx -(\beta/\Gamma)\,d\phi/d\ln r$; so, with the NFW potential where $d\phi/d\ln r \approx 3$ holds the approximation $n(r) \propto r^{-3\beta/\Gamma}$ obtains.

4.3 Issues Arising

We next discuss how the thermal state of the diffuse baryons is originated.

4.3.1 Gravitational Heating

One contribution involves solely the *dynamics* of the hierarchical buildup. In fact, as partner DM halos merge with the main progenitor, the associated baryons shock at about $r \approx R$. At the shock the pointlike particles of the ICP are heated up to (equilibrium) temperatures

$$kT \approx \frac{\mu m_p\, v_1^2}{3}, \qquad (4.8)$$

and are compressed by factors up to 4. This occurs for strong shocks taking place when the merging baryons fall in supersonically, i.e., in a partner with low internal temperature T_v' and high velocity v_1. The latter is set by energy conservation in terms of the DM potential Φ_2 at $r \approx R$, to read

$$v_1 = \sqrt{-2\,\chi\,\Phi_2}\;; \qquad (4.9)$$

the parameter $\chi = 1 - R/R_f$ contains the position R_f where the infall begins. Many numerical simulations converge with the analytical models of Bertschinger (1985) in suggesting the value $\chi \approx 0.37$, closely corresponding to the turnaround radius in the standard spherical collapse model (see § 4.2.1). Adopting this, and considering that $\Phi_2 \approx -5.7\,\sigma^2$ holds for the NFW potential with the concentration $c = 5$ appropriate for rich clusters, Lapi (2004) obtains $v_1 \approx 2.1\,(kT_v/\mu m_p)^{1/2}$. This gratifyingly yields the constant value $\beta = 3\,kT_v/\mu m_p\,v_1^2 \approx 0.7$, close to that observed in rich clusters.

4.3.2 Scale-Invariant Baryons?

The above simple picture of the ICP may be combined with the DM scaling laws recalled in § 4.2.1; these would imply the relations $n \propto \rho$ and $R \propto T^{1/2} \rho^{1/2}$ to hold in Equation (4.1), to yield the scaling $L_X \propto T^2$ for the x-ray luminosities (Kaiser 1986). This is observed for very rich clusters.

As to the z-dependent coefficient, note that the DM scaling laws predict the above correlation to evolve at a given T following $L_X \propto \Delta_c(z)^{1/2} H(z)$, in terms of the threshold for non-linear collapse $\Delta_c(z)$ and of the Hubble parameter $H(z)$. In the concordance cosmology this yields a dependence close to $L_X \propto (1 + z)$ for $z \lesssim 1$ (see Lapi 2004), milder than in the critical universe; this may help to reconcile the discrepancies in the evolution of the $L_X - T$ relation between the observations by Ettori et al. (2004), and those by Markevitch (1998) and Mullis et al. (2004).

However, real surprises arise in moving from rich clusters to poor clusters and groups. These are observed to emit in X rays far *less* than expected; the data (Osmond & Ponman 2004; see Figure 4.8) may be described as an upper envelope close to $L_X \propto T^3$, plus a wide, largely intrinsic scatter downward for poor groups and large galaxies (Mushotzky 2004). So in such structures the plasma is surprisingly underluminous and hence *underdense*, an even more surprising feature on considering that these earlier condensations ought to be denser, if anything; meanwhile, for $kT < 2$ keV the contribution from line emission (see Sarazin 1988) becomes increasingly important, to yield a flatter $L_X \propto T$, if anything (see Lapi 2004). How the observed steep decrease may come about constitutes a widely and hotly debated issue.

4.3.3 Cooling or Heating?

Interesting clues toward a solution are provided on recasting the observed densities and temperatures in terms of the plasma specific entropy s. This is related to the adiabat $K = kT/n^{2/3}$ after $s = k \ln K^{3/2} + \text{const.}$ (Bower 1997; Balogh, Babul & Patton 1999). In turn, the levels of K are linked to those of L_X by the (local) relation

$$K \propto L_X^{-1/3} T^{5/3} \; ; \tag{4.10}$$

the last factor goes over to $T^{4/3}$ for $kT < 2$ keV when lines become important.

The quantity K is telling because it provides a combination of n and T that is invariant under adiabatic processes in the plasma. The basic non-adiabatic process, gravitational heating, by itself would yield the scaling $K \propto T$, but Figure 4.9 shows the data to follow instead a relation closer to $K \propto T^{2/3}$, and to deviate increasingly upwards in moving toward groups.

This points toward additional processes, namely, energy *losses* or *gains*, as important contributors during these structures' merging history.

One view focuses on extensive radiative losses (see Voit & Bryan 2001); these generally operate by removing low-entropy gas and condensing it into stars, a process plainly important *within* galaxies. However, the extensive cooling as needed in the bulk of the IGP to raise K (or depress L_X) to the observed levels would produce too many, unseen stars (e.g., Muanwong et al. 2002, Sanderson & Ponman 2003). On the other hand, cooling triggers catastrophic instabilities unless closely checked and offset by other processes feeding energy back to baryons (e.g., Blanchard, Valls-Gabaud & Mamon 1992; Borgani et al. 2002); so energy *additions* $\Delta E > 0$ are mandatory anyway. Thus an alternative view (Cavaliere, Menci, & Tozzi 1997, 1999; Valageas & Silk 1999; Menci & Cavaliere 2000; Wu, Fabian & Nulsen 2000; Cavaliere, Lapi & Menci 2002) focuses on energy injections affecting the plasma equilibrium. The inputs are provided when the baryons in member galaxies condense to form massive stars (possibly in starbursts) then exploding as type II supernovae (SNe), or when they accrete onto central supermassive black holes (BHs) energizing AGNs. Such feedback actions *preheat* the gas exterior to newly forming structures, hindering its flow into the DM potential wells. When strong enough, they also *deplete* the plasma density from the inside by causing thermal outflow or even dynamical blowout.

We next discuss how the latter view explains not only the entropy levels K, but also the entropy profiles $K(r)$ both in clusters and in groups. Recall the context provided by the relation of $K(r)$ to the (local) polytropic index $\Gamma = 5/3 + d \ln K / d \ln n$; values of Γ ranging from $5/3$ to 1 imply profiles from flat to rising from a deep central minimum. The observed Γ and $K(r)$ strike an intermediate course closer to the isothermal limit, if anything. The alert reader will perceive that such conditions are not straightforward, and actually indicate large, balanced entropy production (e.g., by shocks) and/or smearing out (e.g., by enhanced diffusion).

4.4 Preheating by SNe

As to preheating, obvious energy sources are constituted by type II SN explosions, the final events in the life of massive stars. These provide energies $E_{SN} \approx 10^{51}$ ergs with occurrence $\eta_{SN} \lesssim 5 \times 10^{-3}/M_\odot$, i.e., per solar mass condensed into the stellar IMF. Such outputs may be coupled to the surrounding gas at maximal levels $f_{SN} \approx 1/2$ when starbursts induce cooperative propagation of SN remnants and drive subsonic galactic winds (see Matteucci 2003). Then the integrated thermal input attains the level

$$k\Delta T = f_{SN}\, E_{SN}\, \eta_{SN}\, m_\star \, \frac{2\,\mu m_p}{3\,m} \approx \frac{1}{4}\, \text{keV/particle} \qquad (4.11)$$

(see Cavaliere et al. 2002) in groups with stellar to gas mass ratios $m_\star/m \approx 1/2$; somewhat lower levels obtain in clusters where $m_\star/m$ is smaller.

4.4.1 How Preheating Affects the Density Run

The plasma response to such energy inputs comprises two *de-amplification* factors in groups relative to clusters.

The first factor stems from the hydrostatic equilibrium of the plasma in the DM potential wells. Recall from § 4.3.1 that $T_2 = T_v/\beta_{\text{grav}}$ holds, with $\beta_{\text{grav}} \approx 0.7$ in rich clusters. Energy inputs with thermalization efficiency $\zeta \sim 1$ (to be discussed next) will clearly lead to a form $T_2 = T_v/\beta_{\text{grav}} + \zeta\,\Delta T$ (see Cavaliere et al. 2002); so the β parameter will read

$$\beta = \frac{\beta_{\text{grav}}}{1 + \beta_{\text{grav}}\,\zeta\,\Delta T/T_v} \ . \tag{4.12}$$

At cluster scales where $T_v \gg \Delta T$ holds, nearly constant $\beta \approx \beta_{\text{grav}}$ applies. Moving toward groups the prevailing smaller values of T_v lead to reduced values of β, so to flatter density profiles. But with the limited input from SNe, the outcome is easily offset by the increased concentration, that is, by the increasing depth of the normalized $\Delta\phi$; in the balance, the density profiles are little changed down to poor clusters. Quantitatively, the central luminosity scales proportionally to $n^2 \propto e^{2\beta\,\Delta\phi}$; this factor does not appreciably decrease the luminosities of poor compared to rich clusters with preheating values around $1/4$ keV per particle; see Figures 4.5 and 4.8. We shall show in the next section that preheating levels approaching 1 keV per particle are called for, and these are out of question with SNe.

The second de amplification factor is provided by the sensitivity to ΔT of the *boundary* condition n_2, which sets the normalization of the density profiles. This is addressed next.

4.4.2 How Preheating Changes the Boundary Condition

At the accretion shocks, the standard Rankine–Hugoniot conditions (see Landau & Lifshitz 1959) may be recast to yield the density jump in the form

$$\frac{n_2}{n_1} = 2\left(1 - \frac{1}{\theta}\right) + \sqrt{4\left(1 - \frac{1}{\theta}\right)^2 + \frac{1}{\theta}} \simeq 4\left(1 - \frac{15}{16\,\theta}\right) , \tag{4.13}$$

with the last approximation holding in the strong shock limit. Meanwhile, we (Cavaliere et al. 2002) find the temperature jump $\theta = T_2/T_1$ to be given for all kinds of shocks by the basic relation

$$\theta = \frac{5}{16}\mathcal{M}^2 + \frac{7}{8} - \frac{3}{16}\frac{1}{\mathcal{M}^2} \simeq \frac{5}{16}\mathcal{M}^2 + \frac{7}{8} , \tag{4.14}$$

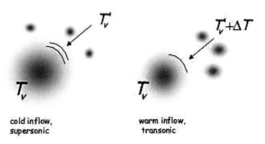

cold inflow,
supersonic

warm inflow,
transonic

Figure 4.6. To illustrate the valve mechanism of the accretion shocks. In rich clusters preheating is negligible, and strong shocks take place in supersonic inflows; so gravitational heating enforces scale-invariant behavior of the ICP. Toward groups the preheating becomes increasingly relevant, the inflows are less supersonic, and the shocks weaken; the scale-invariant trend is broken.

in terms of the inflow Mach number $\mathcal{M} \equiv (3\,\mu m_p\,\tilde{v}_1^2/5\,kT_1)^{1/2}$. Shock heating $(\theta > 1)$ clearly requires the flow to be supersonic in the shock rest frame; i.e., $\mathcal{M} > 1$.

The accretion shocks that constitute our scope here actually occur across a finite transitional layer at $r \approx R$. The above simple relations hold as long as the thickness δ of this layer is longer than a Coulomb collisional mean free path to ensure equipartition with the electrons of the energy primarily gained by the ions (see Ettori & Fabian 1998); δ is also required to be narrow compared with R so that the gravitational forces in the layer can be neglected. The latter condition holds in the inflow of merging partners with masses $M' \lesssim 10^{-1}\,M$, as discussed by Lapi, Cavaliere & Menci (2004). Note that few major clumps do not stop within the layer but rather penetrate deep into the structure (as observed by Mazzotta et al. 2002); they stir up turbulence (see Inogamov & Sunyaev 2003), and conceivably contribute to repeated acceleration of the relativistic electrons energizing the radiohalos (see Feretti et al. 2000).

We find the inflow velocity $\tilde{v}_1$ (used in Equation (4.14)) in the accretion *shock* rest frame to be related to the infall velocity v_1 (appearing in § 4.3.1) by

$$\tilde{v}_1 = \frac{2}{3}\,v_1\left[1 + \left(1 + \frac{15}{4}\,\frac{kT_1}{\mu m_p\,v_1^2}\right)^{1/2}\right] \simeq \frac{4\,v_1}{3}\left(1 + \frac{15\,kT_1}{16\,\mu m_p\,v_1^2}\right) \; ; \; (4.15)$$

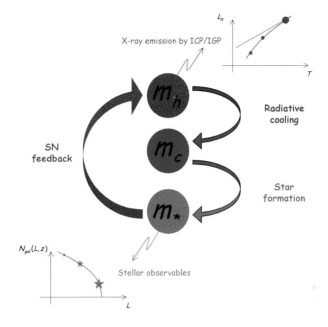

Figure 4.7. The cycling of the baryonic components during structure formation: $m_\star$ is the mass condensed into stars, m_c is the mass radiatively cooled, and m_h is the diffuse component at the virial temperature, i.e., the ICP. The SAM models basing on the DM merging histories treat in detail such a cycling, and yield not only stellar quantities such as the galaxy luminosity functions $N_{\text{gal}}(L, z)$ and the cosmic SFR, but also the x-ray emission L_X associated with m_h: see Menci & Cavaliere (2000).

here the kinetic energy downstream is assumed small. Note in passing that for strong shocks with $\mathcal{M}^2 \gtrsim 3$ one can easily improve upon Equation (4.8) to obtain $kT_2 \simeq \mu m_p v_1^2/3 + 3\, kT_1/2$; the result not only validates the form of Equation (4.12) but also provides the value $\zeta = 3/2$.

4.4.3 The Overall Outcome, Hierarchical Preheating

To capture the full import of the second deamplification factor, one has to consider that the effective value of n_2/n_1 for a structure of mass M at redshift z is given by the average over its merging history. Specifically, one has to average (following Menci & Cavaliere 2000) over the shocks caused in all previous progenitors of mass $M' \lesssim M$ (weighting with their number) when they in turn accreted smaller clumps (weighting with the associated merging rates, see Lacey & Cole 1993). This straightforward if laborious procedure is carried out in the so-called Semi-Analitic Models or SAMs (see

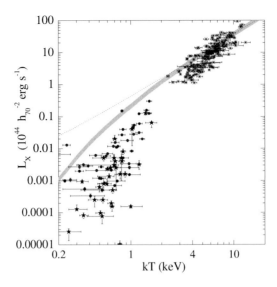

Figure 4.8. Integrated x-ray luminosity L_X vs. x-ray temperature T. Data for clusters (*crosses*) are from Horner (2001), for groups (*circles*) from Osmond & Ponman (2004), and for early-type galaxies (*stars*) from O'Sullivan, Ponman & Collins (2003). The *dotted line* represents the outcome expected from pure gravitational heating, with line-emission included. The *strip* (with 2-sigma width provided by the merging histories) illustrates our result for SN preheating ($k\Delta T = 1/4$ keV per particle) of the plasma with $\Gamma = 1.2$ in the NFW potential.

Figure 4.7), and yields the light strips in Figs. 4.8 and 4.9; these illustrate the variance (at 96% probability level) around the mean value induced by the merging stochasticity.

Two model-independent points are to be stressed. First, the SN feedback acts while a structure and its gaseous content are being built up hierarchically through merging events with a range of partners; in the process, the effective progenitors are those with $M' \lesssim M/3$. This is because these contribute about half the final DM mass in the main progenitor, and half the plasma mass likewise. Meanwhile, being generally colder, they cause stronger shock compressions with Mach numbers actually set by the higher temperature,

$$T_1 = \max\left[T'_v , \Delta T\right] , \tag{4.16}$$

between the virial T'_v in the partner and the preheated value ΔT.

The second point is that SNe make optimal use of their energy in that they produce *hierarchical* preheating of the ICP in groups or clusters. SNe actually affect more the smaller progenitors of such structures; these have

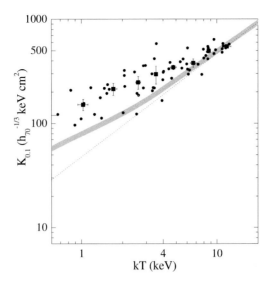

Figure 4.9. Central entropy levels K (at $r \approx 10^{-1} R$) vs. the x-ray temperature T. Data for clusters and groups are from Ponman, Sanderson & Finoguenov (2003): *circles* mark individual systems and *squares* refer to binned data. The *dotted line* represents the outcome expected from pure gravitational heating. The *strip* (with 2-sigma width provided by the merging histories) illustrates our result for SN preheating as in Figure 4.8.

shallower potential wells and are the sites of higher star formation rates, so they are more effective both at producing star-related energy and at using it for heating/expelling the associated gas. As the structure hierarchy develops, such preheated gas will be partly hindered from flowing back into the next forming structures and from contributing to their hot plasma (Menci & Cavaliere 2000). Thus lower densities are propagated some steps up the hierarchy.

In the specific terms of shocks, preheating operates as does a *valve* (see Figure 4.6) that weakens the accretion shocks in smaller structures, and so acts to lower both the boundary density (by decreasing n_2/n_1) and the central one (by decreasing β); from clusters to groups these two deamplification factors progressively break down the scale-invariant behavior. However, in the NFW potential with $\Gamma = 1.2$ the actual depressions of L_X or enhancements of K resulting from the mild extra-heating by SNe turn out to be limited both in amplitude and scatter, as shown by the light strips in Figures 4.8 and 4.9. So higher preheating levels are called for.

An additional issue concerns the entropy *profiles*. In clusters (see Tozzi & Norman 2001; Voit et al. 2003) the entropy deposition increases by the stronger accretion shocks during the last formation stages; this provides rising outer profiles $K(r) \propto r^{1.1}$ equivalent to values $\Gamma \approx 1.1$, in agreement with the cluster data (see § 4.2.2). But in groups any preheating scheme intended to explain the lower baryonic fractions in terms of weaker, nearly isentropic accretion shocks implies the related entropy profiles to be flat, in disagreement with the data (see Pratt & Arnaud 2003; Rasmussen & Ponman 2004). So additional feedback processes are called for.

We address these issues in turn.

4.5 Feedback from AGNs

We now focus on the stronger feedback provided by quasars and AGNs (see Valageas & Silk 1999; Wu et al. 2000; Yamada & Fujita 2001; Nath & Roychowdhury 2002; Cavaliere et al. 2002). These are kindled when the cold galactic gas driven by interactions of the host galaxy with group companions is funneled to form circumnuclear starbursts, but partly trickles down to the very nucleus (see Menci et al. 2004) and accretes onto a central supermassive BH.

AGNs are observed to interact with the surrounding medium in various ways: ripples, bubbles, jets, and shocks; see Fabian et al. (2003), Kraft et al. (2003), Forman et al. (2004), and Fujita et al. (2004). Jets and shocks appear particularly suitable to transport kinetic energy far away from the source. However, the former are highly collimated, and primarily drill narrow tunnels through the ambient medium; the latter are instead more isotropic and more apt to blow considerable fractions of the ambient gas beyond the edge of the structure.

4.5.1 External Preheating by AGNs

Given the BH mass $M_\bullet$, the integrated AGN input comes to

$$k\Delta T = f\,\eta\,m_\star c^2\,\frac{M_\bullet}{4\,M_b}\,\frac{2\,\mu m_p}{3\,m} \approx \frac{1}{2}\ \text{keV/particle}\ . \qquad (4.17)$$

Here we have used the standard mass-energy conversion efficiency $\eta \approx 10^{-1}$ and the locally observed ratio $M_\bullet/M_b \approx 2 \times 10^{-3}$ of BH to galactic bulge masses (Merritt & Ferrarese 2001); the factor $1/4$ accounts for the ratio of the bulge mass observed in blue light compared to that integrated over the star formation history (Fabian 2004). The input above amounts to only a few times the SN's on adopting average values $f \approx 5 \times 10^{-2}$ for the fractional AGN output coupled to the surrounding gas. In fact, small values $f \approx v_w/2\,c \approx 10^{-1}$ are indicated for the average coupling by wind speeds up to $v_w \approx 0.4\,c$ associated with covering factors of order 10^{-1} (see Chartas, Brandt, & Gallagher 2003; Pounds et al. 2003).

Considering that the AGN activity closely parallels the star formation in spheroids (Franceschini et al. 1999; Granato et al. 2004; Umemura 2004), Lapi et al. (2004) add the AGN energy injections to SNs to obtain preheating energies to $k\Delta T \approx 3/4$ keV per particle. Such a *combined* value is sufficient to provide a sizeable step toward the locus of the data, as shown in Figures 4.11 and 4.12 by the heavy strips. While providing a detailed fit in the range from rich to poor clusters (consistent with the argument at the end of § 4.4.1), these strips still fail to comprise the bulk of the data concerning groups and galaxies. Moreover, as anticipated in § 4.4.3 preheating enhanced by AGNs still has the drawback of producing flat entropy profiles in groups.

4.5.2   Internal Impacts of Quasars

But right in groups and galaxies the *impulsive* inputs by powerful quasars take over, providing from *inside* an additional impact on the plasma. For this to occur, the energy $\Delta E \approx 2 \times 10^{62} f \, (M_\bullet/10^9 \, \mathrm{M_\odot}) \, (1 + z)^{-3/2}$ erg discharged over the host dynamical time $t_d \approx 10^{-1}$ Gyr by a quasar (or a number of quasars) while accreting the overall mass $M_\bullet$, has to compete with the energy $E \approx 2 \times 10^{61} \, (kT/\mathrm{keV})^{5/2} \, (1 + z)^{-3/2}$ erg of the plasma in equilibrium; see Lapi, Cavaliere, & De Zotti (2003). The relevant ratio

$$\frac{\Delta E}{E} \approx 0.5 \, \frac{f}{5 \times 10^{-2}} \, \frac{M_\bullet}{10^9 \, \mathrm{M_\odot}} \left(\frac{kT}{\mathrm{keV}}\right)^{-5/2} \tag{4.18}$$

is small in clusters but increases toward groups, approaching unity in poor groups and attaining a few in large galaxies. Then, over distances of order 10^2 pc the quasar wind acts as an efficient piston (see Lamers & Cassinelli 1999; Granato et al. 2004) to drive through the plasma a blastwave terminating into a leading shock; see Figure 4.10.

To describe these blasts, Cavaliere et al. (2002) develop realistic self-similar solutions for the hydrodynamic flow perturbing an equilibrium plasma density approximated as $n(r) \propto r^{-\omega}$ ($2 \leq \omega < 2.5$; see end of § 4.2.2), under the push of the energy $\Delta E(t) \propto t^{2\,(5-2\,\omega)/\omega}$ continuously added over t_d. These solutions include the restraints set to gas dynamics by a finite initial pressure $p(r) \propto r^{2\,(1-\omega)}$ and also by the DM gravity; thus they can describe the full *range* of blast strengths, from weak in clusters to strong in galaxies.

Correspondingly, the ratio of the kinetic to the thermal energy ranges from 10^{-1} to 2; so not only *outflow* but also *ejection* of considerable plasma amounts outwards of the virial radius R are caused in a large galaxy or in a poor group. The parameter $\Delta E/E$ and the related Mach number $\mathcal{M}$ of the leading shock turn out to be independent of r and t during the blast transit; in the simple case $\omega = 2$ they are related by $\mathcal{M}^2 \simeq 1+(\Delta E/E)$ for $\Delta E/E \lesssim$

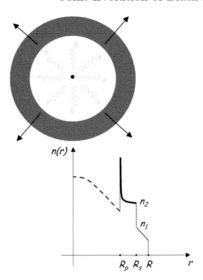

Figure 4.10. To illustrate the density distribution during the propagation of a quasar-driven blast throughout the plasma. By the *dashed* line we outline the initial density run, by the *thick solid* line the flow perturbed by the running blast, and by the *thin solid* line the outer initial density. The perturbed flow is confined between the leading shock at R_s and the trailing piston at $R_p = \lambda R_s$; the thickness $\lambda \approx 0.8 \div 0.5$ decreases as the blast strength increases.

2, and the fractional mass ejected or flowed out is well approximated by $\Delta m/m \simeq \Delta E/2\,E$; see Table 4.6.

Similar results hold independently of model details; in particular, we compare two opposite extreme cases of ejection/outflow. In the first, we take Δm to be the mass in the blast driven out of the virial radius R at $t = t_d$ by the blast *kinetic* energy. In the second, we adopt a constant pressure at $r = R$ as boundary condition, to obtain new densities $n' = n\,T/T'$; now Δm is the mass flowed out of the structure due to the excess *thermal* energy deposited by the blast. For both models, we find $\Delta m/m \simeq \Delta E/2\,E$ to hold, similarly governed by the basic parameter $\Delta E/E$.

After the passage of the blast the plasma recovers hydrostatic equilibrium, described by Equation (4.3). But now all new densities n' will be *depleted* by the factor $1-\Delta m/m$ below the initial value (already affected by the preheating from SNe and AGNs). The resulting $L_X \propto (1 - \Delta m/m)^2$ is shown by the solid and dashed lines in Figure 4.11 for both models (ejection and outflow) discussed above.

Entropy is added by these quasar-driven blasts. While they sweep through the plasma, a moderate jump $K_2/K_1 = T_2\,n_1^{2/3}/T_1\,n_2^{2/3}$ is pro-

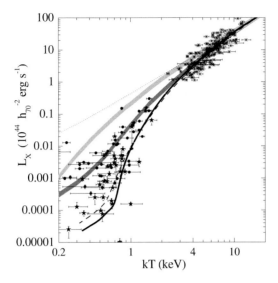

Figure 4.11. Integrated x-ray luminosity L_X vs. x-ray temperature T. Data as in Figure 4.8. The *dotted line* represents the outcome expected from pure gravitational heating, with line-emission included. The *light shaded strip* is for preheating by SNe alone ($k\Delta T = 1/4$ keV per particle, as in Figure 4.8), and the *heavy shaded strip* includes the AGN contribution (a total $k\Delta T = 3/4$ keV per particle). The internal impact from quasars is illustrated by the *solid* (ejection model) and *dashed* (outflow model) lines; see text for details. The coupling level of the quasar output to the ambient medium is taken at $f = 5 \times 10^{-2}$.

duced (see the basic Equations (4.13), (4.14)) across the leading shock. In addition, in the equilibrium recovered with densities depleted by ejection or outflow, the entropy levels are further enhanced as given by

$$K_2'/K_2 = (1 - \Delta m/m)^{-2/3} \ . \tag{4.19}$$

For example, in a group with $kT = 3/4$ keV the combined preheating by SNe and AGNs yields entropy *levels* corresponding to 100 keV cm^2. This is raised to 180 keV cm^2 by the quasar-driven blast arising when a BH accretes 10^9 M$_\odot$, and its output is coupled at levels $f \approx 5 \times 10^{-2}$ to produce $\Delta E/E \approx 1$. The entire K–T correlation is shown by the solid and dashed lines in Figure 4.12 for both the ejection and the outflow models.

We stress two points. From groups to clusters the energy ΔE injected over the related, and only slightly longer t_d, cannot keep pace with the substantially larger E; see Equation 4.18. In galaxies, at the other end, the luminosity deficit or the entropy excess saturate because here $\Delta E/E$

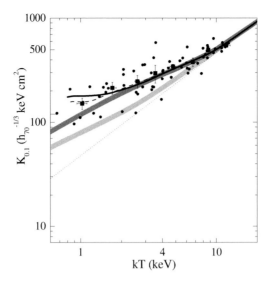

Figure 4.12. Central entropy levels K (at $r \approx 10^{-1} R$) vs. the x-ray temperature T. Data as in Figure 4.9. The *dotted line* represents the outcome expected from pure gravitational heating. *Light shaded strip* is for preheating by SNe alone ($k\Delta T = 1/4$ per particle, as in Figure 4.9), and *heavy shaded strip* includes the AGN contribution (a total $k\Delta T = 3/4$ keV per particle). The internal impact from quasars is illustrated by the *solid* (ejection model) and *dashed* (outflow model) lines; see text for details. As before the coupling level of the quasar output to the ambient medium is taken at $f = 5 \times 10^{-2}$.

is not to exceed a few, lest the gas contained within kpcs is blown away and the accretion it feeds is cut down (see Silk & Rees 1998). The pivotal value $\Delta E/E \approx 1$ yields

$$M_\bullet = 6 \times 10^8 \, M_\odot \left(\frac{f}{5\%} \right)^{-1} \left(\frac{\sigma}{300 \, \text{km s}^{-1}} \right)^5. \qquad (4.20)$$

This is written in terms of the DM velocity dispersion σ, which in turn is found to correlate as $\sigma \propto \sigma_\star^{4/5}$ with the dispersion of the galactic bulge (Ferrarese 2002). Then, for the same value $f \approx 5 \times 10^{-2}$ indicated by the x-ray data, the above relation pleasingly agrees with the observations of relic black holes in the bulges of many galaxies (see Tremaine et al. 2002).

As to the entropy profiles, we (Lapi et al. 2004) consider in closer detail the entropy produced *inside* the structure by the quasar-driven blasts. This is clearly piled up toward the leading shock; we find the distribution $K(m) \propto m^{4/3}$ in terms of the plasma mass $m(< r)$ within the blast. In the residual plasma, adiabatically readjusted to equilibrium after the blast passage, this leaves a strong imprint in the form of a *steep* profile $K'(r) \propto r^{1.3}$, consistent with the data.

4.6 Enhanced Sunyaev–Zel'dovich Effects

The ICP can be also probed through the SZ effect (Sunyaev & Zel'dovich 1972). This arises when some of the CMB photons crossing the structure are Compton upscattered by the hot ICP electrons; then the pure black body spectrum is tilted toward higher energies.

In the μwave band the tilt mimics a diminution of the CMB temperature by $\Delta T_{\mu w} \approx -5.5\, y \sim 1/2$ mK in terms of the Comptonization parameter $y \propto n\, T\, R$. This is evenly contributed by the electron density n and the temperature T; in fact, what matters is the electron pressure $p = n\, k\, T$ integrated along a line of sight at a distance s from the cluster center (see Figure 4.14):

$$y(s) = 2\, \frac{\sigma_T}{m_e c^2} \int_0^{\ell_{\max}} d\ell\, p(r) \ . \tag{4.21}$$

where $\ell_{\max} \equiv \sqrt{R^2 + s^2}$. Since $y \propto E/R^2$ holds, the SZ effect emulates a *calorimeter* (Birkinshaw 1999) probing the ICP total electron thermal energy $E \propto p\, R^3$.

To now, SZ signals have been measured in many rich clusters at levels $y \approx 10^{-4}$ or $\Delta T_{\mu w} \approx -0.5$ mK (see Zhang & Wu 2000; Reese et al. 2002; Birkinshaw 2004), consistent with the ICP densities and temperatures derived from x-rays (see Figure 4.13). SZ measurements in groups are challenging at present, and still missing.

In groups at equilibrium, where the luminosity $L_X \propto T^3$ is depressed well below the gravitational scaling $L_X \propto T^2$ and the adiabat $K \propto T^{2/3}$ is *enhanced* well above the gravitational scaling $K \propto T$, Cavaliere & Menci (2001) predict a related *deficit* of y

$$y \propto L_X^{1/2}\, T^{1/2} \propto K^{-3/2}\, T^3 \propto T^2 \tag{4.22}$$

relative to the baseline value $y \propto T^{3/2}$ that would hold for a constant m/M ratio (Cole & Kaiser 1988). But if such depressions are caused by substantial energies ΔE added by AGNs to the large-scale gravitational energy E, we also expect transient, *enhanced* SZ signals during and soon after the source activity (see also Aghanim, Balland & Silk 2000; Platania et al. 2002). In Lapi et al. (2003) we make realistic predictions for the SZ enhancements when these are originated by quasar-driven blasts with limited strengths $\Delta E/E \lesssim 3$, consistent with the $M_\bullet - \sigma$ correlation, as discussed in the previous section.

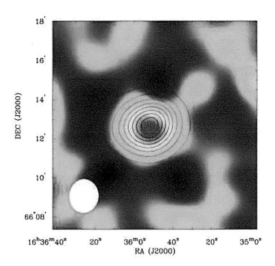

Figure 4.13. SZ image of A2218 taken by BIMA (Reese et al. 2002); x-ray contours from *ROSAT* are superimposed.

In computing how y is *enhanced* during the transit of the quasar-driven blast, we focus on the value $\bar{y} \propto (2/R^2) \int ds\, s\, y(s)$ averaged over the structure area (see Figure 4.14); this is because an early group or galaxy will subtend small angles $\lesssim 1'$. Specifically, we find

$$\frac{\bar{y}}{\bar{y}_{eq}} = \frac{\langle p \rangle}{p_{eq}} \frac{1 - \lambda^3}{3} \tag{4.23}$$

for the parameter $\bar{y}$ averaged over the shell at $R_s \approx R$, a position that maximizes the transit time and optimizes the observability. This is written in terms of the mean pressure enhancement $\langle p \rangle / p_1$ corresponding to $\Delta E / E$

Table 4.1. Relevant quantities for quasar-driven blasts

| | $n \propto r^{-2}$, $L = \text{const}$ | | | $n \propto r^{-2.4}$, $L \propto t^{-5/6}$ | |
$\Delta E / E$	$\mathcal{M}$	$\langle p \rangle / p_1$	$1 - \Delta m/m$	$\mathcal{M}$	$\langle p \rangle / p_1$
0.3.....	1.2	3.6	0.92	2.1	17.8
1........	1.5	4.6	0.58	3.0	21.7
3........	1.9	6.3	~ 0	4.7	26.2

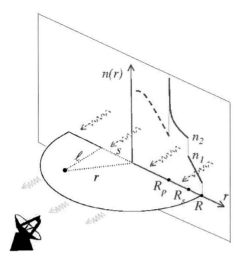

Figure 4.14. The geometry underlying Equation (4.23). For a point in the structure r is the radial coordinate, s is its projection on the plane of the sky, and ℓ is the coordinate along the line of sight. On the vertical plane we outline the initial density run, and the flow perturbed by the quasar-driven blastwave.

as given in Table 4.1.

Our numerical results are represented in Figure 4.15 vs. the depth kT_v of the host potential well. The square illustrates the *minimal* (for $\omega = 2$) enhancement we expect from an early group with $kT_v = 1$ keV at $z = 1.5$. With radii $R \approx 250$ kpc, the angular sizes $\approx 1'$ are close to their minimum in the concordance cosmology. The circles represent our results for a massive ($\sigma = 300$ km s^{-1}, $R \approx 100$ kpc) galaxy at $z = 2.5$. The related angular sizes are around $0'.5$; with resolution fixed at $\approx 1'$, the signals will be diluted by a factor $\approx 1/4$ and scaled down to $\Delta T_{\mu w} \approx -15\,\mu$K.

To evaluate the corresponding statistics, we insert the blue luminosities $L = \Delta E/10\, f\, t_d$ (with a bolometric correction 10) in the quasar luminosity function observed by Croom et al. (2004), and discussed by Cavaliere & Vittorini (2000). In terms of the cumulative fraction of bright galaxies hosting a type 1 quasar brighter than L, shape and evolution of the luminosity function are described by

$$N(L)\,L \approx 2\,10^{-2}\,(1+z)^{1.6}\left(\frac{L_b}{L}\right)^{2.2}, \qquad (4.24)$$

beyond an approximate break at $L_b = 5 \times 10^{45}\,[(1+z)/3.5]^{2.8}$ ergs s^{-1}. This implies that fractions around 10^{-2} and 10^{-1} of bright galaxies will show the signals represented as filled and empty circles in Figure 4.15, respectively.

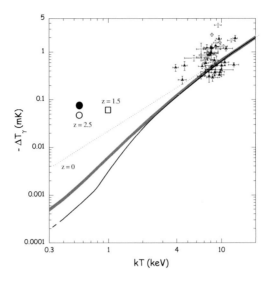

Figure 4.15. Predicted SZ signals as a function of T_v in galaxies, groups, and clusters. Data from Zhang & Wu (2000) and Reese et al. (2002). *Strip* and *lines* as in Figure 4.11. *Square*: area-averaged, undiluted signal from a group at $z = 1.5$, driven by quasar activity with $M_\bullet = 10^9 \ M_\odot$ and $f = 5 \times 10^{-2}$, corresponding to $\Delta E/E = 0.5$. *Circles*: same from a massive galaxy at $z = 2.5$, for $M_\bullet = 6 \times 10^8 \ M_\odot$ or $\Delta E = E$ (*open*) and $M_\bullet = 2 \times 10^9 \ M_\odot$ or $\Delta E = 3 E$ (*filled*).

The same luminosity function, interpreted in terms of interactions of the host galaxy with its group companions, yields a few signals per 10 poor groups, with the strength represented by the square in Figure 4.15.

If a galaxy happens to grow a large black hole in times shorter than t_d, the quasar will inject an energy $\Delta E \gtrsim E$ impulsively; we expect this to stop accretion and cause the quasar to fade or quench. We describe such conditions using our models with steep density gradients $\omega > 2$ and impulsive but fading power injections $L(t) \propto (5 - 2\,\omega)\,\omega^{-1}\,t^{5\,(2-\omega)/\omega}$ (see Table 4.1); these imply strongly enhanced SZ signals during the subsequent transient, but also imply eventual ejection of a substantial gas fraction. We remark that these events are likely to bring star formation to an early end at $z \sim 2$; such may have been the case in some of the EROs (see Cimatti et al. 2002; also Granato et al. 2004).

4.7 Discussion and Conclusions

We have reviewed the main physical processes (summarized in Table 4.2) relevant to the x-ray observations of the plasma pervading not only clusters but also groups of galaxies. Here we summarize their actions on referring to the terms in the expression for the x-ray emission

$$L_X \propto T^2 \left(\frac{n_2}{n_1}\right)^2 \left(1 - \frac{\Delta m}{m}\right)^2 \int dr\, r^2\, e^{2\beta\,\Delta\phi} . \tag{4.25}$$

The factor T^2 stems from the DM scaling laws for the density $n \propto \rho$ and the virial radius R (see § 4.3.2). The factor $(n_2/n_1)^2$ is the (squared) density jump produced by accretion shocks at the structure boundary close to R (see § 4.4.2). The factor $(1 - \Delta m/m)^2$ accounts for hot gas fractions smaller in groups than the cosmic value $m \approx 0.15\, M$ (see § 4.5.2). The last term is the shape factor (in the simple isothermal approximation, see Equation 4.1 and § 4.2.2) that includes the classic parameter $\beta = T_v/T_2$ for the plasma in hydrostatic equilibrium within the DM potential well $\Delta\phi$ (see § 4.4.1).

In *clusters*, we have argued that the AGN external preheating exceeds SN's; the two actually combine to yield a level approaching 1 keV per particle. In moving from rich to poor clusters, the increasing preheating ratio $\Delta T/T_v$ weakens the accretion shocks, leading to a lower value of β and to a jump factor $(n_2/n_1)^2$ decreased from about 10 to 1, once averaged over the merging histories. All that boils down to a relation close to $L_X \propto T^3$ for the x-ray emission and to a parallel relation $K \propto T^{2/3}$ for the entropy levels; these trends are fitting for poor clusters but are still marginal for groups. There, in addition, all preheating schemes produce by the same token *weak* accretion shocks and flat entropy profiles, and the latter conflict with the data.

Just in *groups*, on the other hand, the quasar feedback increasing with the relative energy input $\Delta E/E$ takes over, and acts from the inside to produce *strong* blasts and eject considerable amounts of plasma $\Delta m/m \approx \Delta E/2\,E$, so depleting the residual baryon fraction $(1 - \Delta m/m)$. This further *depresses* the x-ray luminosities $L_X \propto (1 - \Delta m/m)^2$, while *enhancing* the entropy levels K and also producing non-isentropic, *steep* profiles.

In *galaxies*, however, such internal effects must saturate; $\Delta E/E$ is constrained not to exceed a few, lest the impulsive quasar feedback ejects so much galactic gas as to utterly stop the accretion (see Silk & Rees 1998; also King 2003). Such a saturation is reflected in the lower *knees* of the solid and dashed lines in Figure 4.11. By the same token, the saturation condition $\Delta E/E \approx 1$ directly yields $M_\bullet \approx 6 \times 10^8\, M_\odot\, (f/5 \times 10^{-2})^{-1} (\sigma/300\,\mathrm{km\,s^{-1}})^5$ in terms of the DM velocity dispersion σ, as argued by Cavaliere et al. (2002). In turn, σ is found to correlate less than

Table 4.2. Processes affecting the ICP/IGP

Physical process	Affected scales	Modus operandi	Comments
Gravitational heating[a]	gals. grs., cls.	Accretion shocks	Preserves scale invariance
Radiative cooling[b]	gals., grs.	Condenses low-K gas	Excess SF
SN feedback[c]	gals., grs.	Preheating modulates shocks	Low energetics, marginal K levels
AGN feedback[d]	gals., grs. poor cls.	Enhanced pre-heating, blowout	Fits L_X, K; relates to $M_\bullet$; enhances SZ

Useful (if far from exhaustive) references include: [a]Kaiser (1986); [b]Voit & Bryan (2001); [c]Cavaliere et al. (2002); [d]Lapi et al. (2003, 2004).

linearly with the velocity dispersion $\sigma_\star$ of the host galactic bulge (Ferrarese 2002; Pizzella et al. 2004); so the result is consistent with the $M_\bullet - \sigma_\star$ relation observed in active and inactive galactic nuclei (Tremaine et al. 2002).

In terms of increasingly sporadic quasar impacts one can also understand why the intrinsic *scatter* in the x-ray data (adding to observational uncertainties) should widen toward smaller systems such as poor groups or massive galaxies. This we trace back to the increasing *variance* in the occurrence of strong quasar events or in their coupling level f, that concur to sensitively modulate the plasma ejection $\Delta m/m \propto f\, M_\bullet$. As the hierarchical clustering proceeds toward the recent clusters, instead, the evolution of the quasars cuts down most internal effects; so the impulsive contributions to ΔE within a dynamical time cannot match the large equilibrium energy E in such massive and late structures.

Thus diverse pieces of data fit together on considering both the *external* preheating from AGNs and the *internal* impact from quasars. Specifically, an overall coupling level $f \approx 5 \times 10^{-2}$ of the quasar outputs to the ambient plasma yields remarkable agreement with the large-scale x-ray emissions; the same value also yields agreement with the mainly optical measurements of nuclear BH masses vs. galactic velocity dispersions. The AGN feedback may be also relevant (see Ruszkowski & Begelman 2002; Fabian 2004) to the puzzling "cool cores" at the very centers of many clusters (Molendi & Pizzolato 2001).

Moreover, in the μwave and submm bands Lapi et al. (2003) have pointed out that real-time evidence of the quasar impacts in action may be caught from resolved Sunyaev–Zel'dovich signals enhanced by the overpressure in running blastwaves. Such signals from early galaxies and groups are challenging to detect at present, but will soon emerge in surveys planned with single-dish radiotelescopes equipped with the new multibeam technology, and will be pinpointed with the upcoming generation of dedicated interferometers. These will culminate with the Atacama Large Millimeter Array (ALMA) that will be able between 10 mm and 350 microns to resolve signals on the scale of $10''$ (see also http://www.alma.nrao.edu/).

Such resolved detections will catch single episodes of quasar feedback in the *act*; the observations should be aimed at sites marked by an x-ray point source (signaling a fully active quasar), and/or by strong IR emissions (signaling a nascent quasar enshrouded by dust; see also Granato et al. 2004). In fact, such events (at strong variance with the outcomes from major mergers; see Ricker & Sarazin 2001) are expected to correlate with *pointlike* x-rays from an AGN, while causing no enhancement of *extended* emission $L_X \propto m(r) \, T^{1/2} \, n(r)$ from the plasma mass $m(r)$ swept up outwards to lower densities $n(r)$.

Thus we expect that AGN and quasar energy outputs up to 10^{62} ergs with coupling $f \sim 5 \times 10^{-2}$ should leave 3 kinds of consistent *relics*: the depressed x-ray luminosities L_X, and the excess entropy levels K with steep distributions $K(r)$ on supergalactic scales; the $M_\bullet - \sigma$ relation on subgalactic scales (Equation 4.20). In addition, on the intermediate scales of large galaxies/poor groups we expect (Figure 4.15) *transient* SZ signals standing out of a generally depressed landscape.

We conclude our discussion with two summations highlighting the trends from clusters to groups.

• Outflow and ejection beat inflow and merging. The latter processes dominantly affect the ICP in clusters; the former prevail in the IGP of groups on account of the relatively larger energy input $\Delta E/E$. Actually in the IGP the two kinds of processes cumulate to a high level of *complexity*.

• Strong conquers weak gravity. It takes the energy liberated by baryons condensed under *strong* gravity to break down in groups the cluster scale-invariance striven for by the DM *weak* gravity. Strong gravity actually cooperates with other interactions in the stellar cores that eventually explode as type II SNe; but it runs wild near the horizon of the accreting supermassive BHs that energize the stronger feedback from AGNs and quasars.

Acknowledgments

We thank the organizers for the stimulating school. We are grateful to Gianfranco De Zotti and Nicola Menci for fruitful collaborations, and to Giancarlo Setti and Francesca Matteucci for helpful discussions. Work partially supported by grants from ASI and MIUR.

References

Aghanim N, Balland C and Silk J 2000 *A&A* **357** 1

Allen S W and Fabian A C 1994 *MNRAS* **269** 409

Balogh M L, Babul A and Patton D R 1999 *MNRAS* **307** 463

Bennett C L, et al 2003 *ApJS* **148** 1

Bertschinger E 1985 *ApJS* **58** 39

Birkinshaw M 1999 *Phys. Rept.* **310** 97

Birkinshaw M, *Carnegie Observatories Astrophysics Series* **Vol. 3** (2004 Cambridge: Cambridge Univ. Press)

Blanchard A, Valls-Gabaud D and Mamon G A 1992 *A&A* **264** 365

Borgani S, et al 2002 *MNRAS* **336** 409

Bower R G 1997 *MNRAS* **288** 355

Bullock J S, et al 2001 *MNRAS* **321** 559

Cavaliere A and Fusco-Femiano R 1976 *A&A* **49** 137

Cavaliere A and Fusco-Femiano R 1978 *A&A* **70** 677

Cavaliere A and Menci N 2001 *MNRAS* **327** 488

Cavaliere A and Vittorini V 2000 *ApJ* **543** 599

Cavaliere A, Gursky H and Tucker W A 1971 *Nature* **231** 437

Cavaliere A, Lapi A and Menci N 2002 *ApJ* **581** L1

Cavaliere A, Menci N and Tozzi P 1997 *ApJ* **484** L21

Cavaliere A, Menci N and Tozzi P 1999 *MNRAS* **308** 599

Chartas G, Brandt W N and Gallagher S C 2003 *ApJ* **595** 85

Cimatti A, et al 2002 *A&A* **381** L68

Cole S and Kaiser N 1988 *MNRAS* **233** 637

Croom S M, et al 2004 *MNRAS* **349** 1397

De Grandi S and Molendi S 2002 *ApJ* **567** 163

Ettori S and Fabian A C 1998 *MNRAS* **293** L33

Ettori S and Fabian A C 1999 *MNRAS* **305** 834

Ettori S, Tozzi P, Borgani S and Rosati P 2004 *A&A* **417** 13

Fabian A C, et al 2003 *MNRAS* **344** L43

Fabian A C, *Carnegie Observatories Astrophysics Series* **Vol. 1** (Cambridge: Cambridge Univ. Press)

Feretti L, Brunetti G, Giovannini G, Govoni F and Setti G, in *IAP 2000 Conf. "Constructing the Universe with Clusters of Galaxies"* eds. Florence Durret and Daniel Gerbal p. 37

Ferrarese L 2002 *ApJ* **578** 90

Forman W, et al 2004 *ApJ* submitted (preprint astro-ph/0312576

Franceschini A, Hasinger G, Miyaji T and Malquori D 1999 *MNRAS* **310** L5

Fujita Y, et al 2004 *ApJ* in press (preprint astro-ph/0407596

Granato G L, De Zotti G, Silva L, Bressan A and Danese L 2004 *ApJ* **600** 580

Horner D J 2001 *X-ray Scaling Laws for Galaxy Clusters and Groups* Ph.D. Thesis Univ. of Maryland

Inogamov N A and Sunyaev R A 2003 *Astron. Lett.* **29** 791

Jones C and Forman W 1984 *ApJ* **276** 38

Kaiser N 1986 *MNRAS* **222** 323

King A R 2003 *ApJ* **596** L27

King I R 1972 *ApJ* **174** L123

Kraft R P et al 2003 *ApJ* **592** 129

Lacey C and Cole S 1993 *MNRAS* **262** 627

Lamers H J G L M and Cassinelli J P, *Introduction to stellar winds* (1999 Cambridge: Cambridge University Press)

Landau L D and Lifshitz E M, *Fluid Mechanics* (1959 London: Pergamon Press)

Lapi A 2004 *The Energy Budget of Cosmic Baryons* Ph.D. Thesis Univ. "Tor Vergata" Rome

Lapi A, Cavaliere A and De Zotti G 2003 *ApJ* **597** L93

Lapi A, Cavaliere A and Menci N 2004 *ApJ* submitted

Makino N, Sasaki S and Suto Y 1998 *ApJ* **497** 555

Markevitch M 1998 *ApJ* **504** 27

Matteucci F, *The Chemical Evolution of the Galaxy* 2003 Dordrecht: Kluwer Academic Publishers)

Mazzotta P, Markevitch M, Vikhlinin A and Forman W R 2002 *ASP Conf. Proc.* **257** 173

Menci N and Cavaliere A 2000 *MNRAS* **311** 50

Menci N, Cavaliere A, Fontana A, Giallongo E, Poli F and Vittorini V 2004 *ApJ* **604** 12

Merritt D and Ferrarese L 2001 *MNRAS* **320** L30

Molendi S and Pizzolato F 2001 *ApJ* **560** 194

Muanwong O, Thomas P A, Kay S T and Pearce F R 2002 *MNRAS* **336** 527

Mullis C R, et al 2004 *ApJ* **607** 175

Mushotzky R F, *Carnegie Observatories Astrophysics Series* **Vol. 3** (2004 Cambridge: Cambridge Univ. Press)

Nath B B and Roychowdhury S 2002 *MNRAS* **333** 145

Navarro J F, Frenk C S and White S D M 1997 *ApJ* **490** 493

Norman M L 2004 *PASP Conf. Ser.* **301** in press (preprint astro-ph/0403079

Osmond J P F and Ponman T J 2004 *MNRAS* **350** 1511

O'Sullivan E, Ponman T J and Collins R S 2003 *MNRAS* **340** 1375

Padmanabhan T, *Theoretical Astrophysics* **Vol. III** (2003 Cambridge: Cambridge Univ. Press)

Peebles P J E, *Principles of Physical Cosmology* (1993 Princeton: Princeton Univ. Press)

Pizzella A, dalla Bontà E, Corsini E M and Bertola F 2004 *MNRAS* submitted

Platania P, Burigana C, De Zotti G, Lazzaro E and Bersanelli M 2002 *MNRAS* **337** 242

Ponman T J, Sanderson A J R and Finoguenov A 2003 *MNRAS* **343** 331

Pounds K A, King A R, Page K L and O'Brien P T 2003 *MNRAS* **346** 1025

Pratt G W and Arnaud M 2003 *A&A* **408** 1

Press W and Schechter P 1974 *ApJ* **187** 425

Rasmussen J and Ponman T J 2004 *MNRAS* **349** 722

Reese E D et al *ApJ* 2002 **581** 53

Ricker P M and Sarazin C L 2001 *ApJ* **561** 621

Ruszkowski M and Begelman M C 2002 *ApJ* **581** 223

Sanderson A J R and Ponman T J 2003 *MNRAS* **345** 1241

Sarazin C L, *X-ray Emission from Clusters of Galaxies* (1988 Cambridge: Cambridge Univ. Press)

Sheth R K and Tormen G 2002 *MNRAS* **329** 61

Silk J and Rees M J 1998 *A&A* **331** 1

Sunyaev R A and Zel'dovich Ya B 1972 *Comm. Astroph. Sp. Sc.* **4** 173

Tozzi P and Norman C 2001 *ApJ* **546** 63

Tremaine S, et al 2002 *ApJ* **574** 740

Umemura M, *Carnegie Observatories Astrophysics Series* **Vol. 1** (2004 Cambridge: Cambridge Univ. Press)

Valageas P and Silk S 1999 *A&A* **350** 725

Voit G M and Bryan G L 2001 *Nature* **414** 425

Voit G M, Balogh M L, Bower G B, Lacey C G and Bryan G L 2003 *ApJ* **593** 272

White S D M, Navarro J F, Frenk C S and Evrard A E 1993 *Nature* **366** 429

Wu K K S, Fabian A C and Nulsen P E J 2000 *MNRAS* **318** 889

Yamada M and Fujita Y 2001 *ApJ* **553** L145

Zhang T and Wu X 2000 *ApJ* **545** 141

Chapter 5

The Formation of Primordial Luminous Objects

Emanuele Ripamonti
Kapteyn Astronomical Institute
University of Groningen
Groningen, The Netherlands

Tom Abel
Kavli Institute for Astroparticle Physics and
Cosmology
Stanford Linear Accelerator
Menlo Park, California

5.1   Introduction

The scientific belief that the universe evolves in time is one of the legacies of the theory of the Big Bang.

The concept that the universe has a history started to attract the interest of cosmologists soon after the first formulation of the theory: already Gamow (1948; 1949) investigated how and when galaxies could have been formed in the context of the expanding universe.

However, the specific topic of the formation (and of the fate) of the *first* objects dates to two decades later, when no objects with metallicities as low as those predicted by primordial nucleosynthesis ($Z \lesssim 10^{-10} \sim 10^{-8} Z_\odot$) were found. Such concerns were addressed in two seminal papers by Peebles & Dicke (1968; hereafter PD68) and by Doroshkevich, Zel'Dovich & Novikov (1967; hereafter DZN67),[1] introducing the idea that some objects

---

[1] This paper is in Russian and we base our comments on indirect knowledge (e.g., from the Novikov & Zel'Dovich 1967 review).

could have formed before the stars we presently observe.

(i) Both PD68 and DZN67 suggest a mass of $\sim 10^5 M_\odot$ for the first generation of bound systems, based on the considerations on the cosmological Jeans length (Gamow 1948; Peebles 1965) and the possible shape of the power spectrum.

(ii) They point out the role of thermal instabilities in the formation of the protogalactic bound object, and of the cooling of the gas inside it; in particular, PD68 introduces H_2 cooling and chemistry in the calculations about the contraction of the gas.

(iii) Even if they do not specifically address the occurrence of fragmentation, these papers make two very different assumptions: PD68 assumes that the gas will fragment into "normal" stars to form globular clusters, while DZN67 assumes that fragmentation *does not* occur, and that a single "super-star" forms.

(iv) Finally, some feedback effects as considered (e.g., Peebles & Dicke considered the effects of supernovae).

Today most of the research focuses on the issues of when fragmentation may occur, what objects are formed, and how they influence subsequent structure formation.

In these notes we will leave the discussion of feedback to lecture notes by Ferrara & Salvaterra and by Madau & Haardt in this same book and focus only on the aspects of the formation of the first objects.

The advent of cosmological numerical hydrodynamics in particular allows a fresh new look at these questions. Hence, these notes will touch on aspects of theoretical cosmology to chemistry, computer science, hydrodynamics, and atomic physics. For further reading and more references on the subject we refer the reader to other relevant reviews such as Barkana & Loeb 2001, and more recently Ciardi & Ferrara 2004, Glover 2004, and Bromm & Larson 2004.

In these notes, we try to give a brief introduction to only the most relevant aspects. We will start with a brief overview of the relevant cosmological concepts in §5.2, followed by a discussion of the properties of primordial material (with particular emphasis on its cooling and its chemistry) in §5.3. We will then review the technique and the results of numerical simulations in §§5.4 and 5.5: the former will deal with detailed 3D simulations of the formation of gaseous clouds that are likely to transform into luminous objects, while the latter will examine results (mostly from 1D codes) about the modalities of such transformation. Finally, in §5.6 we will critically discuss the results of the previous sections, examining their consequences, and comparing them to our present knowledge of the universe.

5.2 Physical Cosmology

In the following we will adopt the modern physical description of the universe (dating back to at least 1920), based upon the "cosmological principle," which affirms that on cosmological scales the distributions of matter and energy should be homogeneous and isotropic, whose metric is the Robertson–Walker metric. Although derived from mainly philosophical arguments, the cosmological principle is also supported by observations such as the isotropy of the Cosmic Microwave Background (CMB) (e.g., Wu et al. 1999).

We will also make some additional, general assumptions that are quite common in present-day cosmology, and which are believed to be the best explanations for a series of observations. That is:

(i) The cosmological structures we observe at present (galaxies, clusters of galaxies, etc.) formed because of gravitational instability of preexisting, much shallower fluctuations;

(ii) Most of the matter in the universe is in the form of cold dark matter (CDM), that is, of some kind of elementary particle (or particles) that has not been discovered at present. CDM particles are assumed to have a negligibly small cross-section for electromagnetic interactions (i.e., to be *dark*), and to move at non-relativistic speeds (i.e., to be *cold*).

5.2.1 Fluctuations in the Early Universe

5.2.1.1 *Inflation*

Inflation is a mechanism which was first proposed by Guth (1981) and (in a different way) by Starobinsky (1979, 1980), and has since been included in a number of different "inflationary theories" (see Guth 2004 for a review upon which we base this paragraph).

The basic assumption of inflationary models is the existence of states with negative pressure; a typical explanation is that some unidentified kind of scalar field (commonly referred to as *inflaton*) temporarily keeps part of the universe in a "false vacuum" state, in which the energy density must be approximately constant (that is, it can not "diluted" by expansion) at some value $\rho_f c^2$, implying a negative pressure $p_f = -\rho_f c^2$. Inserting these two expressions in the first Friedmann cosmological equation

$$\ddot{a}(t) = -\frac{4\pi}{3} G \left(\rho + 3\frac{p}{c^2} \right) a(t) \tag{5.1}$$

(where $a(t)$ is the scale factor at cosmic time t) it is easy to derive that the considered region expands exponentially: $a(t) \propto e^{t/t_f}$ with $t_f = (8\pi G \rho_f/3)^{-1/2}$; the epoch in which the universe undergoes such exponential expansion is called *inflation*.

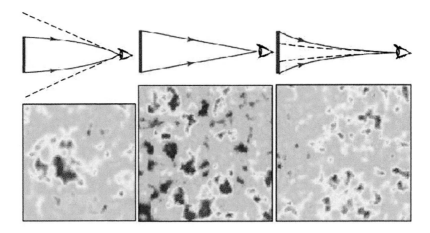

Figure 5.1. Effect of universe geometry on the observed angular size of fluctuations in the CMBR. If the universe is closed (left panel) "hot spots" appear larger than actual size; if the universe is flat (middle panel), "hot spots" appear with their actual size; if the universe is open, "hot spots" appear smaller than their actual size.

If ρ_f is taken to be at about the grand unified theory scale, we get $t_f \sim 10^{-38}$ s, corresponding to a Hubble length of $ct_f \sim 10^{-28}$ cm; if the inflationary phase is long enough (a lower limit for this is about 65 e-folding times, corresponding to an expansion by a factor $\sim 10^{28}$), it smooths the metric, so that the expanding region approaches a de Sitter flat space, regardless of initial conditions. Furthermore, when inflation ends, the energy stored in the inflaton field is finally released, thermalizes, and leads to the hot mixture of particles assumed by the standard Big Bang picture.

Inflation helps explain several fundamental observations, which were just assumed as initial conditions in non-inflationary models:

(i) The Hubble expansion: repulsive gravity associated with a false vacuum is exactly the kind of force needed to set up a motion pattern in which every two particles are moving apart with a velocity proportional to their distance.

(ii) The homogeneity and isotropy: in "classical" cosmology the remarkable uniformity of the CMB cannot be "explained," because different regions of the present CMB sky never were causally connected. Instead, in inflationary models the whole CMB sky was causally connected *before* inflation started, and uniformity can be established at that time.

(iii) The flatness problem: a flat Friedmann–Robertson–Walker model universe (i.e., with $\Omega(t) \equiv \rho_{tot}(t)/\rho_{crit}(t) = 1$, where $\rho_{tot}(t)$ is the cosmological mean density, including the "dark energy" term, and $\rho_{crit}(t) = 3H(t)^2/8\pi G$, $H(t) \equiv \dot{a}(t)/a(t)$ being the Hubble parameter) always remains flat, but if at an early time Ω was just slightly different from 1, the difference should have grown as fast as $(\Omega - 1) \propto t^{2/3}$. All the observational results (e.g., the Bennet et al. 2003 Wilkinson Microwave Anisotropy Probe (WMAP) result $\Omega_0 = 1.02 \pm 0.02$) show that at present Ω is quite close to 1, implying that at the Planck time ($t \sim 10^{-43}$ s) the closeness was really amazing. Inflation removes the need for this incredibly accurate fine tuning, since during an inflationary phase Ω is driven towards 1 as $(\Omega - 1) \propto e^{-2Ht}$.

(iv) The absence of magnetic monopoles: grand unified theories, combined with classical cosmology, predict that the universe should be largely dominated by magnetic monopoles, which instead have never been observed. Inflation provides an explanation, since it can dilute the monopole density to completely negligible levels.

(v) the *anisotropy* properties of the CMB radiation: inflation provides a prediction for the power spectrum of fluctuations, which should be generated by quantum fluctuations and nearly scale-invariant, a prediction in close agreement with the WMAP results (see the next subsection).

A peculiar property of inflation is that most inflationary models predict that inflation does not stop everywhere at the same time, but just in localized "patches" in a succession which continues eternally; since each patch (such as the one we would be living in) can be considered a whole universe, it can be said that inflation produces an infinite number of universes.

5.2.1.2 Primordial Fluctuation Evolution – Dark Matter

Inflation predicts the statistical properties of the density perturbation field, defined as

$$\delta(\mathbf{x}) \equiv \frac{\rho(\mathbf{r}) - \bar{\rho}}{\bar{\rho}} \tag{5.2}$$

where $\mathbf{r}$ is the proper coordinate, $\mathbf{x} = \mathbf{r}/a$ is the comoving coordinate, and $\bar{\rho}$ is the mean matter density.

In fact, if we look at the Fourier components in terms of the comoving wave-vectors $\mathbf{k}$

$$\delta(\mathbf{k}) = \int \delta(\mathbf{x}) \, e^{-i\mathbf{k}\mathbf{x}} \, d^3\mathbf{x} \tag{5.3}$$

the inflationary prediction is that the perturbation field is a Gaussian random field, that the various $\mathbf{k}$ modes are independent, and that the power spectrum $P(k)$ (where $k \equiv |\mathbf{k}|$) is close to scale-invariant, i.e., it is given

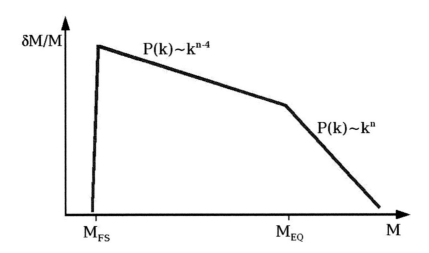

Figure 5.2. Schematic shape of the DM power spectrum after accounting for the effects of free streaming and of the processing that takes place for fluctuations entering the horizon before t_{eq}.

by a power law

$$P(k) \equiv <|\delta_{\mathbf{k}}|^2> \propto k^{n_s}, \qquad \text{with } n_s \simeq 1. \qquad (5.4)$$

This prediction applies to the *primordial* power spectrum; in order to make comparisons with observations, we need to include effects that subsequently modified its shape:

(i) Free streaming: the dark matter particles are in motion, and since they are believed to interact only through gravity, they freely propagate from overdense to underdense regions, wiping out the density perturbations. However, at any given cosmic time t, this will affect only wavelengths smaller than the *free streaming* length, i.e., the proper distance $l_{FS}(t)$ that a dark matter particle can have travelled in time t. It can be shown (see e.g., Padmanabhan 1993, §4.6) that this scale depends primarily on the epoch t_{nr} when the dark matter particles become non-relativistic: before that epoch they move at the speed of light, and can cover a proper distance $l_{FS}(t_{nr}) \sim 2ct_{nr}$, corresponding to a comoving distance $\lambda_{FS} = (a_0/a_{nr})2ct_{nr}$; after becoming non-relativistic, particle motions become much slower than cosmological expansion, and the evolution after t_{nr} only increases λ_{FS} by a relatively small factor. In turn, t_{nr} depends primarily on the mass m_{DM}

of dark matter particles, so that

$$\lambda_{\mathrm{FS}} \sim 5 \times 10^{-3} \, (\Omega_{\mathrm{DM}} h^2)^{1/3} \left(\frac{m_{\mathrm{DM}}}{1 \, \mathrm{GeV}} \right)^{-4/3} \mathrm{pc} \qquad (5.5)$$

corresponding to a mass scale

$$M_{\mathrm{FS}} \sim 6 \times 10^{-15} \, (\Omega_{\mathrm{DM}} h^2) \left(\frac{m_{\mathrm{DM}}}{1 \, \mathrm{GeV}} \right)^{-4} \mathrm{M}_\odot. \qquad (5.6)$$

Since the most favored candidates for the DM particles are Weakly Interacting Massive Particles (WIMPs) with mass between 0.1 and 100 GeV, we probably have $M_{\mathrm{FS}} \lesssim 10^{-8} \, \mathrm{M}_\odot$. Some super-symmetric theories (e.g., Schwartz et al. 2001, and more recently Green, Hofmann, & Schwarz 2004) instead point towards $M_{\mathrm{FS}} \sim 10^{-6} \mathrm{M}_\odot$, and Diemand, Moore, & Stadel (2005) used this result in order to argue that the first structures which formed in the early universe were Earth-mass dark-matter halos (but see also Zhao et al. 2005 for criticism about this result).

(ii) Growth rate changes: in general, perturbations grow because of gravity; however, the details of the process change with time, leaving an imprint on the final processed spectrum. The time t_{eq} when the matter density becomes larger than the radiation density is particularly important: before t_{eq}, perturbations with λ larger than the Hubble radius $c/H(t)$ grow as $\delta \propto a^2$, while the growth of smaller perturbations is almost completely suppressed; after t_{eq} both large and small fluctuations grow as $\delta \propto a$. Because of these differences, the size of the Hubble radius at t_{eq}, $\lambda_{\mathrm{eq}} \simeq 13 \, (\Omega h^2)^{-1} \mathrm{Mpc}$ (in terms of mass, $M_{\mathrm{eq}} \simeq 3.2 \times 10^{14} (\Omega h^2)^{-2} \, \mathrm{M}_\odot$) separates two different spectral regimes. In the wavelength range $\lambda_{\mathrm{FS}} \leq \lambda \leq \lambda_{\mathrm{eq}}$ the growth of fluctuations pauses between the time they enter the Hubble radius and t_{eq}. As a result the slope of the processed spectrum is changed, and $P(k) \propto k^{n_s - 4} \simeq k^{-3}$. Instead, at scales $\lambda > \lambda_{\mathrm{eq}}$ all the fluctuations keep growing at the same rate at all times, and the shape of the power spectrum remains similar to the primordial one, $P(k) \propto k^{n_s} \simeq k^1$.

WMAP (Bennett et al. 2003) measured the spectral index, obtaining $n_s = 0.99 \pm 0.04$, and did not detect deviations from Gaussianity, both results in agreement with inflationary predictions.

This kind of spectrum, in which fluctuations are typically larger on small scales, leads naturally to hierarchical structure formation, since small-scale fluctuations are the first to become non-linear (i.e., to reach $\delta \sim 1$), collapse, and form some kind of astronomical object. It is also worth remarking that the very first objects, coming from the highest peaks of $\delta(\mathbf{x})$, are typically located where modes $\delta(\mathbf{k})$ of different wavelength make some kind of "constructive interference": the very first objects are likely to

Figure 5.3. Map of the temperature anisotropies in the CMB as obtained by combining the five bands of the WMAP satellite in order to minimize foreground contamination. (From Bennett et al. 2003.)

be on top of larger ones, and they are likely to be clustered together, rather than uniformly distributed. For this reason, it is also very likely that the halos where these objects formed have long since been incorporated inside larger objects, such as the bulges of M_* galaxies or the cD galaxy at the center of galaxy clusters (see e.g., White & Springel 1999).

5.2.1.3 *Fluctuation Evolution - Baryons*

Before the equivalence epoch t_{eq} the baryons evolve in the same way as dark matter. Instead, in the matter dominated era they behave differently: we mentioned that all the dark matter fluctuations that were not erased by free streaming grow as $\delta \propto a \propto (1+z)^{-1}$, but this does not apply to baryons. In fact, baryons decouple from radiation only at t_{dec}, significantly later than t_{eq} (we remember that $1 + z_{eq} \sim 10^4$, while $1 + z_{dec} \simeq 10^3$). The persistence of the coupling with radiation prevents the growth of baryonic fluctuations on all scales; even worse, on relatively small scales all the fluctuations in the baryonic component are erased through a mechanism similar to free streaming. Such effect takes place below the so-called Silk scale (Silk 1968), which is given by the average distance that the photons (and the baryons coupled with them) can diffuse before $t = t_{dec}$; this translates into a comoving distance

$$\lambda_S \simeq 3.5 \left(\frac{\Omega}{\Omega_b} \right)^{1/2} (\Omega h^2)^{-3/4} \text{ Mpc} \qquad (5.7)$$

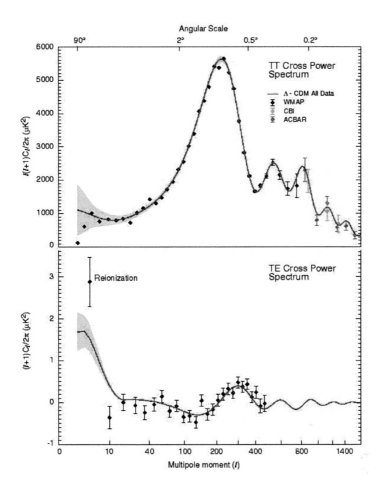

Figure 5.4. Angular CMB power spectrum of temperature (top panel) and temperature-polarization (bottom panel) as obtained by the WMAP satellite (from Bennett et al. 2003). The line shows the best fit with a ΛCDM model, and grey areas represent cosmic variance.

(where Ω_b is the baryonic contribution to Ω) and encloses a mass

$$M_S \simeq 6.2 \times 10^{12} \left(\frac{\Omega}{\Omega_b} \right)^{3/2} (\Omega h^2)^{-5/4} \, M_\odot. \qquad (5.8)$$

This result was a major problem for cosmology before the existence of dark matter started to be assumed: it implies that in a purely baryonic universe there should be no structures on a scale smaller than those of galaxy clusters (if $\Omega = \Omega_b \simeq 0.1$). Furthermore, even fluctuations that were not erased can grow only by a factor $1 + z_{dec}$ between decoupling and present time, and this is not enough to take a typical CMB fluctuation (of amplitude $\delta \sim 10^{-5}$) into the non-linear regime. The introduction of CDM solved this problem, since after recombination the baryons are finally free to fall inside dark matter potential wells, whose growth was unaffected by the photon drag.

It can be found that after decoupling from radiation, the baryonic fluctuations quickly "reach" the levels of dark matter fluctuations, evolving as

$$\delta_b = \delta_{DM} \left(1 - \frac{1+z}{1+z_{dec}} \right), \qquad (5.9)$$

so that the existing dark matter potential wells "regenerate" baryonic fluctuations, including the ones below the Silk scale.

This result is correct as long as pressure does not play a role, that is for objects with a mass larger than the cosmological Jeans mass $M_J \propto T^{3/2} \rho^{-1/2}$. Such mass behaves differently at high and low redshift. Before a redshift $z_{Compton} \simeq 130$ we have that the temperature of the baryons is still coupled to that of the CMB because of Compton scattering of radiation on the residual free electrons; for this reason, $T_b(z) \simeq T_{CMB}(z) \propto (1+z)$, and as $\rho(z) \propto (1+z)^3$ the value of M_J is constant:

$$M_J(z) \simeq 1.35 \times 10^5 \left(\frac{\Omega_m h^2}{0.15} \right)^{-1/2} M_\odot \qquad \text{(for } z_{dec} \gtrsim z \gtrsim z_{Compton}\text{)}$$
$$(5.10)$$

where $\Omega_m = \Omega_b + \Omega_{DM}$ is the total matter density (baryons plus dark matter). At lower redshifts the baryon temperature is no longer locked to that of the CMB and drops adiabatically as $T_b \propto (1+z)^2$. At such redshifts the Jeans mass evolves as

$$M_J(z) \simeq 5.7 \times 10^3 \left(\frac{\Omega_m h^2}{0.15} \right)^{-1/2} \left(\frac{\Omega_b h^2}{0.022} \right)^{-3/5} \left(\frac{1+z}{10} \right)^{3/2} M_\odot \quad (5.11)$$
$$\text{(for } z \lesssim z_{Compton}\text{)}.$$

5.2.2 From Fluctuations to Cosmological Structures

5.2.2.1 Non-Linear Evolution

When the density contrast δ becomes comparable to unity, the evolution becomes non-linear, and the Fourier modes no longer evolve independently. The easiest way to study this phase is through the "real space" $\delta(\mathbf{x})$ (rather than its Fourier components $\delta(\mathbf{k})$), considering the idealized case of the collapse of a spherically symmetric overdensity, and in particular the collapse

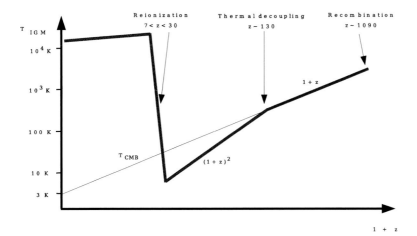

Figure 5.5. Schematic evolution of the temperature of the InterGalactic Medium (IGM) as a function of redshift: after recombination and before thermal decoupling (at z_{Compton}), T_{IGM} is locked to T_{CMB} and evolves as $(1+z)$. After thermal decoupling T_{IGM} is no longer locked to the radiation temperature, and drops adiabatically as $(1+z)^2$, until the first non-linear objects collapse, and emit light which is able to re-ionize the universe, raising T_{IGM} above $\sim 10^4$ K.

of a *top-hat* fluctuation. It is well known that, through some simple further assumption (such as that "spherical shells do not cross"), the time evolution of the radius R of a top-hat perturbation (see e.g., Padmabhan 1993, §8.2 for the treatment of a slightly more general case) can be written down as

$$R(t) = \frac{R_i}{2} \frac{1 + \delta_i}{\delta_i - (\Omega_i^{-1} - 1)} [1 - \cos \theta(t)] \qquad (5.12)$$

where $R_i = R(t_i)$, $\delta_i = \delta(t_i)$ and $\Omega_i = \Omega(t_i)$ are the "initial conditions" for the fluctuation evolution, and θ is defined by the equation

$$[\theta(t) - \sin \theta(t)] \frac{1 + \delta_i}{2H_i \Omega_i^{1/2} [\delta_i - (\Omega_i^{-1} - 1)]^{3/2}} \simeq t \qquad (5.13)$$

where again H_i is the Hubble parameter at t_i, and the last approximate equality is valid only as long as $\delta_i \ll 1$ (that is, a sufficiently early t_i must be chosen). The fluctuation radius R reaches a maximum R_{ta} at the so-called *turn-around* epoch (when $\theta = \pi$) when the overdense region finally detaches itself from the Hubble flow and starts contracting. However, while Equation (5.12) suggests an infinite contraction to $R = 0$ (when $\theta = 2\pi$), the violent relaxation process (Lynden-Bell 1967) prevents this from happening, leading to a configuration in virial equilibrium at $R_{\text{vir}} \simeq R_{\text{ta}}/2$.

Here, we summarize some well-known, useful findings of this model.

First of all, combining the evolution of the background density evolution and of Equation (5.12) it is possible to estimate the density contrast evolution

$$\delta(t) = \frac{9}{2} \frac{(\theta - \sin\theta)^2}{(1 - \cos\theta)^3} - 1 \tag{5.14}$$

which leads to some noteworthy observations, such as that the density contrast at turn-around is $\delta_{ta} = (9\pi^2/16) - 1 \simeq 4.6$, which at virialization becomes $\delta_{vir} = \Delta_c$, where it can be usually assumed that $\Delta_c \simeq 18\pi^2$ but sometimes higher-order approximations are necessary, such as the one in Bryan & Norman 1998,

$$\Delta_c = 18\pi^2 + 82(1 - \Omega_m^z) - 39(1 - \Omega_m^z)^2 \tag{5.15}$$

with

$$\Omega_m^z = \frac{\Omega_m(1+z)^3}{\Omega_m(1+z)^3 + \Omega_\Lambda + \Omega_k(1+z)^2} \tag{5.16}$$

where Ω_Λ is the dark energy density, and $\Omega_k = 1-\Omega_m-\Omega_\Lambda$ is the curvature. From this, it is possible to estimate the virial radius

$$R_{vir} \simeq 0.784 \left(\frac{M}{10^8 h^{-1} M_\odot}\right)^{1/3} \left(\frac{\Omega_m}{\Omega_m^z} \frac{\Delta_c}{18\pi^2}\right)^{-1/3} \left(\frac{1+z}{10}\right)^{-1} h^{-1} \text{ kpc}, \tag{5.17}$$

the circular velocity for such a halo

$$V_{circ} \simeq 23.4 \left(\frac{M}{10^8 h^{-1} M_\odot}\right)^{1/3} \left(\frac{\Omega_m}{\Omega_m^z} \frac{\Delta_c}{18\pi^2}\right)^{1/6} \left(\frac{1+z}{10}\right)^{1/2} \text{ km s}^{-1}, \tag{5.18}$$

and the virial temperature

$$T_{vir} \simeq 19800 \left(\frac{M}{10^8 h^{-1} M_\odot}\right)^{2/3} \left(\frac{\Omega_m}{\Omega_m^z} \frac{\Delta_c}{18\pi^2}\right)^{1/3} \left(\frac{1+z}{10}\right) \left(\frac{\mu}{0.6}\right) \text{ K.} \tag{5.19}$$

5.2.2.2 The Press–Schechter Formalism

The simple top-hat model is at the core of the so-called Press–Schechter formalism (Press & Schechter 1974, but see also the contribution by Sommerville in this book), which predicts the density of virialized halos of a given mass at a given redshift. This model assumes that the distribution of the smoothed density field δ_M (where M is the mass scale of the smoothing) at a certain redshift z_0 is Gaussian with a variance σ_M, so that the probability of having δ_M larger than a given δ_{crit} is

$$P(\delta_M > \delta_{crit}) = \int_{\delta_{crit}}^{\infty} 1/(2\pi)^{1/2} \sigma_M e^{-x^2/2\sigma_M^2} \, dx; \tag{5.20}$$

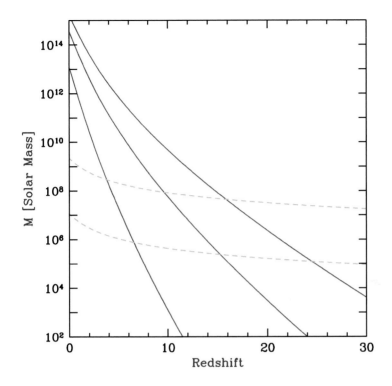

Figure 5.6. Characteristic mass of 1σ (bottom solid curve), 2σ (middle solid curve), and 3σ (top solid curve) collapsing halos as a function of redshift. These were obtained from the Eisenstein & Hu (1999) power spectrum, assuming an $\Omega_\Lambda = 0.7$, $\Omega_m = 0.3$ cosmology. The dashed curves show the minimum mass that is required for the baryons to be able to cool and collapse (see next section) in the case of pure atomic cooling (upper curve) and of molecular cooling (lower curve). (From Barkana & Loeb 2001.)

a common choice is $z_0 = 0$, requiring δ_M to be estimated through a purely linear evolution of the primordial power spectrum.

The Press–Schechter model then chooses a $\delta_{\text{crit}} = \delta_{\text{crit}}(z)$ (but it is also possible to assume a constant δ_{crit} and make σ_M a function of redshift; see e.g., Viana & Liddle 1996) and assumes that this probability (multiplied by a factor of 2 - see Bond et al. 1991 for an explanation of this extra factor) also gives the fraction mass which at a redshift z is inside virialized halos of mass M or larger. This can be differentiated over M in order to

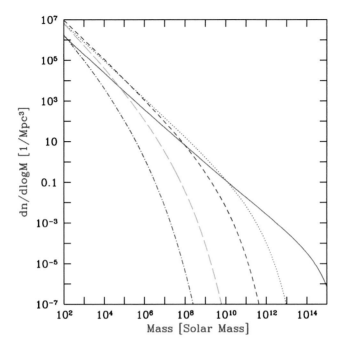

Figure 5.7. Halo mass functions at several redshifts (from bottom left to bottom right, $z = 30$, $z = 20$, $z = 10$, $z = 5$, and $z = 0$, respectively). The assumed power spectrum and cosmology are the same as in Figure 5.6. (From Barkana & Loeb 2001.)

get the mass distribution at each given redshift

$$\frac{dn}{dM} = \frac{2}{(2\pi)^{1/2}} \frac{\rho_m}{M} \frac{dln(1/\sigma_M)}{dM} \frac{\delta_{\text{crit}}(z)}{\sigma_M} e^{-\frac{\delta_{\text{crit}}(z)^2}{2\sigma_M^2}} \qquad (5.21)$$

In this way, the abundance of halos is completely determined through the two functions $\delta_{\text{crit}}(z)$ and σ_M. The first one is commonly written as $\delta_{\text{crit}}(z) = \delta_0/D(z)$, where $D(z)$ is the growth factor ($D(z) \simeq (1 + z)^{-1}$ for Einstein–de Sitter models; see Peebles 1993 for a more general expression), coming from cosmology; instead δ_0 is usually taken to be 1.686, since the top-hat model predicts that an object virializes at the time when the linear theory estimates its overdensity at $\delta = 1.686$. Instead σ_M depends on the power spectrum; for example, Figures 5.6 and 5.7 are based on the Eisenstein & Hu (1999) results.[2]

---

[2] The authors of this paper also provide some very useful codes for dealing

5.3 Primordial Gas Properties

5.3.1 Cooling

The typical densities reached by the gas after virialization (of the order of $n_B \equiv \rho_B/m_H \sim 0.01\Omega_b[(1 + z_{vir})/10]^3 cm^{-3}$ are far too low for the gas to condense and form an object like a star. The only way to proceed further in the collapse and in the formation of luminous objects is to remove the gas thermal energy through radiative cooling.

For this reason, cooling processes are important in determining where, when, and how the first luminous objects will form.

In Figure 5.8 it is possible to see that the cooling of primordial (i.e., metal-free) *atomic* gas at temperatures below $\sim 10^4$ K is dramatically inefficient, because in that temperature range the gap between the fundamental and the lowest excited levels ($\simeq 10.2$ eV for H atoms) is so much larger than the thermal energy $\sim k_B T \lesssim 1$ eV that very few atoms get collisionally excited, and very few photons are emitted through the corresponding de-excitations.

This is important because in all hierarchical scenarios the first objects to virialize are the smallest ones, and such halos have the lowest virial temperatures. If the primordial gas were completely atomic, the first luminous objects would probably form relatively late ($z \lesssim 10 - 15$), in moderately massive halos ($M \sim 10^8$ M$_\odot$) with $T_{vir} \gtrsim 10^4$ K.

However, it is also possible to see from Figure 5.8 that the presence of molecules in small amounts ($f_{H_2} \equiv 2n_{H_2}/(n_H + 2n_{H_2}) \gtrsim 5 \times 10^{-4}$; the dashed curve in Figure 5.8 was obtained assuming $f_{H_2} = 10^{-3}$) can dramatically affect the cooling properties of primordial gas at low temperatures, making low mass halos virializing at high redshift ($z \gtrsim 20$) the most likely sites for the formation of the first luminous objects.

5.3.2 Molecular Cooling

In the current scenario for the formation of primordial objects, the most relevant molecule is H_2. The only reason for this is the high abundance of H_2 when compared with all other molecules. In fact, the radiating properties of an H_2 molecule are very poor: because of the absence of a dipole moment, radiation is emitted only through weak quadrupole transitions. In addition, the energy difference between the H_2 ground state and the lowest H_2 excited roto-vibrational levels is relatively large ($\Delta E_{01}/k_B \gtrsim 200$ K, between the fundamental and the first excited level; however, such transition is prohibited by quadrupole selection rules, and the lowest energy gap for a quadrupole transition is $\Delta E_{02}/k_B \simeq 510$ K), further reducing the cooling efficiency at low temperatures.

with the power spectrum and the Press–Schechter formalism at the Web page http://background.uchicago.edu/~whu/transfer/transferpage.html.

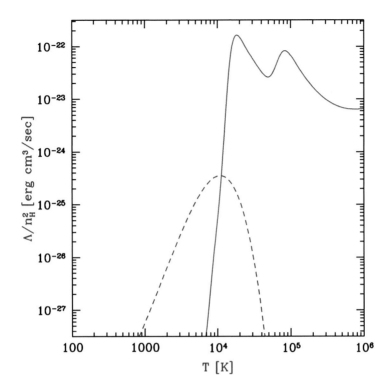

Figure 5.8. Cooling rate per atomic mass unit of metal-free gas, as a function of temperature. The solid line shows the gas assumed to be completely atomic (the two peaks correspond to H and He excitations, while the high temperature tail is dominated by free-free processes) and drops to about zero below $T \sim 10^4$ K; the dashed line shows the contribution of a small ($f_{H_2} = 10^{-3}$) fraction of molecular hydrogen, which contributes extra cooling in the range 100 K $\lesssim T \lesssim 10^4$ K. (From Barkana & Loeb 2001.)

Apart from H_2 the most relevant molecular coolants are HD and LiH. In the following, we will briefly list the cooling rates (mainly taken from Galli & Palla 1998, hereafter GP98; see also Hollenbach & McKee 1979, Lepp & Shull 1983, 1984, and Martin et al. 1996, Le Bourlot et al. 1999, Flower et al. 2000) of H_2 and of the other two possibly relevant species.[3]

We note that Bromm et al. 1999 included HD cooling in some of their

---

[3] In Figure 5.10 the cooling rate for H_2^+ is shown, too. But while it is still marginally possible that HD or LiH cooling can play some kind of role in the primordial universe, this is much more unlikely for H_2^+, mainly because its underabundance with respect to H_2 is always much larger than the difference in the cooling rates; for this reason we choose to omit a detailed discussion of H_2^+ cooling.

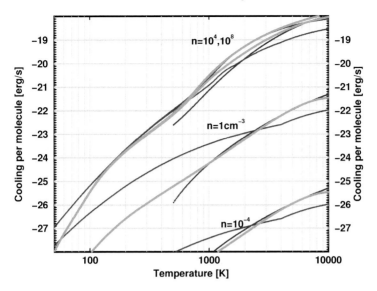

Figure 5.9. Cooling rate per H_2 molecule as computed by different authors (Lepp & Shull 1983, Martin et al. 1996, GP98) as a function of temperature and for different densities; note that for $n \gtrsim 10^4$ cm^{-3} the cooling rate is almost independent of density.

simulations but found that it never accounted for more than $\sim 10\%$ of H_2 cooling; however, they did not completely rule out the possibility that an n_{HD}/n_{H_2} ratio substantially larger than the equilibrium value (close to $n_D/n_H \sim 10^{-5}$) could change this conclusion. Also, in present-day simulations of primordial star formation the gas temperature never goes below a few hundred degrees: in case it did, even a tiny amount of LiH could be enough to dominate the cooling rate.

5.3.2.1 H_2 Cooling Rate

The H_2 cooling rate *per molecule* $\Lambda_{H_2}(\rho, T)$ can be conveniently expressed in the form:

$$\Lambda_{H_2}(\rho, T) = \frac{\Lambda_{H_2,\text{LTE}}(T)}{1 + \frac{\Lambda_{H_2,\text{LTE}}(T)}{n_H \Lambda_{H_2,\rho \to 0}(T)}} \qquad (5.22)$$

where $\Lambda_{H_2,\text{LTE}}(T)$ and $n_H \Lambda_{H_2,\rho \to 0}$ are the high and low density limits of the cooling rate (which apply at $n \gtrsim 10^4$ cm^{-3} and at $n \lesssim 10^2$ cm^{-3}, respectively).

The high-density (or LTE, from the Local Thermal Equilibrium assumption which holds in these conditions) limit of the cooling rate per H_2

molecule is given by Hollenbach & McKee (1979):

$$\Lambda_{H_2,LTE}(T) = \frac{9.5 \times 10^{-22} T_3^{3.76}}{1 + 0.12\, T_3^{2.1}} e^{-\left(\frac{0.13}{T_3}\right)^3} + 3 \times 10^{-24} e^{-0.51/T_3} +$$

$$+6.7 \times 10^{-19} e^{-5.86/T_3} + 1.6 \times 10^{-18} e^{-11.7/T_3} \;\; erg\, s^{-1} \quad (5.23)$$

where $T_3 \equiv T/(1000\ K)$. Note that the first row in the formula accounts for rotational cooling, while the second row accounts for the first two vibrational terms.

For the low-density limit, GP98 found that in the relevant temperature range ($10\ K \leq T \leq 10^4\ K$) the cooling rate $\Lambda_{H_2,\rho\to0}$ is independent of density, and is well approximated by

$$\log \Lambda_{H_2,\rho\to0}(T) \simeq -103.0 + 97.59 \log T - 48.05(\log T)^2 +$$
$$10.80(\log T)^3 - 0.9032(\log T)^4 \quad (5.24)$$

where T and $\Lambda_{H_2,\rho\to0}$ are expressed in K and $erg\, s^{-1}\, cm^3$, respectively.

Note that even if both $\Lambda_{H_2,LTE}$ and $\Lambda_{H_2,\rho\to0}$ do not depend on density, $\Lambda_{H_2}(\rho, T)$ is independent of ρ only in the high density limit.

5.3.2.2 HD and LiH Cooling Rates

The cooling rates of HD and of LiH are more complicated (see Flower et al. 2000 for HD and Bogleux & Galli 1997 for LiH), but in the low-density limit (and in the temperature range $10\ K \leq T \leq 1000\ K$) it is possible to use the relatively simple expressions given by GP98.

For HD, we have that the low density limit of the cooling rate per molecule, $\Lambda_{HD,\rho\to0}$, is:

$$\Lambda_{HD,\rho\to0}(T) \simeq 2\gamma_{10}E_{10}e^{-E_{10}/k_B T} + (5/3)\gamma_{21}E_{21}e^{-E_{21}/k_B T} \quad (5.25)$$

where E_{10} and E_{21} are the energy gaps between HD levels 1 and 0 and levels 2 and 1, respectively; they are usually expressed as $E_{10} = k_B T_{10}$ and $E_{21} = k_B T_{21}$, with $T_{10} \simeq 128\ K$ and $T_{21} \simeq 255\ K$, γ_{10} and γ_{21} are the approximate collisional de-excitation rates for the 1–0 and 2–1 transitions, and are given by

$$\gamma_{10} \simeq 4.4 \times 10^{-12} + 3.6 \times 10^{-13} T^{0.77} \quad (5.26)$$
$$\gamma_{21} \simeq 4.1 \times 10^{-12} + 2.1 \times 10^{-13} T^{0.92} \quad (5.27)$$

where we use the numerical value of T (in Kelvin) and the rates are expressed in $cm^3\, s^{-1}$.

For LiH instead we have that the same density limit of the cooling rate per molecule, $\Lambda_{LiH,\rho\to0}$, can be fitted by:

$$\log_{10}(\Lambda_{LiH,\rho\to0}) = c_0 + c_1(\log_{10} T) + c_2(\log_{10} T)^2 +$$
$$c_3(\log_{10} T)^3 + c_4(\log_{10} T)^4 \quad (5.28)$$

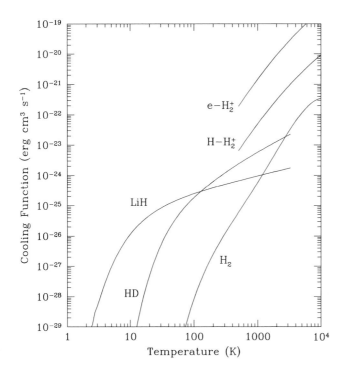

Figure 5.10. Comparison of the low-density $(n \lesssim 10 \text{ cm}^{-3})$ cooling rates per molecule of several molecular species, in particular H_2, HD, and LiH (from GP98). Note that at $T \sim 100$ K both HD and LiH molecules are more than 10^3 times more efficient coolants than H_2 molecules, but this difference is believed to be compensated by the much higher H_2 abundance (see e.g., Bromm et al. 2002). The plot also shows the cooling due to H–H_2^+ and e^-–H_2^+ collisions, but these contributions are never important because of the very low H_2^+ abundance.

where $c_0 = -31.47$, $c_1 = 8.817$, $c_2 = -4.144$, $c_3 = 0.8292$, and $c_4 = -0.04996$, assuming that T is expressed in Kelvin, and that $\Lambda_{\text{LiH},\rho \to 0}$ is expressed in $\text{erg cm}^3 \text{ s}^{-1}$.

5.3.2.3 Cooling at High Densities: Collision-Induced Emission

During the formation of a protostar an important role is played by the so-called Collision-Induced Emission (CIE; very often known as Collision-Induced Absorption, or CIA), a process that requires pretty high densities $(n = \rho/m_{\text{H}} \gtrsim 10^{13} - 10^{14} \text{ cm}^{-3})$ to become important (Lenzuni et al. 1991; Frommold 1993; Ripamonti & Abel 2004). In fact, CIE takes place when

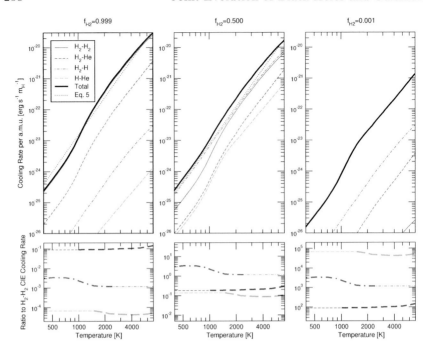

Figure 5.11. CIE cooling for different collisions and different molecular fractions(0.999, 0.5, 0.001). Top panels show the cooling rates per unit mass: the thick solid line is the total CIE cooling, while thin lines show the various components: H_2–H_2 (solid), H_2–He (short dashed), H_2-H (dot dashed) and H-He (long dashed); the dotted line shows the results of the approximate formula given in the text. In the bottom panels the ratios of the various components of H_2–H_2 CIE is shown. All quantities were calculated assuming $n = 10^{14}$ cm^{-3} and $X = 0.75$. (From Ripamonti & Abel 2004.)

a collision between two H_2 molecules (or H_2 and H, or H_2 and He, or even H and He) takes place: during the collision, the interacting pair briefly acts as a "super-molecule" with a nonzero electric dipole, and a much higher probability of emitting (CIE) or absorbing (CIA) a photon than an unperturbed H_2 molecule (whose dipole is 0). Because of the very short durations of the collisions ($\lesssim 10^{-12}$ s), such a mechanism can compete with "normal" emission only in high density environments. Furthermore, because of the short durations of the interactions, collision-induced lines become very broad and merge into a continuum: the H_2 CIE spectrum only shows broad peaks corresponding to vibrational bands, rather than a large number of narrow roto-vibrational lines. This is also important because self-absorption is much less relevant than for line emission, and in primordial proto-stars simulations CIE cooling can be treated with the optically thin approximation up to $n \sim 10^{16}$ cm^{-3} (for H$_2$ lines, the optically thin approximation breaks at about $n \sim 10^{10}$ cm^{-3}). In Figure 5.11 we show the CIE cooling rate for a gas with $n = 10^{14}$ cm^{-3}, as a function of temperature and for different chemical compositions; at temperatures between 400 and 7000 K. For H$_2$ abundances $f_{H_2} \equiv 2n_{H_2}/(n_{H^+} + n_H + 2n_{H_2}) \gtrsim 0.5$ the total CIE cooling rate can be approximated by the simple expression (see Ripamonti & Abel 2004)

$$L_{CIE}(\rho, T, X, f_{H_2}) \simeq 0.072 \rho T^4 X f_{H_2} \text{ erg g}^{-1}\text{s}^{-1} \qquad (5.29)$$

where the density is in g cm^{-3}, the temperature is in K, and $X \simeq 0.75$ is the hydrogen fraction (by mass).

5.3.3   Chemistry

Since molecules play such an important role in the formation of the first luminous objects, it is important to include a proper treatment of their abundance evolution. However, a full treatment should take into account ~ 20 different species and ~ 100 different reactions. For instance, GP98 give a chemical network of 87 reactions, which includes e^-, H, H$^+$, H$^-$, D, D$^+$, He, He$^+$, He^{++}, Li, Li$^+$, Li$^-$, H$_2$, H$_2^+$, HD, HD$^+$, HeH$^+$, LiH, LiH$^+$, H$_3^+$, and H$_2$D$^+$; even their *minimal model* is too complicated to be described here, so we will just describe the most basic processes involved in H$_2$ formation. However, we note that the papers by Abel et al. (1997)[4] and by GP98 provide a much more accurate description of primordial chemistry.

5.3.3.1   *Atomic Hydrogen and Free Electrons*

Apart from molecule formation (see below), the main reactions involving hydrogen (and helium, to which we can apply all the arguments below) are ionizations and recombinations.

   Ionizations can be produced both by radiation (H $+ h\nu \rightarrow$ H$^+ + e^-$) and by collisions, mainly with free electrons (H $+ e^- \rightarrow$ H$^+ + 2e^-$) but also with other H atoms (H $+$ H $\rightarrow$ H$^+ e^- +$ H). Photoionizations dominate over collisions as long as UV photons with energy above the H ionization threshold are present (see e.g., Osterbrock 1989), but this is not always the case before the formation of the first luminous objects, when only the CMB radiation is present; even after some sources of radiation have appeared, it is likely that their influence will be mainly local, at least until reionization.

---

[4] The collisional rate coefficients given in the Abel et al. (1997) paper can be readily obtained through a FORTRAN code available on the Web page http://www.astro.psu.edu/users/tabel/PGas/LCA-CM.html; also note that on Tom Abel's Web site (http://www.tomabel.com/) it is possible to find much useful information about the primordial universe in general.

However, in the primordial universe the electron temperature is low, and collisional ionizations are relatively rare.

Recombinations ($H^+ + e^- \to H + h\nu$) have much higher specific rates, since they do not have a high energy threshold. They probably dominate the evolution of free electrons, which can be relatively abundant as residuals from the recombination era (e.g., Peebles 1993), and are important for H_2 formation (see below).

Simple approximations for the reaction rates of collisional ionizations and recombinations are given in Table 5.1.

Table 5.1. Reaction rates for some of the most important reactions in primordial gas, plus the main reactions involved in the formation of HD. In the formulae, T_γ is the temperature of the radiation field (for our purposes, the temperature of the CMB radiation) and temperatures need to be expressed in Kelvin; $j(\nu_{\mathrm{LW}})$ is the radiation flux (in $\mathrm{erg\,s^{-1}\,cm^{-2}}$) at the central frequency ν_{LW} of the Lyman–Werner bands, $h\nu_{\mathrm{LW}} = 12.87$ eV. The rates come from compilations given by Tegmark et al. 1997 (reactions 1–7), Palla et al. 1983 (reactions 8–13), Abel et al. 1997 (reaction 14) and Bromm et al. 2002 (reactions 15–19).

Reaction			Rate
$H^+ + e^-$	$\to$	$H + h\nu$	$k_1 \simeq 1.88 \times 10^{-10} T^{-0.644} \ \mathrm{cm^3\,s^{-1}}$
$H + e^-$	$\to$	$H^- + h\nu$	$k_2 \simeq 1.83 \times 10^{-18} T^{0.88} \ \mathrm{cm^3\,s^{-1}}$
$H^- + H$	$\to$	$H_2 + e^-$	$k_3 \simeq 1.3 \times 10^{-9} \ \mathrm{cm^3\,s^{-1}}$
$H^- + h\nu$	$\to$	$H + e^-$	$k_4 \simeq 0.114 T^{2.13} e^{-8650/T_\gamma} \ \mathrm{cm^3\,s^{-1}}$
$H^+ + H$	$\to$	$H_2^+ + h\nu$	$k_5 \simeq 1.85 \times 10^{-23} T^{1.8} \ \mathrm{cm^3\,s^{-1}}$
$H_2^+ + H$	$\to$	$H_2 - +H^+$	$k_6 \simeq 6.4 \times 10^{-10} \ \mathrm{cm^3\,s^{-1}}$
$H_2^+ + h\nu$	$\to$	$H^+ + H$	$k_7 \simeq 6.36 \times 10^5 e^{-71600/T_\gamma} \ \mathrm{cm^3\,s^{-1}}$
$H + H + H$	$\to$	$H_2 + H$	$k_8 \simeq 5.5 \times 10^{-29} T^{-1} \ \mathrm{cm^6\,s^{-1}}$
$H_2 + H$	$\to$	$H + H + H$	$k_9 \simeq 6.5 \times 10^{-7} T^{-1/2} e^{-52000/T} \times$
			$\times (1 - e^{-6000/T}) \ \mathrm{cm^6\,s^{-1}}$
$H + H + H_2$	$\to$	$H_2 + H_2$	$k_{10} \simeq k_8/8$
$H_2 + H_2$	$\to$	$H + H + H_2$	$k_{11} \simeq k_9/8$
$H + e^-$	$\to$	$H^+ + e^- + e^-$	$k_{12} \simeq 5.8 \times 10^{-11} T^{1/2} e^{-158000/T} \ \mathrm{cm^3\,s^{-1}}$
$H + H$	$\to$	$H^+ + e^- + H$	$k_{13} \simeq 1.7 \times 10^{-4} k_{12}$
$H_2 + h\nu$	$\to$	$H^+ + H$	$k_{14} \simeq 1.1 \times 10^8 j(\nu_{\mathrm{LW}}) \ \mathrm{s^{-1}}$
$D^+ + e^-$	$\to$	$D + h\nu$	$k_{15} \simeq 8.4 \times 10^{-11} T^{-0.5} \times$
			$\times (\frac{T}{10^3})^{-0.2} [1 + (\frac{T}{10^6})^{0.7}]^{-1} \ \mathrm{cm^3 s^{-1}}$
$D + H^+$	$\to$	$D^+ + H$	$k_{16} \simeq 3.7 \times 10^{-10} T^{0.28} e^{-43/T} \ \mathrm{cm^3 s^{-1}}$
$D^+ + H$	$\to$	$D + H^+$	$k_{17} \simeq 3.7 \times 10^{-10} T^{0.28} \ \mathrm{cm^3 s^{-1}}$
$D^+ + H_2$	$\to$	$H^+ + HD$	$k_{18} \simeq 2.1 \times 10^{-9} \ \mathrm{cm^3 s^{-1}}$
$HD + H^+$	$\to$	$H_2 + D$	$k_{19} \simeq 1.0 \times 10^{-9} e^{-464/T} \ \mathrm{cm^3 s^{-1}}$

5.3.3.2 H_2 *Formation and Disruption*

At present, H_2 is commonly believed to form mainly through reactions taking place on the surface of dust grains (but see e.g., Cazaux & Spaans 2004). Such a mechanism could not work in the primordial universe, when the metal (and dust) abundance was negligible. The first two mechanisms leading to formation of H_2 in a primordial environment were described by McDowell (1961) and by Saslaw & Zipoy (1967); soon after the publication of this latter paper, H_2 started to be included in theories of structure formation (such as the PD68 paper about globular cluster formation through H_2 cooling). Both these mechanisms consist of two stages, involving either e^- or H^+ as catalyzers. The first (and usually most important) goes through the reactions

$$H + e^- \rightarrow H^- + h\nu \tag{5.30}$$

$$H^- + H \rightarrow H_2 - +e^- \tag{5.31}$$

while the second proceeds as

$$H^+ + H \rightarrow H_2^+ + h\nu \tag{5.32}$$

$$H_2^+ + H \rightarrow H_2 - +H^+. \tag{5.33}$$

In both cases, H_2 production can fail at the intermediate stage if a photodissociation occurs ($H^- + h\nu \rightarrow H + e^-$, or $H_2^+ + h\nu \rightarrow H^+ + H$). The rates of all these reactions are listed in Table 5.1

By combining the reaction rates of these reactions, it is possible (see Tegmark et al. 1997; hereafter T97) to obtain an approximate evolution for the ionization fraction $x \equiv n_{H^+}/n$ and the H_2 fraction $f_{H_2} \equiv 2n_{H_2}/n$ (here n is the total density of protons, $n \simeq n_H + n_{H^+} + 2n_{H_2}$):

$$x(t) \simeq \frac{x_0}{1 + x_0 n k_1 t} \tag{5.34}$$

$$f_{H_2}(t) \simeq f_0 + 2\frac{k_m}{k_1}\ln(1 + x_0 n k_1 t) \qquad \text{with} \tag{5.35}$$

$$k_m = \frac{k_2 k_3}{k_3 + k_4/[n(1-x)]} + \frac{k_5 k_6}{k_6 + k_7/[n(1-x)]} \tag{5.36}$$

where the various k_i are the reaction rates given in Table 5.1, and x_0 and f_0 are the initial fractions of ionized atoms and of H_2 molecules.

Another mechanism for H_2 formation, which becomes important at (cosmologically) very high densities $n_H \gtrsim 10^8$ cm^{-3} are the so-called three-body reactions described by Palla et al. (1983). While the previous mechanisms are limited by the small abundance of free electrons (or of H^+), reactions such as

$$H + H + H \rightarrow H_2 + H \tag{5.37}$$

can rapidly convert all the hydrogen into molecular form, provided the density is high enough. For this reason, they are likely play an important role during the formation of a primordial protostar.

Finally, H_2 can be dissociated through collisional processes such as reactions 9 in Table 5.1, but probably the most important process is its photodissociation by photons in the Lyman–Werner bands (11.26–13.6 eV; but Abel et al. 1997 found that the most important range is between 12.24 and 13.51 eV). These photons are below the H ionization threshold, therefore they can diffuse to large distances from their source. So, any primordial source emitting a relevant number of UV photons (e.g., a $\sim 100 M_\odot$ star) is likely to have a major feedback effect, since it can strongly reduce the amount of H_2 in the halo where it formed (and also in neighboring halos), inhibiting further star formation.

5.3.3.3   *Approximate Predictions*

The above information can be used for making approximate predictions about the properties of the halos hosting the first luminous objects. Such kinds of predictions were started by Couchman & Rees (1986), and especially by T97 and their example was followed and improved by several authors (see below).

The basic idea is to have a simplified model for the evolution of the H_2 fraction and the cooling rate inside spherical top-hat fluctuations, in order to check whether after virialization the baryons are able to cool (and keep collapsing), or they just "settle down" at about the virial radius. Such approximate models are fast to compute, and it is easy to explore the gas behavior on a large range of virialization redshifts and fluctuation masses (or, equivalently, virial temperatures). In this way, for each virialization redshift it is possible to obtain the minimum molecular fraction that is needed for the collapse to proceed, and the minimum size of halos where such abundance is achieved. The results interestingly point out that the molecular fraction threshold separating collapsing and non-collapsing objects has an almost redshift-independent value of $f_{H_2} \sim 5 \times 10^{-4}$ (see Figure 5.12). Instead, the minimum halo mass actually evolves with redshift (see Figure 5.13).

Predictions about the ability of the baryons inside each kind of halo to keep collapsing after virialization can then be combined with Press–Schechter predictions about the actual abundances of halos. For instance, in Figure 5.13 the solid black (almost vertical) line shows where the masses of 3σ fluctuations lie, as a function of redshift. So, if we decide to neglect the rare fluctuations at more than 3σ from the average, that figure tells us that the first luminous objects can start forming only at $z \lesssim 30$, in objects with a total mass $\gtrsim 2 \times 10^6 \ M_\odot$.

Such a result is subject to a number of uncertainties, both about the

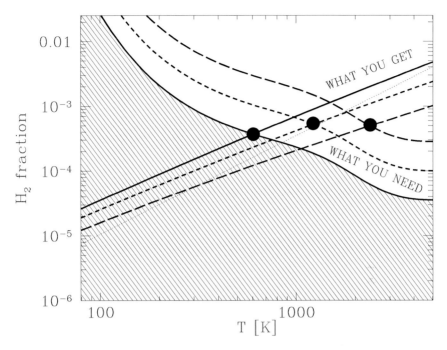

Figure 5.12. Comparison of the H_2 fraction needed for a halo to collapse and H_2 fraction that can be formed inside the halo in an Hubble time (from T97). This is shown as a function of halo virial temperature and for three different virialization redshifts ($z = 100$: solid; $z = 50$: short dashes; $z = 25$: long dashes). The three dots mark the minimum H_2 abundance which is needed for collapse at the three considered redshift, and it can be seen that they all are at $f_{H_2} \sim 5 \times 10^{-4}$.

"details" of the model and about the processes it neglects; here are some of the more interesting developments:

(i) Abel et al. (1998) found that the minimum mass is strongly affected by the uncertainties in the adopted H_2 cooling function, with differences that could reach a factor ~ 10 in the minimum mass estimate.
(ii) Fuller & Couchman (2000) used numerical simulations of single, high-σ density peaks in order to improve the spherical collapse approximation.
(iii) Machacek et al. (2001) and Kitayama et al. (2001) investigated the influence of background radiation.
(iv) Yoshida et al. (2003) used larger-scale numerical simulations and found that the merging history of an halo also could play a role, since frequent mergings heat the gas and prevent or delay the collapse of $\sim 30\%$ of the halos.

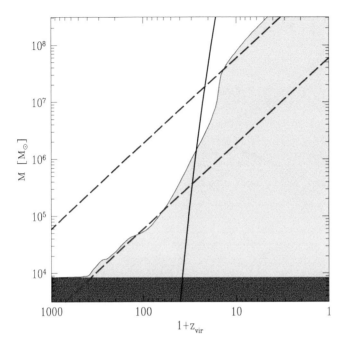

Figure 5.13. Evolution with virialization redshift of the minimum mass which is able to cool, collapse, and possibly form stars. Only the halos whose (z_{vir}, M) fall outside the filled region are able to cool fast enough. The filled region is made of a red part, where CMB radiation prevents the cooling, and a yellow part where the cooling is slow because of a dearth of H_2 molecules. The two parallel dashed lines correspond to virial temperatures of 10^4 K (the highest one) and 10^3 K (the lowest one), while the almost vertical line in the middle of the plot corresponds to 3σ peaks in a SCDM ($\Omega_m = 1$, $\Omega_\Lambda = 0$) cosmology.

As a result of the improved modeling (and of a different set of cosmological parameters as ΛCDM has substituted SCDM), in the most recent papers the value of the minimum halo mass for the formation of the first luminous objects is somewhat reduced to the range 0.5–1×10^6 $M_\odot$, with a weak dependence on redshift and a stronger dependence on other parameters (background radiation, merging history etc.).

5.4 Numerical Cosmological Hydrodynamics

Numerical simulations are an important tool for a large range of astrophysical problems. This is especially true for the study of primordial stars, given the absence of direct observational data about these stars, and the relatively scarce indirect evidence we have.

5.4.1 Adaptive Refinement Codes (ENZO)

The two problems that immediately emerge when setting up a simulation of primordial star formation are the dynamical range and the required accuracy in solving the hydrodynamical equations.

When studying the formation of objects in a cosmological context, we need both to simulate a large enough volume (with a box size of at least 100 comoving kpc), and to resolve objects of the size of a star ($\sim 10^{11}$ cm in the case of the Sun), about 11 orders of magnitude smaller.

This huge difference in the relevant scales of the problem is obviously a problem. It can be attenuated by the use of Smoothed Particle Hydrodynamics (SPH; see e.g., Monaghan 1992), whose Lagrangian nature has some benign effects, as the simulated particles are likely to concentrate (i.e., , provide resolution) in the regions where the maximum resolution is needed. However, even if this kind of method can actually be employed (see Bromm et al. 1999, 2002), it has at least two important drawbacks. First of all, the positive effects mentioned above cannot bridge in a completely satisfactory way the extreme dynamical range we just mentioned, since the mass resolution is normally fixed once and for all at the beginning of the simulation. Second, SPH is known to have poor shock resolution properties, which casts doubts on the results when the hydrodynamics becomes important.

The best presently available solution for satisfying both requirements is the combination of an Eulerian approach (which is good for the hydrodynamical part) and an adaptive technique (which can extend the dynamical range). This is known as Adaptive Mesh Refinement (AMR; see e.g., Berger & Colella 1989; Norman 2004), and basically consists in having the simulation volume represented by a hierarchy of nested *grids* (*meshes*) which are created according to resolution needs.

In particular, the simulations we are going to describe in the following paragraphs were made with the code ENZO[5] (see O'Shea et al. 2004 and references therein for a full description). Briefly, ENZO includes the treatment of gravitational physics through N-body techniques, and the treatment of hydrodynamics through the Piecewise Parabolic Method (PPM) of Woodward & Colella 1984 (as modified by Bryan et al. 1995 in order to adapt to cosmological simulations). ENZO can optionally include the treatment of gas cooling (in particular H_2 cooling, as described in the preceding sections), of primordial non-equilibrium chemistry (see the preceding sections) and of UV background models (e.g., the ones by Haardt & Madau 1996)

---

[5] The ENZO code can be retrieved at the Web site http://cosmos.ucsd.edu/enzo/

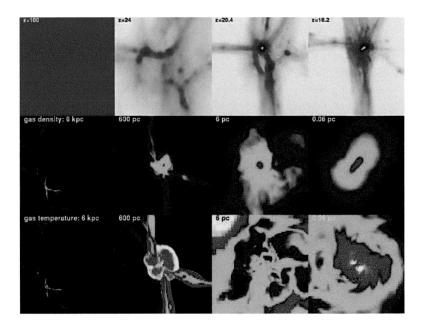

Figure 5.14. Overview of the evolution leading to the formation of a primordial star. The top row shows the gas density, centered at the pre-galactic object within which the star is formed. The four projections are labeled with their redshifts. Pre-galactic objects form from very small density fluctuations and continuously merge to form larger objects. The middle and bottom rows show thin slices through the gas density and temperature at the final simulation stage. The four pairs of slices are labelled with the scale on which they were taken, starting from 6 (proper) kpc (the size of the simulated volume) and zooming in down to 0.06 pc (12,000 AU). In the left panels, the larger scale structures of filaments and sheets are seen. At their intersections, a pre-galactic object of $\sim 10^6$ M$_\odot$ is formed. The temperature slice (second panel, bottom row) shows how the gas shock heats as it falls into the pre-galactic object. After passing the accretion shock, the material forms hydrogen molecules and starts to cool. The cooling material accumulates at the center of the object and forms the high-redshift analogue to a molecular cloud (third panel from the right), which is dense and cold ($T \sim 200 \, K$). Deep within the molecular cloud, a core of ~ 100 M$_\odot$, a few hundred K warmer, is formed (right panel) within which a ~ 1 M$_\odot$ fully molecular object is formed (yellow region in the right panel of the middle row). (From ABN02.)

and heuristic prescriptions (see Cen & Ostriker 1992) for star formation in larger-scale simulations.

5.4.2 Formation of the First Star

The use of AMR codes for cosmological simulations of the formation of the first objects in the universe was pioneered by Abel et al. (2000) and further refined by Abel et al. (2002; hereafter ABN02), where the dynamic range covered by the simulations was larger than 5 orders of magnitude in scale length, i.e., the wide range between an almost cosmological ($\gtrsim 100$ comoving kpc, i.e., $\gtrsim 5$ proper kpc) and an almost stellar scale ($\lesssim 1000$ AU $\sim 10^{-2}$ pc). In the whole process, the AMR code kept introducing new (finer resolution) meshes whenever the density exceeded some thresholds, or the Jeans length was resolved by less than 64 grid cells.

The simulations were started at a redshift $z = 100$ from cosmologically consistent initial conditions,[6] and the code also followed the non-equilibrium chemistry of the primordial gas, and included an optically thin treatment of radiative losses from atomic and molecular lines, and from Compton cooling.

The main limitation of these simulations was the assumption that the cooling proceeds in the optically thin limit; such assumption breaks down when the optical depth inside H_2 lines reaches unity (corresponding to a Jeans length $\sim 10^3$ AU ~ 0.01 pc). However, the simulations were halted only when the optical depth at line centers became larger than 10 (Jeans length of about ~ 10 AU ($\sim 10^{-4}$ pc), since it was unclear whether Doppler shifts could delay the transition to the optically thick regime.

5.4.2.1 *Summary of the Evolution: Radial Profiles*

In Figures 5.14 and 5.15 we show the evolution of gas properties both in pictures and in plots of spherically averaged quantities, as presented in Abel et al. (2002). From these figures, and in particular from the local minima in the infall velocity (Figure 5.15e) it is possible to identify four characteristic mass scales:

(i) The mass scale of the pre-galactic halo, of $\sim 7 \times 10^5 M_\odot$ in total, consistent with the approximate predictions discussed in the previous sections.

(ii) The mass scale of a "primordial molecular cloud," ~ 4000 $M_\odot$: the molecular fraction in this region is actually very low ($\lesssim 10^{-3}$), but it is

---

[6] Such conditions were taken from an SCDM model with $\Omega_\Lambda = 0$, $\Omega_m = 1$, $\Omega_b = 0.06$, $H_0 = 50$ km/s Mpc^{-1} which is quite different from the "concordance" ΛCDM model. However, the final results are believed to be only marginally affected by differences in these cosmological parameters.

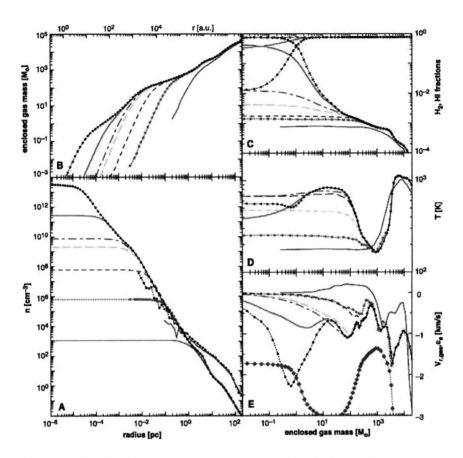

Figure 5.15. Radial mass-weighted averages of physical quantities at seven different simulation times. (A) Particle number density in cm^{-3} as a function of radius; the bottom line corresponds to $z = 19$, and moving upwards the "steps" from one line to the next are of 9×10^6 yr, 3×10^5 yr, 3×10^4 yr, 3000 yr, 1500 yr, and 200 yr, respectively; the uppermost line shows the simulation final state, at $z = 18.181164$. The two lines between 0.01 and 200 pc give the DM mass density (in $GeV\,cm^{-3}$) at $z = 19$ and the final time, respectively. (B) Enclosed gas mass. (C) Mass fractions of H and H_2. (D) Temperature. (E) Radial velocity of the baryons; the bottom line in (E) shows the negative value of the local speed of sound at the final time. In all panels the same output times correspond to the same line styles. (From ABN02.)

enough to reduce the gas temperature from the virial value ($\gtrsim 10^3$ K) to ~ 200 K.

(iii) The mass scale of a "fragment," ~ 100 $M_\odot$, which is determined by the change in the H_2 cooling properties at a density $n \sim 10^4$ cm^{-3} (for a complete discussion, see Bromm et al. 1999, 2002), when local thermal equilibrium is reached and the cooling rate dependence on density flattens to $\Lambda \propto n$ (from $\Lambda \propto n^2$); this is also the first mass scale where the gas mass exceeds the Bonnor–Ebert mass (Ebert 1955; Bonnor 1956) $M_{\mathrm{BE}}(T, n, \mu) \simeq 61$ $M_\odot T^{3/2} n^{-1/2} \mu^{-2}$ (with T in Kelvin and n in cm^{-3}), indicating an unstable collapse.

(iv) The mass scale of the "molecular core," ~ 1 $M_\odot$ is determined by the onset of three-body reactions at densities in the range $n \sim 10^8 - 10^{10}$ cm^{-3}, which leads to the complete conversion of hydrogen into molecular form; at this stage, the infall flow becomes supersonic (which is the precondition for the appearance of an accretion shock and of a central hydrostatic flow); the increase in the H_2 abundance due to the formation of this molecular core also leads to the transition to the optically thick regime.

5.4.2.2 Angular Momentum

In Figure 5.16 we show the evolution of the radial distribution of average specific angular momentum (and related quantities). It is remarkable that in ABN02 simulations rotational support does not halt the collapse at any stage, even if this could be a natural expectation.

There are two reasons for such an (apparently odd) fact:

(i) As can be seen in panel A of Figure 5.16, the collapse starts with the central gas having much less specific angular momentum than the average (this is typical of halos produced by gravitational collapse; see e.g., Quinn & Zurek 1988). That is, the gas in the central regions starts with little angular momentum to lose in the first place.

(ii) Second, some form of angular momentum transport is clearly active, as is demonstrated by the decrease in the central specific angular momentum. Turbulence is likely to be the explanation: at any radius, there will be both low and high angular momentum material, and redistribution will happen because of pressure forces or shock waves: lower angular momentum material will selectively sink inwards, displacing higher angular momentum gas. It is notable that this kind of transport will be suppressed in situations in which the collapse occurs on the dynamical time scale, rather than the longer cooling time scale: this is the likely reason why this kind of mechanism has not been observed in simulations of present day star formation (see e.g., . Burkert & Bodenheimer 1993).

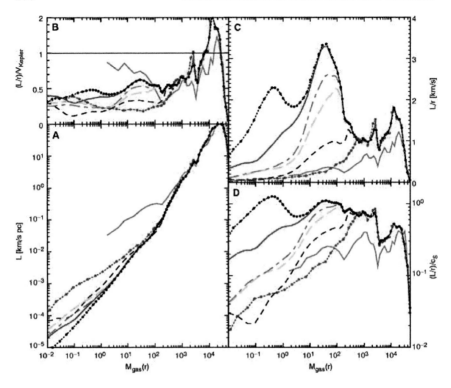

Figure 5.16. Radial mass weighted averages of angular momentum-related quantities at different times (the same as in Figure 5.15). (A) specific angular momentum L. (B) Rotational speed in units of Keplerian velocity $v_{\mathrm{Kep}} \equiv (GM_r/r)^{1/2}$. (C) Rotational speed (L/r). (D) Rotational speed in units of the sound speed c_s (From ABN02.)

5.4.3 SPH Results

Bromm et al. (1999, 2002) have performed simulations of primordial star formation using an SPH code that included essentially the same physical processes as the simulations we just described. Apart from the numerical technique, the main differences were that they included deuterium in some of their models, and that their initial conditions were not fully cosmological (e.g., , since they chose to study isolated halos, the angular momenta were assigned as initial conditions, rather than generated by tidal interactions with neighboring halo).

In Figures 5.17 and 5.18 we show the results of one of their simulations, which gives results in essential agreement with those we have previously discussed.

An interesting extra-result of this kind of simulation is that the authors are able to assess the mass evolution of the various gaseous clumps in the hypothesis that feedback is unimportant (which could be the case if the clumps directly form intermediate mass black holes without emitting much radiation, or if fragmentation to a quite small stellar mass scale happens). They find that a significant fraction (~ 0.5) of the halo gas should end up inside one of the gas clumps, although it is not clear at all whether this gas will form stars (or some other kind of object). Furthermore, they find that clumps are likely to increase their mass on a timescale of about 10^7 yr (roughly corresponding to the initial time scale of the simulated halo), both because of gas accretion and of mergers, and they could easily reach masses $\gtrsim 10^4$ $M_\odot$. Obviously, this result is heavily dependent on the not very realistic assumed lack of feedback.

5.5 Protostar Formation and Accretion

Full three-dimensional simulations (such as the ones of ABN02) are not able to reach the stage when a star is really formed. They are usually stopped at densities ($n \sim 10^{10} - 10^{11}$ cm^{-3}) which are much lower than typical stellar densities ($\rho \sim 1$ g cm^{-3}, $n \sim 10^{24}$ cm^{-3}). In fact, at low densities the gas is optically thin at all frequencies, and it is not necessary to include radiative transfer in order to estimate the gas cooling. Instead, at densities $n \gtrsim 10^{10}$ cm^{-3} some of the dominant H_2 roto-vibrational lines become optically thick and require the treatment of radiative transfer.

For this kind of problem, this is a prohibitive computational burden, and at present the actual formation of a protostar can not be fully investigated through self-consistent 3D simulations. In order to proceed further, it is necessary to introduce some kind of simplification in the problem.

5.5.1 Analytical Results

Historically, there were several studies based on analytical arguments (i.e., stability analysis) or single-zone models, leading to different conclusions about the properties of the final object.

An early example is the paper by Yoneyama (1972), in which it was argued that fragmentation takes place until the opacity limit is reached. However, Yoneyama looked at the opacity limit of the entire "cloud" (roughly corresponding to one of the 10^5–10^6 $M_\odot$ mini-halos we consider at present; originally, this was mass the scale proposed by PD68) rather than the putative fragments and arrived at the conclusion that fragmentation stopped for masses $\lesssim 60$ $M_\odot$.

More recently, Palla et al. 1983 pointed to the increase of the cooling rate at $n \gtrsim 10^8$ cm^{-3} (due to H_2 fast formation through three-body reactions) as a possible trigger for instability, leading to fragmentation on mass

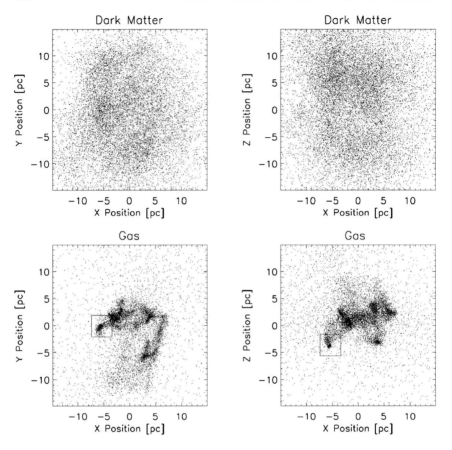

Figure 5.17. Typical result of the SPH simulations by Bromm et al. 2002. The figure shows the morphology of a simulated 2×10^6 M$_\odot$ halo virializing at $z \sim 30$ just after the formation of the first clump of mass 1400 M$_\odot$ (which is likely to produce stars). The two top row panels shows the distribution of the dark matter (which is undergoing violent relaxation). The two bottom panels show the distribution of gas, which has developed a lumpy morphology and settled at the center of the potential well.

scales of ~ 0.1 M$_\odot$; such instability has actually been observed in the simulations of Abel et al. (2003), but it does not lead to fragmentation because its growth is too slow (see also Sabano & Yoshii 1977, Silk 1983, Omukai & Yoshii 2003, and Ripamonti & Abel 2004 for analytical fragmentation criteria and their application to this case).

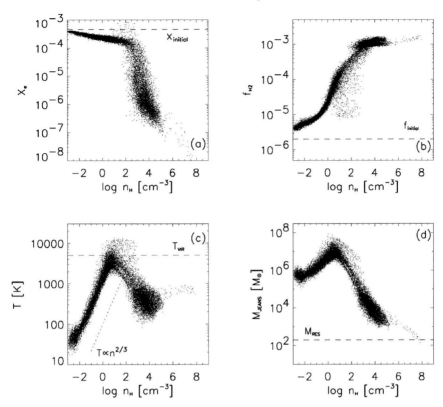

Figure 5.18. Gas properties in the SPH simulations by Bromm et al. 2002. The figure shows the properties of the particles shown in Figure 5.17. Panel (a) shows the free electron abundance x_e as a fraction of H number density. Panel (b) shows the H$_2$ abundance f_{H_2} (note that even the highest density particles have $n_H \lesssim 10^8$ cm^3, so three-body reactions are unimportant and $f_{H_2} \lesssim 10^{-3}$). Panel (c) shows the gas temperature (low-density gas gets to high temperatures because H$_2$ cooling is inefficient). Panel (d) shows the value of the Jeans mass which can be obtained by using each particle temperature and density. All the panels have the hydrogen number density n_{rmH} on the x-axis.

5.5.2 Mono-Dimensional Models

5.5.2.1 The Formation of the Protostellar Core

A less simplified approach relies on 1D studies, such as those of Omukai & Nishi (1998) and Ripamonti et al. (2002): in such studies the heavy price of assuming that the collapsing object always remains spherical (which makes it impossible to investigate the effects of angular momentum, and prevents fragmentation) is compensated by the ability to properly treat

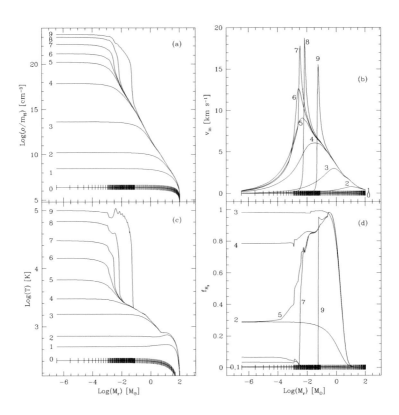

Figure 5.19. Protostellar collapse, formation of an hydrostatic core, and start of the protostellar accretion phase as can be found with 1D simulations. The four panels show the profiles of density (top left; (a)), infall velocity (top right; (b)), temperature (bottom left; (c)) and H_2 abundance (bottom right; (d)) as a function of enclosed mass at 10 different evolutionary stages (0 = initial conditions; 9 = final stage of the computation). The most relevant phases are the rapid formation of H_2 (2–3) and the formation of a shock on the surface of the hydrostatic core (5–6–7), followed by the onset by accretion. Also note the almost perfect power-law behavior of the density profile before core formation. (From Ripamonti et al. 2002.)

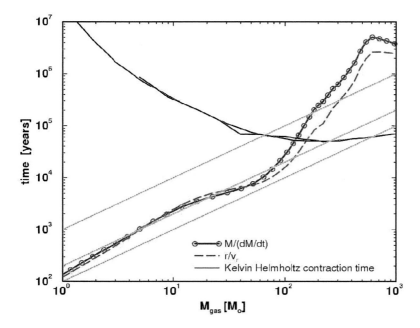

Figure 5.20. Comparison of the accretion and the Kelvin–Helmoltz time scales for primordial protostars, as a function of protostellar mass. The Kelvin–Helmoltz contraction time (obtained using the ZAMS luminosity as given by Schaerer 2002) is shown as the solid black line, while the dashed line and the solid line with circles show the time that is needed to accrete each mass of gas; they are based on the results of ABN02 and differ slightly in the way in which they were obtained. The dotted lines mark three constant accretion rates of 10^{-2} (bottom line), 5×10^{-3} (middle line), and 10^{-3} $M_{\odot}/yr$ (top line).

all the other physical processes that are believed to be important: first of all, the detailed radiative transfer of radiation emitted in the numerous H_2 resonance lines, then the transition to continuum cooling (at first from molecular CIE, and later from atomic processes) and finally the non-ideal equation of state which becomes necessary at high densities, when the gas becomes increasingly hot and largely ionized.

Such studies find that the collapse initially proceeds in a self-similar fashion (in good agreement with the solution found by Larson 1969 and Penston 1969); at the start of this phase (stages 1 and 2 in Figure 5.19, to be compared with the profiles at corresponding densities in Figure 5.15) their results can be compared with those of 3D simulations, and the agreement is good despite the difference in the assumed initial conditions: this is likely due to the self-similar properties. At later stages the comparison is not as

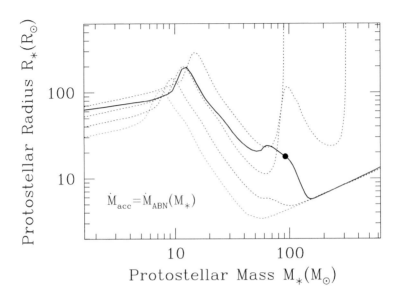

Figure 5.21. Evolution of the protostellar radius as a function of protostellar mass in the models of Omukai & Palla (2003). The models differ in the assumed accretion rates. The dotted curves show the results of models where $\dot{M}_{\rm core}$ is assumed to be constant at values of 1.1×10^{-3} $M_\odot$/yr (dotted curve starting at $R_{\rm core} \simeq 30$ $R_\odot$), 2.2×10^{-3} $M_\odot$/yr (starting at $R_{\rm core} \simeq 40$ $R_\odot$), 4.4×10^{-3} $M_\odot$/yr ("fiducial" model; starting at $R_{\rm core} \simeq 50$ $R_\odot$) and 8.8×10^{-3} $M_\odot$/yr (starting at $R_{\rm core} \simeq 65$ $R_\odot$). The solid thick curve shows the evolution of a model in which the accretion rate was obtained from the extrapolation of the ABN02 data.

good, but this is presumably due to the differences in the cooling rate (the 3D models at $n \gtrsim 10^{10}$ cm^{-3} only include the optically thin cooling rate).

It is noticeable that the transition to optically thick cooling does not stop the self-similar phase, contrary to theoretical expectations (see e.g., Rees 1976). The reason is that the reduction in the cooling rate is smooth, and is further delayed by the onset of CIE cooling and by the start of H_2 dissociation, which acts as a heat well (each H_2 dissociation absorbs 4.48 eV of thermal energy) and provides an alternative "disposal" mechanism for the excess thermal energy.

The self-similar phase proceeds until the central density is $n \gtrsim 10^{21}$ cm^{-3} $\sim 10^{-3}$ g cm^{-3}, when a small ($M_{\mathrm{core},0} \sim 3 \times 10^{-3}$ M$_\odot$) hydrostatic core eventually forms. Such a core starts accreting the surrounding gas at a very high rate ($\dot{M}_{\mathrm{core}} \sim 0.01 - 0.1$ M$_\odot$ yr^{-1}), comparable to the theoretical predictions of Stahler et al. 1980

$$\dot{M}_{\mathrm{core}} \sim c_s^3/G \simeq 4 \times 10^{-3} \left(\frac{T}{1000\,\mathrm{K}}\right)^{3/2} \left(\frac{\mu}{1.27}\right)^{-3/2} \mathrm{M}_\odot\,\mathrm{yr}^{-1} \quad (5.38)$$

where $c_s = [k_{\mathrm{B}}T/(\mu m_{\mathrm{H}})]^{1/2}$ is the isothermal sound speed of a gas with mean molecular weight μ.

Although the Omukai & Nishi (1998) and the Ripamonti et al. (2002) studies predict that such a high accretion rate will decline (approximately as $\dot{M}_{\mathrm{core}} \propto t^{-0.3}$), it is clear that, if the accretion proceeds unimpeded, it could lead to the formation of a star with a mass comparable to the whole proto-stellar cloud where the proto-star is born, that is, about 100 M$_\odot$; such a process would take a quite short time of about $10^4 - 10^5$ yr.

This scenario could be deeply affected by feedback effects: even not considering the radiation coming from the interior of the protostar, the energy released by the shocked material accreting onto the surface can reach very high values, comparable with the Eddington luminosity, and is likely to affect the accretion flow.

Figure 5.20 compares the accretion timescale (which can be extrapolated from the data at the end of the ABN02 simulations) to the Kelvin–Helmoltz timescale,

$$t_{\mathrm{KH}} = \frac{GM^2}{RL} \quad (5.39)$$

where L is the luminosity of the protostar. This plot tells us that the accretion timescale is so fast that the stellar interior has very little space for a "re-adjustment" (which could possibly stop the accretion) before reaching a mass of $\sim 10 - 100$ M$_\odot$.

However, it can be argued that such a re-adjustment is not necessary, since in the first stages most of the protostellar luminosity comes from the accretion process itself:

$$L_{\mathrm{acc}} \simeq \frac{GM_{\mathrm{core}}\dot{M}_{\mathrm{core}}}{R_{\mathrm{core}}}$$

$$\simeq 8.5 \times 10^{37} \left(\frac{M_{\mathrm{core}}}{\mathrm{M}_\odot}\right) \left(\frac{\dot{M}_{\mathrm{core}}}{0.01\,\mathrm{M}_\odot/yr}\right) \left(\frac{R_{\mathrm{core}}}{10^{12}\,\mathrm{cm}}\right)^{-1} \mathrm{erg\,s}^{-1} \quad (5.40)$$

where we have inserted realistic values for M_{core}, $\dot{M}_{\mathrm{core}}$, and R_{core}. This luminosity is quite close to the Eddington luminosity ($L_{\mathrm{Edd}} \simeq 1.3 \times 10^{38}$ $(M/\mathrm{M}_\odot)$ erg s^{-1}), and radiation pressure could have a major effect.

Ripamonti et al. (2002) found that this luminosity is not able to stop the accretion, but they do not properly trace the internal structure of the core, so their results cannot be trusted except in the very initial stages of accretion (say, when $M_{core} \lesssim 0.1$ M$_\odot$).

A better suited approach was used in studies by Stahler et al. (1986) and, more recently, by Omukai & Palla (2001, 2003), who assumed that the accretion can be described as a sequence of steady-state accretion flows onto a growing core. The core is modeled hydrostatically, as a normal stellar interior (including the treatment of nuclear burning), while the accreting envelope is modeled through the steady-state assumption, in conjunction with the condition that outside the "photosphere" (the region where the optical depth for a photon to escape the cloud is $\gtrsim 1$) the accreting material is in free-fall. As shown in Figure 5.21, Omukai & Palla (2003) find that even if feedback luminosity deeply affects the structure of the accreting envelope, it is never able to stop the accretion before the protostellar mass reaches $60 - 100$ M$_\odot$ as a minimum; after that, the final mass of the protostar depends on the assumed mass accretion rate $\dot{M}_{core}$: for $\dot{M}_{core} \lesssim 4 \times 10^{-3}$ M$_\odot$ yr^{-1}, the accretion can proceed unimpeded until the gas is completely depleted (or the star explodes as a supernova); otherwise, the radiation pressure is finally able to revert the accretion; with high accretion rates this happens sooner, and the final stellar mass is relatively low, while for accretion rates only slightly above the critical value of $\sim 4 \times 10^{-3}$ M$_\odot$ yr^{-1} the stellar mass can reach about 300 M$_\odot$.

If such predictions are correct, the primordial Initial Mass Function (IMF) is likely to be much different from the present one, reflecting mainly the mass spectrum of the gas fragments from which the stars originate, and a mass of ~ 100 M$_\odot$ could be typical for a primordial star. However, we note that these important results could change as a result of better modeling. For example, deep modifications of the envelope structure are quite possible if a frequency-dependent opacity (rather than the mean opacity used in the cited studies) is included.

5.6 Discussion

The previous sections show that, although not certain at all (because of the big uncertainty about feedback effects), numerical simulations tend to favor the hypothesis that primordial stars had a larger typical mass than present-day stars.

If this is true, it could indeed solve some observational puzzles, such as why we have never observed a single zero-metallicity star (answer: they have exploded as supernovae and/or transformed into compact objects at a very high redshift), and maybe help explain the relatively high metallicities ($Z \gtrsim 10^{-4} Z_\odot$ even in low column density systems) measured in the Lyman α forest (see e.g., Cowie & Songaila 1998), or the high ($\sim$ solar) metallicities we observe in the spectra of some quasars already at redshift 6 (Fan et al. 2000, 2001, 2003, 2004; Walter et al. 2003).

However, a top-heavy primordial IMF also runs into a series of problems, which we will discuss in the remainder of this section.

5.6.1 UV Radiation Feedback

First of all, if a moderately massive star (even $M \gtrsim 20 - 30$ M$_\odot$ is likely to be enough) actually forms in the center of a halo in the way shown by ABN02, it will emit copious amounts of UV radiation, which will produce important feedback effects. Indeed, the scarcity of metals in stellar atmospheres (see Schaerer 2002, and also Tumlinson & Shull 2000, Bromm Kudritzki & Loeb 2001) results in UV fluxes that are significantly larger than for stars of the same mass but with "normal" metallicity. Furthermore, this same scarcity of metals is likely to result in a negligibly small density of dust particles, further advancing the UV fluxes at large distances from the sources when compared to the present-day situation.

Massive primordial objects are also likely to end up in black holes (either directly or after having formed a star[7]), which could be the seeds of present-day super-massive black holes (see e.g., Volonteri, Haardt & Madau 2003). If accretion can take place onto these black holes (also known as *mini-quasars*), they are likely to emit an important amount of radiation in the UV and the x-rays (this could also be important in explaining the WMAP result about Thomson scattering optical depth; see e.g., Madau et al. 2003).

UV photons can have a series of different effects. First of all, we already mentioned in the section about chemistry that Lyman–Werner photons (11.26 eV $\leq h\nu \leq 13.6$ eV) can dissociate H_2 molecules, preventing further star formation. Such photons are effective even at large distances from their sources, and even in a neutral universe, since their energy is below the H ionization threshold; the only obstacles they find on their way are some of the Lyman transitions of H, which can scatter these photons or remove them from the band; this results in a "sawtooth" modulation (see Figure 5.22) of the spectrum. Haiman, Rees & Loeb (1997) argue that this negative feedback could conceivably bring primordial star formation to an early stop; however, other authors found that this is not necessarily the case (Ciardi et al. 2000; Omukai 2001; Glover & Brand 2001), or that the feedback effect on H_2 abundance can be moderated by the effects of x-ray photons coming from mini-quasars (Haiman Abel and Rees 2000; Glover

---

[7] Here we think of a star as an object where quasi-hydrostatic equilibrium is reached because of stable nuclear burning; according to this definition, a black hole can be formed without passing through a truly stellar phase, even if it is very likely to emit significant amounts of radiation in the process.

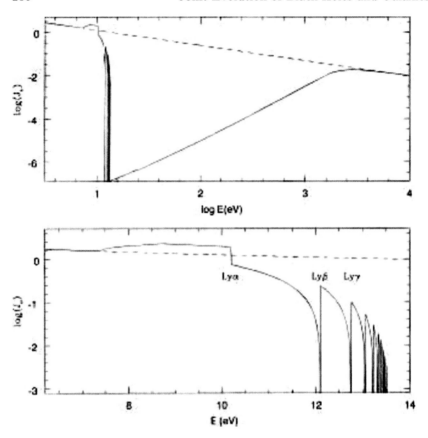

Figure 5.22. Average flux (in units of $\mathrm{erg\,cm^{-2}\,s^{-1}\,sr^{-1}\,Hz^{-1}}$) at an observation redshift $z_{\mathrm{obs}} = 25$ coming from sources turning on at $z = 35$. The top panel shows the effects of absorption of neutral hydrogen and helium, strongly reducing the flux between ~ 13.6 eV and ~ 1 keV (the solid and dashed lines show the absorbed and unabsorbed flux, respectively). The bottom panel shows the same quantities in a much smaller energy range, in order to illustrate the sawtooth modulation due to line absorption below 13.6 eV. (From Haiman, Rees & Loeb 1997.)

& Brand 2003).

A second obvious effect from UV photons is to ionize the interstellar and the intergalactic medium; it is unclear whether these kinds of objects are important in the reionization history of the universe, but they will definitely form HII regions in their immediate neighborhoods. In Figure 5.23 we show the results of Whalen, Abel & Norman 2004 about the evolution of an HII region produced by a 200 M$_\odot$ star inside an halo with the same

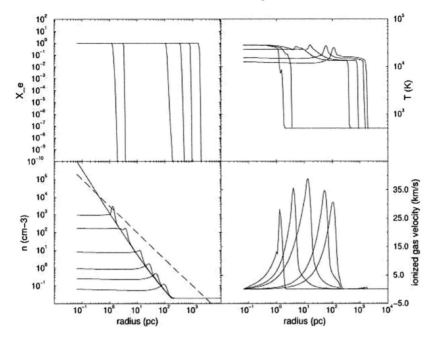

Figure 5.23. Dynamical evolution of the HII region produced by a 200 $M_\odot$ star. Panels refer to ionization fraction profiles (top left), temperature distributions (top right), densities (bottom left), and velocities (bottom right). The dashed line in the density panel is for the Strömgren density (the density required to form a Strömgren sphere and therefore initially bind the ionization front) at a given radius. Profiles are given at times (from top to bottom in the density panel; from left to right in the others) 0 (density panel), 63 kyr (all panels), 82 kyr (ionization and temperature panels), 95 kyr (ionization panel), 126.9 kyr (all panels), 317 kyr (all panels), 1.1 Myr (temperature, density and velocity panels), and 2.2 Myr (all panels), which is the approximate main sequence lifetime of a 200 $M_\odot$ star. (From Whalen, Abel & Norman 2004).

density profile as found in the ABN02 paper: in the beginning the ionization front is D-type (that is, it is "trapped" because of the density) and expands because a shock forms in front of it; later (after about 70 kyr from the start of the UV emission) the ionization front reaches regions of lower densities and becomes R-type (radiation driven), expanding much faster than the shock. However, the shock keeps moving at speeds of the order of 30 km s^{-1} even after the source of ionizing photons is turned off. Such a speed is much larger than the rotational velocities of the mini-halos where primordial star formation is supposed to be taking place ($\lesssim$ 10 kms^{-1}), so these shocks are likely to expel a very large fraction of the original gas content of the mini-halo. UV emission could even lead to the *photo-evaporation* of neighboring mini-halos, similar to what Barkana & Loeb 1999 (see also Shapiro, Iliev & Raga 2004) found in the slightly different context to cosmological reionization.

Since the dynamical timescale of a mini-halo is $\gtrsim 10^7$ yr and is longer than the timescale for this kind of phenomena (definitely shorter than the $\sim$ 1–10 Myr main sequence lifespan of massive stars, and very likely to be of the order of $\sim 10^4$–10^5 yr), the star formation in one mini-halo is likely to stop almost immediately after the first (massive) object forms. This means that, unless two or more stars form at exactly the same time (within $\sim$ 1% of the mini-halo dynamical time), each mini-halo will form *exactly one* massive star.

5.6.2   Supernovae Feedback and Metallicities

After producing plenty of UV photons during their lives primordial massive stars are likely to explode as supernovae. This must be true for some fraction of them, otherwise it would be impossible to pollute the IGM with metals, as mass loss from zero-metal stars is believed to be unimportant (this applies to stars with $M \lesssim 500$ M$_\odot$; see Baraffe, Heger & Woosley 2001, and also Heger et al. 2002); however, it is clearly possible that some fraction of primordial stars directly evolve into black holes.

In Figure 5.24 we show the results about the fate of zero metallicity stars, as obtained by Heger & Woosley 2002 (but see also Heger et al. 2003), indirectly confirming this picture and suggesting that pair-instability supernovae could play a major role if the primordial IMF is really skewed towards masses $\gtrsim 100$ M$_\odot$.

This has a series of consequences. First of all, supernova explosions are one of the very few phenomena directly involving primordial stars that we can realistically hope to observe. Wise & Abel (2004; see also Marri & Ferrara 1998 and Marri, Ferrara & Pozzetti 2000 for the effects of gravitational lensing) investigated the expected number of such supernovae by means of a semi-analytic model combining Press–Schechter formalism, an evolving minimum mass for star forming halos and negative feedbacks, finding the rates shown in Figure 5.25; if such objects are pair-instability supernovae with masses $\gtrsim 175$ M$_\odot$, they should be detectable by future missions such as JWST;[8] if some of them are associated with gamma ray bursts, the recently launched Swift[9] satellite has a slim chance of observing them at redshifts $\lesssim 30$ (see Gou et al. 2004).

Supernova explosions obviously have hydrodynamical effects on the gas of the mini-halo where they presumably formed, especially in the case of the particularly violent pair-instability supernovae (see e.g., Ober, El Elid

---

[8] James Webb Space Telescope; more information at http://www.jwst.nasa.gov.
[9] More information at http://swift.gsfc.nasa.gov.

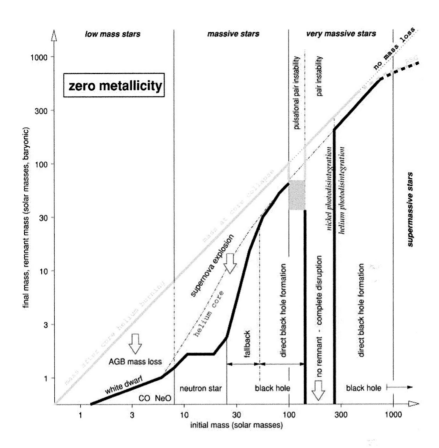

Figure 5.24. The "final mass function" of non-rotating metal-free stars as a function of their initial mass, as given by from Heger & Woosley 2002. The thick black curve gives the final mass of the collapsed remnant, and the thick gray curve and the mass of the star at the start of the event that produces that remnant (mass loss, SN etc.); for zero-metal stars this happens to be mostly coincident with the dotted line, corresponding to no mass loss. Below initial masses of 5–10 $M_\odot$ white dwarfs are formed; above that, initial masses of up to ~ 25 $M_\odot$ lead to neutron stars. At larger masses, black holes can form in different ways (through fall-back of SN ejecta or directly). Pair instability starts to appear above ~ 100 $M_\odot$, and becomes very important at ~ 140 $M_\odot$: stars with initial masses in the 140–260 $M_\odot$ range are believed to completely disrupt the star, leaving no remnant; instead, above this range pair instability is believed to lead to a complete collapse of the star into a black hole.

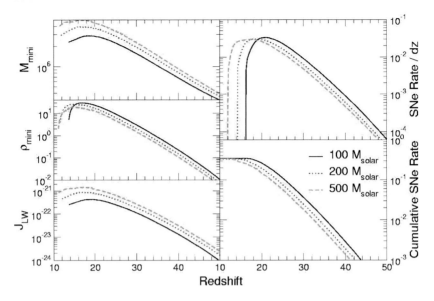

Figure 5.25. Primordial supernova properties as reported by Wise & Abel 2004. The right panels show the differential rate of primordial supernovae per unit redshift (top) and the cumulative rate (bottom); both are per year of observation and per square degree. The left panels show the critical halo mass (in $M_\odot$) for primordial star formation (top), the comoving number density (in Mpc^{-3}) of halos above the critical mass (middle) and the predicted specific intensity of the soft UV background in the Lyman–Werner band (in $erg\ s^{-1}\ cm^{-2}\ Hz^{-1}\ sr^{-1}$, bottom). The three lines in each panel refer to fixed primordial stellar masses of 100 $M_\odot$ (solid), 200 $M_\odot$ (dotted), and 500 $M_\odot$ (dashed).

& Fricke 1983), and they are likely to expel it even if it was not removed by the effects of UV radiation (see e.g., MacLow & Ferrara 1999 and Ferrara & Tolstoy 2000), further reducing the probability of having more than 1 star per mini-halo. The similarity with UV effects extends to the influence on neighboring halos that can be wiped out by SN explosions (Sigward, Ferrara & Scannapieco 2004).

Finally, supernovae provide a mechanism for spreading metals into the IGM, as discussed e.g., by Madau, Ferrara & Rees 2001. In turn, these metals will modify the cooling properties of the gas, and lead to star formation with a normal (Salpeter-like) IMF. If this is true, and if pair-instability supernovae dominate the yields from primordial stars, we should be able to distinguish the peculiar pair-supernova abundance pattern (described in Heger & Woosley 2001) as a "signature" of primordial origin.

Searches for low-metallicity stars have a long history (see e.g., Beers 1999, 2000), and their inability to find stars with metallicities $Z \leq 10^{-4} Z_\odot$

led to speculation that this metallicity marked the transition to a normal IMF (see e.g., Bromm et al. 2001, Schneider et al. 2002). However, Christlieb et al. 2002 finally reported the discovery of an extremely metal poor star (HE0107-5240) with $[Fe/H] = -5.3$[10], a level compatible with a "second generation" star (a star formed from material enriched by very few supernovae; for comparison, Wise & Abel 2004 find that primordial SNe could enrich the IGM to $[Fe/H] \sim -4.1$, although this is probably an upper limit). The abundance patterns in this star are quite strange (for example, the carbon abundance is slightly less than 1/10 solar), but a much better fit is provided by supernova yields of moderately massive stars (15–40 M$_\odot$) rather than from yields predicted for pair-instability supernovae coming from very massive stars. However, at the moment no model can satisfactorily fit all the observed abundances (see Bessel, Christlieb & Gustafsson 2004).

Even if the primordial nature of this star must still be established, it represents a cautionary tale about the currently favored predictions of a large number of massive or very massive primordial stars, and a reminder that better theoretical modeling is still needed, with particular regard to feedback effects during the stellar accretion phase.

Acknowledgments

We are grateful to the SIGRAV School for providing the opportunity to write these lecture notes. We thank M. Mapelli and M. Colpi for assistance and useful comments about the manuscript. This work was started when both authors were at the Astronomy Department of Pennsylvania State University. E.R. gratefully acknowledges support from NSF CAREER award AST-0239709 from the U.S. National Science Foundation, and from the Netherlands Organization for Scientific Research (NWO) under project number 436016.

---

[10] $[Fe/H] = \log_{10}(N_{Fe}/N_H) - \log_{10}(N_{Fe}/N_H)_\odot$.

References

Abel T, Anninos P, Norman M L and Zhang Y 1998 *ApJ* **508** 518–529

Abel T, Anninos P, Zhang Y and Norman M 1997 *New Astr.* **2** 181

Abel T, Bryan G L and Norman M L 2000 *ApJ* **540** 39

Abel T, Bryan G L and Norman M L 2002 *Science* **295** 93–98 [ABN02]

Baraffe I, Heger A and Woosley S E 2001 *ApJ* **550** 890

Barkana R and Loeb A 1999 *ApJ* **523** 54

Barkana R and Loeb A 2001 *Physics Reports* **349** 125-238

Beers T C 1999a in *The Third Stromlo Symposium: The Galactic Halo*, eds. Gibson B K Axelrod T S & Putman M E, *Astron. Soc. Pacif. Conf. Ser.* **165** 202

Beers T C 1999b *The First Stars*, proceedings of the MPA/ESO Workshop held at Garching, Germany, 4-6 August 1999, eds. A. Weiss, T. Abel & V. Hill (Springer), p. 3

Bennett C L et al. 2003 *ApJS* **148** 1

Berger M J and Colella P 1989 *J. Comp. Phys.* **82** 64

Bessel M S, Christlieb N and Gustafsson B 2004 *ApJL* **612** 61

Bogleux E and Galli D 1997 *MNRAS* **288** 638

Bond J R, Cole S, Efstathiou G and Kaiser N 1991 *ApJ* **379** 440

Bonnor W B 1956 *MNRAS* **116** 351

Bromm V and Larson R B 2004 *ARAA* **42** 79

Bromm V, Coppi P and Larson R B 1999 *ApJL* **527** L5

Bromm V, Coppi P and Larson R B 2002 *ApJ* **564** 23

Bromm V, Ferrara A, Coppi P and Larson R B 2001 *MNRAS* **328** 969

Bromm V, Kudritzki R P and Loeb A 2001 *ApJ* **552** 464

Bryan G L and Norman M 1998 *ApJ* **495** 80

Bryan G L, Norman M L, Stone J M, Cen R and Ostriker J P 1995 *Comp. Phys. Comm.* **89** 149

Burkert A and Bodenheimer P 1993 *MNRAS* **264** 798

Cazaux S and Spaans M 2004 *ApJ* **611** 40

Cen R and Ostriker P 1992 *ApJL* **399** L113

Christlieb N, Bessel M S, Beers T C, Gustafsson B, Korn A, Barklem P S, Karlsson T, Mizuno-Wiedner M and Rossi S 2002 *Nature* **419** 904

Ciardi B and Ferrara A 2004 *Space Science Reviews* accepted [astro-ph/0409018]

Ciardi B, Ferrara Governato F and Jenkins A *MNRAS* **314** 611

Coles P and Lucchin F 1995 *Cosmology. The origin and evolution of cosmic structure* (Chichester, England: John Wiley & Sons)

Couchman H M and Rees M J 1986 *MNRAS* **221** 53

Cowie L L and Songaila A 1998 *Nature* **394** 44

Diemand J, Moore B and Stadel J 2005 *Nature* **433** 389 [astro-ph/0501589]

Doroshkevich A G, Zel'Dovich Y B and Novikov I D 1967 *Astronomicheskii Zhurnal* **44** 295 [DZN67]

Ebert R. 1955 *Z. Astrophys.* **37** 217

Fan X et al. 2000 *AJ* **120** 1167

Fan X et al. 2001 *AJ* **122** 2833

Fan X et al. 2003 *AJ* **125** 2151

Fan X et al. 2004 *AJ* **128** 1649

Ferrara A and Salvaterra 2005 **THIS BOOK**

Ferrara A and Tolstoy E 2000 *MNRAS* **313** 291

Flower D R, Le Bourlot J, Pineau de Forêts G and Roueff E 2000 *MNRAS* **314** 753–758

Frommhold L 1993 *Collision-induced absorption in gases* (Cambridge, UK: Cambridge University Press)

Fuller T M and Couchman H M P 2000 *ApJ* **544** 6–20

Galli D and Palla F 1998 *A&A* **335** 403–420 [GP98]

Gamow G 1948 *Physical Review* **74** 505

Gamow G 1949 *Rev. Mod. Phys.* **21**, 367

Glover S C O 2004 *Space Science Reviews* accepted [astro-ph/0409737]

Glover S C O and Brand P W J L 2001 *MNRAS* **321** 385

Glover S C O and Brand P W J L 2003 *MNRAS* **340** 210

Gou L, Meszaros P, Abel T and Zhang B 2004 *Apj* **604** 508

Green A M, Hofmann S and Schwarz D J 2004 *MNRAS* **353** 23

Guth A H 1981 *Phys. Rev. D* **23** 247

Guth A H 2004 to be published in *Carnegie Observatories Astrophysics Series, Vol. 2: Measuring and Modeling the Universe*, ed. W L Freedman (Cambridge: Cambridge University Press), [astro-ph/0404546]

Haardt F and Madau P 1996 *ApJ* **461** 20

Haiman Z, Abel T and Rees M J 2000 *ApJ* **534** 11

Haiman Z, Rees M J and Loeb A 1997 *ApJ* **476** 458

Heger A and Woosley S E 2002 *ApJ* **567** 532

Heger A, Fryer C L, Woosley S E, Langer N and Hartmann D H 2003 *ApJ* **591** 288

Heger A, Woosley S, Baraffe I and Abel T 2002 in *Lighthouses in the Universe; The Most Luminous Celestial Objects and Their Use for Cosmology*, proceedings of the MPA/ESO Conference held at Garching (Germany), eds. M. Gilfanov, R. A. Siunyaev and E. Curazov (Berlin: Springer), 369 [astro-ph/0112059]

Hollenbach D and McKee C F 1979 *ApJS* **41** 555–592

Kitayama T, Susa H, Umemura M and Ikeuchi S 2001 *MNRAS* **326** 1353

Larson R B 1969 *MNRAS* **145** 271

Le Bourlot J, Pineau de Forêts G and Flower D R 1999 *MNRAS* **305** 802

Lenzuni P, Chernoff D F and Salpeter E E 1991 *ApJS* **76** 759

Lepp S and Shull J M 1983 *ApJ* **270** 578–582

Lepp S and Shull J M 1984 *ApJ* **280** 465–469

Lynden-Bell D 1967 *MNRAS* **136** 101

Mac Low M M and Ferrara A 1999 *ApJ* **513** 142

Machacek M E, Bryan G L and Abel T 2001 *ApJ* **548** 509–521

Madau P and Haardt F 2005 **THIS BOOK**

Madau P, Ferrara A and Rees M J 2001 *ApJ* **555** 92

Madau P, Rees M J, Volonteri M, Haardt F and Oh S P *ApJ* **604** 484

Marri S and Ferrara A 1998 *ApJ* **509** 43

Marri S, Ferrara A and Pozzetti L 2000 *MNRAS* **317** 265

Martin P G, Schwarz D H and Mandy M E 1996 *ApJ* **461** 265–281

McDowell M R C 1961 *Observatory* **81** 240

Monaghan J J 1992 *ARAA* **30** 543

Norman M L 2004 to be published in *Springer Lecture Notes in Computational Science and Engineering: Adaptive Mesh Refinement - Theory and Applications*, eds. T Plewa T Linde and G Weirs [astro-ph/0402230]

Novikov I D and Zel'Dovich Y B 1967 *ARAA* **5** 627

Ober W W, El Elid M F and Fricke K J 1983 *A&A* **119** 61

Omukai K 2001 *ApJ* **546** 635

Omukai K and Nishi R 1998 *ApJ* **508** 141–150

Omukai K and Palla F 2001 *ApJL* **561** L55–L58

Omukai K and Palla F 2003 *ApJ* **589** 677–687

Omukai K and Yoshii Y 2003 *ApJ* **599** 746

O'Shea B W, Bryan G, Bordner J, Norman M L, Abel T, Harkness R and Kritsuk A 2004 to be published in *Springer Lecture Notes in Computational Science and Engineering: Adaptive Mesh Refinement - Theory and Applications*, eds. T Plewa T Linde and G Weirs [astro-ph/0403044]

Osterbrock D E 1989 *Astrophysics of Gaseous Nebulae and Active Galactic Nuclei* (Mill Valley, California: University Science Books)

Padmanabhan T 1993 *Structure Formation in the Universe* (Cambridge: Cambridge University Press)

Palla F, Salpeter E E and Stahler S W 1983 *ApJ* **271** 632

Peebles P J E 1965 *ApJ* **142** 1317

Peebles P J E 1993 *Principles of Physical Cosmology* (Princeton: Princeton University Press)

Peebles P J E and Dicke R H 1968 *ApJ* **154** 891 [PD68]

Penston M V 1969 *MNRAS* **144** 425

Press W H and Schechter P 1974 *ApJ* **187** 425

Quinn P J and Zurek W H 1988 *ApJ* **331** 1

Rees M J 1976 *MNRAS* **174** 483

Ripamonti E and Abel T 2004 *MNRAS* **348** 1019–1034

Ripamonti E, Haardt F, Ferrara A and Colpi M 2002 *MNRAS* **334** 401–418

Sabano Y and Yoshii Y 1977 *PASJ* **29** 207

Saslaw W C and Zipoy D 1967 *Nature* **216** 976

Schaerer D 2002 *A&A* **382** 28

Schneider R, Ferrara A, Natarajan P and Omukai K 2002 *ApJ* **571** 30

Shapiro P R, Iliev I T and Raga A C 2004 *MNRAS* **348** 753

Sigward F, Ferrara A and Scannapieco E 2004 *MNRAS*, accepted [astro-ph/0411187]

Silk J 1968 *ApJ* **151** 459–472

Silk J 1983 *MNRAS* **205** 705

Sommerville R 2006 Part 3 of this book

Stahler S, Palla F and Salpeter E E 1986 *ApJ* **302** 590

Stahler S W, Shu F H and Taam R E 1980 *ApJ* **241** 637

Starobinsky A A 1979 *Pis'ma Zh. Eksp. Teor. Fiz.* **30** 719 [JETP Lett. 30,682]

Starobinsky A A 1980 *Phys. Lett.* **91B** 99

Tegmark M, Silk J, Rees M J, Blanchard A, Abel T and Palla F 1997 *ApJ* **474** 1–12 [T97]

Tumlinson J and Shull J M 2000 *ApJL* **528** 65

Uehara H and Inutsuka S 2000 *ApJL* **531** L91–L94 (HD)

Viana P T P and Liddle A R 1996 *MNRAS* **281** 323

Volonteri M, Haardt F and Madau P 2003 *ApJ* **582** 559

Walter F, Bertoldi F, Carilli C, Cox P, Lo K Y, Neri R, Fan X, Omont A, Strauss M A and Menten K M 2003 *Nature* **424** 406

Whalen D, Abel T and Norman M L 2004 *ApJ* **610** 14

White S M and Springel V 1999 *The First Stars*, proceedings of the MPA/ESO Workshop held at Garching, Germany, 4-6 August 1999, eds. A. Weiss, T. Abel & V. Hill (Springer), p. 327

Wise J H and Abel T 2004 *in preparation*, [astro-ph/0411558]

Woodward P R and Colella 1984 *J. Comp. Phys.* **54** 174

Wu K K S, Lahav O and Rees M J 1999 *Nature* **397** 225–30

Yoneyama T 1972 *Publ. of the Astr. Soc. of Japan* **24** 87

Yoshida N, Abel T, Hernquist L and Sugiyama N 2003 *ApJ* **592** 645–663

Zhao H S, Taylor J, Silk J and Hooper D 2005, astro-ph/0502049

Chapter 6

The Evolution of Baryons Along Cosmic History

Piero Madau
Astronomy Department
University of California Observatories
Santa Cruz, California

Francesco Haardt
Department of Physics and Mathematics
University of Insubria
Como, Italy

The last decade has witnessed great advances in our understanding of the high redshift universe. The pace of observational cosmology and extragalactic astronomy has never been faster, and progress has been equally significant on the theoretical side. The key idea of currently popular cosmological scenarios, that primordial density fluctuations grow by gravitational instability driven by collisionless, cold dark matter (CDM), has been elaborated upon and explored in detail through large-scale numerical simulations on supercomputers, leading to a hierarchical ("bottom-up") scenario of structure formation. In this model, the first objects to form are on subgalactic scales, and merge to make progressively bigger structures ("hierarchical clustering"). Ordinary matter in the universe follows the dynamics dictated by the dark matter until radiative, hydrodynamic, and star formation processes take over. Perhaps the most remarkable success of this theory has been the prediction of anisotropies in the temperature of the cosmic microwave background (CMB) radiation at about the level subsequently measured by the COBE satellite and most recently by the BOOMERANG, MAXIMA, DASI, CBI, Archeops, and WMAP experiments.

In spite of some significant achievements in our understanding of the formation of cosmic structures, there are still many challenges facing hierarchical clustering theories, and many fundamental questions remain, at best, only partially answered. While quite successful in matching the observed large-scale density distribution (such as e.g., the properties of galaxy clusters, galaxy clustering, and the statistics of the Lyman-α forest), CDM simulations appear to produce halos that are too centrally concentrated compared to the mass distribution inferred from the rotation curves of (dark matter-dominated) dwarf galaxies, and to predict too many dark matter subhalos compared to the number of dwarf satellites observed within the Local Group. Another perceived problem (possibly connected with the "missing satellites") is our inability to predict when, how, and to what temperature the universe was reheated and reionized, i.e., to understand the initial conditions of the galaxy formation process.

While N-body + hydrodynamical simulations have convincingly shown that the intergalactic medium (IGM) – the main repository of baryons at high redshift – is expected to fragment into structures at early times in CDM cosmogonies, the same simulations are much less able to predict the efficiency with which the first gravitationally collapsed objects lit up the universe at the end of the "dark ages". The crucial processes of star formation, preheating, and feedback (e.g., the effect of the heat input from the first generation of sources on later ones), and assembly of massive black holes in the nuclei of galaxies are poorly understood. We know that at least some galaxies and quasars were already shining when the universe was less than 10^9 yr old. But when did the first luminous objects form, what was their nature, and what impact did they have on their environment and on the formation of more massive galaxies? While the excess H I absorption measured in the spectra of $z \sim 6$ quasars in the Sloan Digital Sky Survey (SDSS) has been interpreted as the signature of the trailing edge of the cosmic reionization epoch, the recent detection by the Wilkinson Microwave Anisotropy Probe (WMAP) of a large optical depth to Thomson scattering, $\tau_e = 0.17 \pm 0.04$ suggests that the universe was reionized at higher redshifts, $z_{\rm ion} = 17 \pm 5$. This is of course an indication of significant star-formation activity at very early times.

In this chapter, we will summarize some recent developments in our understanding of the dawn of galaxies and the impact that some of the earliest cosmic structures may have had on the baryonic universe.

6.1   Cosmogonic Preliminaries

In a spatially homogeneous and isotropic universe, the energy density, the pressure, and the scale factor are related by the Friedmann equation:

$$\frac{H^2}{H_0^2} = \Omega_M a^{-3} + \Omega_Q a^{-3(1+w)} + \Omega_R a^{-4} + \Omega_K a^{-2}, \tag{6.1}$$

and the acceleration equation:

$$\frac{\ddot{a}}{a} = -\frac{H_0^2}{2}\left[\Omega_M a^{-3} + (1+3w)\Omega_Q a^{-3(1+w)} + 2\Omega_R a^{-4}\right], \tag{6.2}$$

where H_0 is the today value of the Hubble parameter $H \equiv \dot{a}/a$, and Ω_i are the today densities (relative to the critical density) of matter (M), dark energy (Q), radiation (R), and curvature (K). By definition $\Omega_K = 1 - \Omega_M - \Omega_Q - \Omega_R$. The dark energy equation of state is $P_Q = w\,\rho_Q c^2$, and $w = -1$ defines the special case of a "classical" cosmological constant Λ.

In recent years, through SNIa and CMB data (e.g., Riess et al. 1998; Perlmutter et al. 1999; Kogut et al. 2003), the measurements of cosmological parameters have been dramatically refined, leading to a standard model for the universe in which we live. The values of the main parameters are: $h = 0.72 \pm 0.05$, $1 - \Omega_K = 1.03 \pm 0.05$, and $\Omega_M h^2 = 0.14 \pm 0.02$, where h is the Hubble constant in units of 100 km/s/Mpc. The density of baryons is found to be $\Omega_B h^2 = 0.024 \pm 0.001$, the index of the primordial perturbations is $n = 0.99 \pm 0.04$, and the Thomson opacity of the universe $\tau = 0.166 \pm 0.07$.

6.2 The Dark Ages of the Universe

6.2.1 Basic Physics of Recombination

Hydrogen recombination marks the end of the plasma era, and the beginning of the era of neutral matter. According to Boltzmann statistics, the number density distribution of a non-relativistic species of particle mass m_i in thermal equilibrium is given by

$$n_i = g_i\left(\frac{m_i k_B T}{2\pi\hbar^2}\right)^{3/2}\exp\left(\frac{\mu_i - m_i c^2}{k_B T}\right), \tag{6.3}$$

where g_i and μ_i are the statistical weight and chemical potential of the species. If, further, ionization equilibrium between electrons and protons holds, recalling that $g_e = g_p = 0.5$ and $g_H = 2$, we obtain the Saha equation:

$$\frac{n_e n_p}{n_H n_{HI}} = \frac{x^2}{1-x} = \frac{(2\pi m_e k_B T)^{3/2}}{n_H(2\pi\hbar)^3}\exp\left(-\frac{I_H}{k_B T}\right), \tag{6.4}$$

where $x = n_e/n_H$ is the ionization fraction, and $I_H = 13.6$ eV is the ionization potential of atomic hydrogen. If we define the moment of recombination as the exact instant when $x = 0.5$, then the recombination temperature is $kT_{rec} = 0.323$ eV, corresponding to a recombination redshift $z_{rec} = 1370$. The ionization fraction goes from 0.9 to 0.1 as the redshift decreases from 1475 to 1255. The elapsed time is $\approx 70,000$ yr.

Note that recombination is delayed by the inability to maintain equilibrium, as two-photon decay from the metastable $2s$ level to the ground state $1s$, and losses of Lyman-α photons due to cosmological expansion tend to overproduce H I.

Finally, residual free electrons act as catalysts of the formation of molecular hydrogen, that eventually will be the main source of gas cooling in the first generation of gravitationally bound systems.

6.2.2 Post-Recombination Universe

6.2.2.1 CMB-Matter Coupling

The CMB plays a key role in the early evolution of structures. First, it sets the epoch of decoupling, when matter becomes free to move through the radiation field to form the first generation of gravitationally bound systems. Second, CMB fixes the matter temperature that in turn determines the Jeans scale of the minimum size of the first bound objects.

The net drag force acting on an electron moving at speed $v \ll c$ through the CMB is

$$F = \frac{4}{3}\frac{\sigma_T \alpha_B T^4 v}{c}, \tag{6.5}$$

where σ_T is the Thomson cross-section, and α_B the Stefan–Boltzmann constant. From Equation (6.5) we can obtain the equation of electron streaming motion:

$$\frac{1}{v}\frac{dv}{dt} = -\frac{4}{3}\frac{\sigma_T \alpha_B T^4 x}{m_p c}, \tag{6.6}$$

and then the ratio of the cosmic expansion time $t_H = 2/(3H_0)$ to the dissipation time for streaming motion as

$$\frac{t_H}{v}\frac{dv}{dt} = \frac{1.5 \times 10^{-6} x}{\Omega_M^{1/2} h}(1+z)^{5/2}. \tag{6.7}$$

At recombination this ratio is large ($\simeq 60$ at $z = 1200$, corresponding to $x \simeq 0.3$). Since the characteristic timescale for the gravitational growth of mass density fluctuations is of the order of the expansion timescale, we can conclude that baryonic density fluctuations become free to grow only after decoupling.

Consider now the heat transfer rate in a gas at temperature T_e in a blackbody radiation field at temperature T. Given the rate at which electrons are doing work against the radiation drag force as vF, plasma transfers energy to the radiation field at a mean rate per free electron given by

$$-\frac{d\epsilon}{dt} = <Fv> = \frac{4}{3}\frac{\sigma_T \alpha_B T^4}{c} <v^2> \propto T_e. \tag{6.8}$$

To reach thermal equilibrium, $T = T_e$, this rate has to be balanced by the rate at which photons scatter off electrons, leading to the Compton heating timescale:

$$\frac{dT_e}{dt} = \frac{x}{1+x}\frac{8\sigma_T \alpha_B T^4}{3m_e c}(T - T_e), \tag{6.9}$$

where the factor $x/(1 + x)$ accounts for every ionization producing two particles to share the thermal energy. Finally, the ration of the expansion time to the Compton heating timescale reads

$$\frac{t_H}{T_e}\frac{dT_e}{dt} = \frac{0.0056}{\Omega_M h}(T/T_e - 1)\frac{x}{1+x}(1+z)^{5/2}. \tag{6.10}$$

Assuming a residual ionization after recombination of $x \simeq 10^{-5}\Omega_M^{1/2}/(h\Omega_B)$, we can see that the two timescales are equal at $z_{\text{th}} \approx 180$, implying that residual ionization keeps matter and radiation in thermal equilibrium well after recombination. For $z < z_{\text{th}}$ radiation temperature drops as $T \propto (1 + z)$, while matter temperature as $T_e \propto (1 + z)^2$ due to adiabatic expansion.

6.2.2.2 After Decoupling

In the era of precision cosmology we know that, at a redshift $z_{\text{dec}} = 1088 \pm 1$, exactly $t_{\text{dec}} = (372$
$,pm\,14)\times 10^3$ years after the big bang, the universe became optically thin to Thomson scattering (Spergel et al. 2003), and entered a "dark age" (Rees 2000). At this epoch the electron fraction dropped below 13% (Figure 6.1), and the primordial radiation cooled below 3000 K, shifting first into the infrared and then into the radio. We understand the microphysics of the post-recombination universe well. As seen before, the fractional ionization froze out to the value $\sim 10^{-4.8}\Omega_M/(h\Omega_b)$: these residual electrons were enough to keep the matter in thermal equilibrium with the radiation via Compton scattering until a thermalization redshift $z_t \simeq 800(\Omega_b h^2)^{2/5} \simeq 150$, i.e., well after the universe became transparent (Peebles 1993). Thereafter, the matter temperature decreased as $(1 + z)^2$ due to adiabatic expansion (Figure 6.2) until primordial inhomogeneities in the density field evolved into the non-linear regime. The minimum mass scale for the gravitational aggregation of cold dark matter particles is negligibly small. One of the most popular CDM candidates is the neutralino: in neutralino CDM, collisional damping and free streaming smear out all power of primordial density inhomogeneities only below $\sim 10^{-7}$ M$_\odot$ (Hofmann et al. 2001). Baryons,

however, respond to pressure gradients and do not fall into dark matter clumps below the cosmological Jeans mass (in linear theory this is the minimum mass-scale of a perturbation where gravity overcomes pressure),

$$M_J = \frac{4\pi\bar{\rho}}{3}\left(\frac{5\pi k_B T}{3G\bar{\rho}m_p\mu}\right)^{3/2} \approx 2.5\times 10^5\, h^{-1}\,\mathrm{M_\odot}(aT/\mu)^{3/2}\Omega_M^{-1/2}. \quad (6.11)$$

Here $a = (1+z)^{-1}$ is the scale factor, $\bar{\rho}$ the total mass density including dark matter, μ the mean molecular weight, and T the gas temperature. The evolution of M_J is shown in Figure 6.2. In the post-recombination universe, the baryon–electron gas is thermally coupled to the CMB, $T \propto a^{-1}$, and the Jeans mass is independent of redshift and comparable to the mass of globular clusters, $M_J \approx 10^6\,\mathrm{M_\odot}$. For $z < z_t$, the temperature of the baryons drops as $T \propto a^{-2}$, and the Jeans mass decreases with time, $M_J \propto a^{-3/2}$. This trend is reversed by the reheating of the IGM. The energy released by the first collapsed objects drives the Jeans mass up to galaxy scales (Figure 6.2): previously growing density perturbations decay as their mass drops below the new Jeans mass. In particular, photoionization by the ultraviolet radiation from the first stars and quasars would heat the IGM to temperatures of $\approx 10^4\,\mathrm{K}$ (corresponding to a Jeans mass $M_J \lesssim 10^{10}\,\mathrm{M_\odot}$ at $z \simeq 20$), suppressing gas infall into low-mass halos and preventing new (dwarf) galaxies from forming.

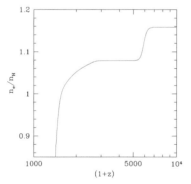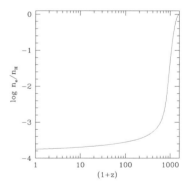

Figure 6.1.   Helium and hydrogen recombination for the WMAP parameters $(\Omega_M, \Omega_\Lambda, \Omega_b, h) = (0.29, 0.71, 0.045, 0.7)$. The step at earlier times in the left panel is due to the recombination of He III into He II. We used the code RECFAST (Seager et al. 1999) to compute the electron fraction.

6.2.3 Linear Growth of Perturbations

The linear evolution of density perturbations in the DM–baryon fluid is governed by two second-order differential equations:

$$\ddot{\delta}_{DM} + 2H\dot{\delta}_{DM} = \frac{3}{2}H^2\Omega_M(a)\left(f_{DM}\delta_{DM} + f_B\delta_B\right), \qquad (6.12)$$

for DM, and

$$\ddot{\delta}_B + 2H\dot{\delta}_B = \frac{3}{2}H^2\Omega_M(a)\left(f_{DM}\delta_{DM} + f_B\delta_B\right) - \frac{c_s^2}{a^2}k^2\delta_B, \qquad (6.13)$$

for baryons, where $\rho = \bar{\rho}(1 + \delta)$, $\delta_{DM}(t, k)$, and $\delta_B(t, k)$ are the Fourier components of the density fluctuations in DM and baryons, f_{DM} and f_B are the DM and baryon mass fractions, c_s is the gas sound speed, and k the (comoving) wavenumber. Note that here $\Omega_M(a)$ is the time-dependent matter density parameter. If DM is gravitationally dominant, then $f_B \simeq 0$, and the cosmic evolution of δ_{DM} can be approximated remarkably well by (Lahav et al. 1991):

$$\delta_{DM} \simeq \frac{5}{2}a\Omega_M(a)\left[\Omega_M^{4/7}(a) - \Omega_\Lambda(a) + \left(1 + \frac{\Omega_M(a)}{2}\right)\left(1 + \frac{\Omega_\Lambda(a)}{70}\right)\right] \tag{6.14}$$

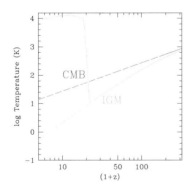
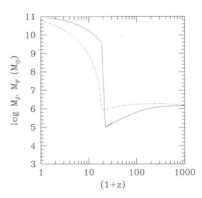

Figure 6.2. Left: Evolution of the radiation (long-dashed line, labeled CMB) and gas (solid line, labeled IGM) temperatures after recombination. The universe is assumed to be reionized by ultraviolet radiation at $z \simeq 20$. The short-dashed line is the extrapolated gas temperature in the absence of any reheating mechanism. Right: Cosmological (gas + dark matter), Jeans (solid line), and filtering (dot-dashed line) mass.

Growing modes actually decrease in density, but not as fast as the average universe. Equation (6.13) shows that, on large scales, (i.e., small k), pressure forces can be neglected, and baryons simply track the DM fluctuations. On the contrary, on small scales (i.e., large k), pressure dominates, and baryon fluctuations will be suppressed relative to DM fluctuations. Gravity and pressure are equal at the characteristic Jeans scale:

$$k_J = \frac{a}{c_s}\sqrt{4\pi G\bar{\rho}}. \tag{6.15}$$

Assuming $f_B \simeq 0$, then

$$\delta_B(t,k) \simeq \frac{\delta_{DM}(t,k)}{1 + k^2/k_J^2}. \tag{6.16}$$

When the Jeans mass itself varies with time, linear gas fluctuations tend to be smoothed on a (filtering) scale that depends on the full thermal history of the gas instead of the instantaneous value of the sound speed (Gnedin & Hui 1998). In general, assuming $a \propto t^{2/3}$, the growth of density fluctuations in the gas is suppressed for comoving wavenumbers $k > k_F$, where the filtering scale k_F is related to the Jeans wavenumber k_J by (Gnedin 2000)

$$\frac{1}{k_F^2(a)} = \frac{3}{a}\int_0^a \frac{da'}{k_J^2(a')}[1 - (a'/a)^{1/2}]. \tag{6.17}$$

This expression for k_F accounts for an arbitrary thermal evolution of the IGM through $k_J(a)$. Corresponding to the critical wavenumber k_F there is a critical (filtering) mass M_F, defined as the mass enclosed in the sphere with comoving radius equal to k_F,

$$M_F = (4\pi/3)\bar{\rho}(2\pi a/k_F)^3, \tag{6.18}$$

in a way completely analogous to the definition of the Jeans mass (Equation (6.11)), starting from the Jeans scale (Equation (6.15)).

It is the filtering mass that is central to calculations of the effects of reheating and reionization on galaxy formation. The filtering mass for a toy model with early photoionization is shown in Figure 6.2: after reheating, the filtering scale is actually smaller than the Jeans scale. Numerical simulations of cosmological reionization confirm that the characteristic suppression mass is typically lower than the linear-theory Jeans mass (Gnedin 2002).

6.3   The Emergence of Cosmic Structures

Linear, and eventual non-linear evolution, of DM-halos and baryonic structures in the benchmark cosmological model lead to well-defined predictions concerning galaxy formation and evolution. Some shortcomings on galactic and subgalactic scales of the currently favored model of hierarchical galaxy formation in a universe dominated by CDM have recently appeared. The significance of these discrepancies is still debated, and "gastrophysical" solutions involving feedback mechanisms may offer a possible way out. Other models have attempted to solve the apparent small-scale problems of CDM at a more fundamental level, i.e., by reducing small-scale power. Although the "standard" ΛCDM model for structure formation assumes a scale-invariant initial power spectrum of density fluctuations, $P(k) \propto k^n$ with $n = 1$, the recent WMAP data favor (but do not require) a slowly varying spectral index, $dn/d\ln k = -0.031^{+0.016}_{-0.018}$, i.e., a model in which the spectral index varies as a function of wavenumber k (Spergel et al. 2003). This running spectral index model predicts a significantly lower amplitude of fluctuations on small scales than standard ΛCDM. The suppression of

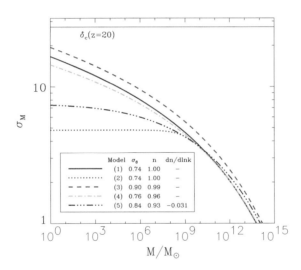

Figure 6.3. The variance of the matter-density field vs. mass M, for different power spectra. All models assume a "concordance" cosmology with parameters $(\Omega_M, \Omega_\Lambda, \Omega_b, h) = (0.29, 0.71, 0.045, 0.7)$. Solid curve: standard ΛCDM with no tilt, cluster normalized. Dotted curve: ΛWDM with a particle mass $m_X = 2\,\mathrm{keV}$, cluster normalized, no tilt. Dashed curve: tilted WMAP model, WMAP data only. Dash-dotted curve: tilted WMAP model, including 2dFGRS and Lyman-α data. Dash-triple dotted curve: running spectral index WMAP model, including 2dFGRS and Lyman-α data. Here n refers to the spectral index at $k = 0.05\,\mathrm{Mpc}^{-1}$. The horizontal line at the top of the figure shows the value of the extrapolated collapse overdensity $\delta_c(z)$ at $z = 20$.

small-scale power has the advantage of reducing the amount of substructure in galactic halos and makes small halos form later (when the universe was less dense) hence less concentrated (Navarro et al. 1997, Zentner & Bullock 2002), relieving some of the problems of ΛCDM. But it makes early reionization a challenge.

Figure 6.3 shows the linearly extrapolated (to $z = 0$) variance of the mass-density field smoothed on a scale of comoving radius R,

$$\sigma_M^2 = \langle (\delta M/M)^2 \rangle = \frac{1}{2\pi^2} \int_0^\infty dk\, k^2 P(k) T^2(k) W^2(kR), \qquad (6.19)$$

for different power spectra. Here $M = H_0^2 \Omega_M R^3 / 2G$ is the mass inside R, $T(k)$ is the transfer function for the matter density field (which accounts for all modifications of the primordial power-law spectrum due to the effects of pressure and dissipative processes), and $W(kR)$ is the Fourier transform of the spherical top-hat window function, $W(x) = (3/x^2)(\sin x/x - \cos x)$. The value of the rms mass fluctuation in a $8\,h^{-1}$ Mpc sphere, $\sigma_8 \equiv \sigma(z = 0, R = 8\,h^{-1}\,\text{Mpc})$, has been fixed for the $n = 1$ models to $\sigma_8 = 0.74$, consistent with recent normalization by the $z = 0$ X-ray cluster abundance constraint (Reiprich & Böhringer 2002).

In the CDM paradigm, structure formation proceeds "bottom-up"; i.e., the smallest objects collapse first, and subsequently merge together to form larger objects. It then follows that the loss of small-scale power modifies structure formation most severely at the highest redshifts, significantly reducing the number of self-gravitating objects then. This, of course, will make it more difficult to reionize the universe early enough. It has been argued, for example, that one popular modification of the CDM paradigm, warm dark matter (WDM), has so little structure at high redshift that it is unable to explain the WMAP observations of an early epoch of reionization (Barkana, Haiman & Ostriker 2001, Spergel et al. 2003). And yet the WMAP running-index model may suffer from a similar problem (Somerville et al. 2003). A look at Figure 6.3 shows that $10^6\,\text{M}_\odot$ halos will collapse at $z = 20$ from $2.9\,\sigma$ fluctuations in a tilted ΛCDM model with $n = 0.99$ and $\sigma_8 = 0.9$, from $4.6\,\sigma$ fluctuations in a running-index model, and from $5.7\,\sigma$ fluctuations in a WDM cosmology. The problem is that scenarios with increasingly rarer halos at early times require even more extreme assumptions (i.e., higher star formation efficiencies and UV photon production rates) in order to be able to reionize the universe by $z \sim 17$ as favored by WMAP. Figure 6.4 depicts the mass fraction in all collapsed halos with masses above the *filtering mass* for a case without reionization and one with reionization occurring at $z \simeq 20$. At early epochs this quantity appears to vary by orders of magnitude in different models.

6.4   The Epoch of Reionization

As discussed above, at epochs corresponding to $z \sim 1000$ the intergalactic medium (IGM) is expected to recombine and remain neutral until sources of radiation and heat develop that are capable of reionizing it. The detection of transmitted flux shortward of the Lyα wavelength in the spectra of sources at $z \gtrsim 5$ implies that the hydrogen component of this IGM was ionized at even higher redshifts. The increasing thickening of the Lyα forest recently measured in the spectra of SDSS $z \sim 6$ quasars (Becker et al. 2001; Djorgovski et al. 2001) may be the signature of the trailing edge of the cosmic reionization epoch. On the other hand, WMAP results indicate that the epoch of reionization could be as early as $z \simeq 17$.

It is clear that substantial sources of ultraviolet photons and mechanical energy, like young star-forming galaxies, were already present back then. The reionization of intergalactic hydrogen at $z \gtrsim 6$ is unlikely to have been accomplished by quasi-stellar sources: the observed dearth of luminous optical and radio-selected QSOs at $z > 3$ (Shaver et al. 1996; Fan et al. 2001), together with the detection of substantial Lyman-continuum flux in a composite spectrum of Lyman-break galaxies at $\langle z \rangle = 3.4$ (Steidel et al. 2001), both may support the idea that massive stars in galactic and subgalactic systems – rather than quasars – reionized the hydrogen component of the IGM when the universe was less than 5% of its current age, and dominate the 1 ryd metagalactic flux at all redshifts greater than 3. An episode of pregalactic star formation may also provide a possible explanation for the

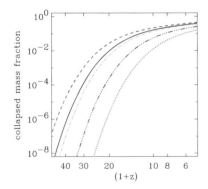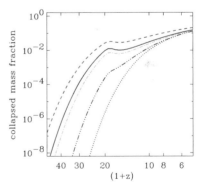

Figure 6.4. Mass fraction in all collapsed halos above the filtering mass M_F as a function of redshift, for different power spectra. Curves are the same as in Figure 6.3. Left panel: filtering mass M_F has been computed in the absence of reionization. Right panel: M_F computed assuming the universe is reionized by ultraviolet radiation at $z \simeq 20$.

widespread existence of heavy elements (such as carbon, oxygen, and silicon) in the IGM. There is some evidence that the double reionization of helium may have occurred later, at a redshift of 3 or so (see Kriss et al. 2001, and references therein): this is likely due to the integrated radiation emitted above 4 ryd by QSOs.

Establishing what ended the dark ages and when is important for determining the impact of cosmological reionization and reheating on several key cosmological issues, from the role reionization plays in allowing protogalactic objects to cool and make stars, to determining the thermal state of baryons at high redshifts and the small-scale structure in the temperature fluctuations of the cosmic microwave background. Conversely, probing the reionization epoch may provide a means for constraining competing models for the formation of cosmic structures: for example, popular modifications of the CDM paradigm that attempt to improve over CDM by suppressing the primordial power-spectrum on small scales, like warm dark matter (WDM), are known to reduce the number of collapsed halos at high redshifts and make it more difficult to reionize the universe (Barkana et al. 2001).

6.4.1 Reionization Equation

Large computer simulations performed in recent years have attacked the problem of modeling reionization from several different sides, while observations of high redshift QSOs and CMB polarization measurements were providing apparently contradicting outcomes. Chen (2003) suggested that a double reionization might have occurred, with a first generation of POPIII stars able to ionize but unable to maintain ionization. Then, when atomic hydrogen cooling became available in more massive galaxies, a second generation of stars took over, doing the job again, and now for good.

To understand the physics of reionization, we can use a simple analytical approach, which was proved to give surprisingly accurate results.

In inhomogeneous reionization scenarios, the history of the transition from a neutral IGM to one that is almost fully ionized can be statistically described by the evolution with redshift of the *volume filling factor* or porosity $Q(z)$ of H II, He II, and He III regions. The radiation emitted by spatially clustered stellar-like and quasar-like sources – the number densities and luminosities of which may change rapidly as a function of redshift – coupled with absorption processes in a medium with a time-varying clumping factor, all determine the complex topology of neutral and ionized zones in the universe. When $Q << 1$ and the radiation sources are randomly distributed, the ionized regions are spatially isolated, every UV photon is absorbed somewhere in the IGM, and the ionization process cannot be described as due to a statistically homogeneous radiation field (Arons & Wingert 1972; Meiksin & Madau 1993). While the final size of the expanding ionized bubbles is only limited by the individual source lifetime, the timescale for full recombination is rather short and relict ionization zones may recombine faster than they can accumulate (cf. Arons & Wingert 1972). As Q grows, the crossing of ionization fronts becomes more and more common, and the neutral phase shrinks in size until the reionization process is completed at the "overlap" epoch, when every point in space is exposed to Lyman continuum radiation and $Q = 1$.

The filling factor of H II regions in the universe, Q_{HII}, is then equal at any given instant t to the integral over cosmic time of the rate of ionizing photons emitted per hydrogen atom and unit cosmological volume by all radiation sources present at earlier epochs, $\int_0^t \dot{n}_{\text{ion}}(t')dt'/\bar{n}_{\text{H}}(t')$, *minus* the rate of radiative recombinations , $\int_0^t Q_{\text{HII}}(t')dt'/\bar{t}_{\text{rec}}(t')$. Differentiating one gets

$$\frac{dQ_{\text{HII}}}{dt} = \frac{\dot{n}_{\text{ion}}}{\bar{n}_{\text{H}}} - \frac{Q_{\text{HII}}}{\bar{t}_{\text{rec}}}. \tag{6.20}$$

It is this simple differential equation – and its equivalent for expanding helium zones – that statistically describes the transition from a neutral universe to a fully ionized one, independently, for a given emissivity, of the complex and possibly short-lived emission histories of individual radiation sources, e.g., on whether their comoving space density is constant or actually varies with cosmic time.

Denoting with $\bar{n}_{\text{H}}$ the mean hydrogen density of the expanding IGM, $\bar{n}_{\text{H}}(0) = 1.7 \times 10^{-7} (\Omega_b h_{50}^2/0.08)$ cm^{-3}, the volume-averaged mean recombination rate is

$$\bar{t}_{\text{rec}} = [(1+2\chi)\bar{n}_{\text{H}}\alpha_B \, C]^{-1} = 0.3\,\text{Gyr} \left(\frac{\Omega_b h_{50}^2}{0.08}\right)^{-1} \left(\frac{1+z}{4}\right)^{-3} C_{30}^{-1}, \tag{6.21}$$

where α_B is the recombination coefficient to the excited states of hydrogen, χ the helium to hydrogen cosmic abundance ratio, $C \equiv \langle n_{\text{HII}}^2 \rangle / \bar{n}_{\text{HII}}^2$ is the ionized hydrogen clumping factor[1], and a gas temperature of 10^4 K has been assumed. Clumps that are dense and thick enough to be self-shielded from UV radiation will stay neutral and will not contribute to the recombination rate. One should point out that the use of a volume-averaged clumping factor in the recombination timescale is only justified when the size of the H II region is large compared to the scale of the clumping, so that the effect of many clumps (filaments) within the ionized volume can be averaged over. This will be a good approximation either at late epochs, when the IGM is less dense and the H II zones have had time to grow (or when overlapping ionized regions from an ensemble of sources are able to properly sample the large-scale density fluctuations), or at earlier epochs if the ionized bubbles are produced by very luminous sources such as quasars or the stars within

---

[1] This may be somewhat lower than the total gas clumping factor if higher density regions are less ionized (Gnedin & Ostriker 1997).

halos collapsing from very high-σ peaks. From Equation (6.46), the mean free path between absorbers having neutral columns $> N_{HI}$ is 0.8 Mpc $h_{50}^{-1} [(1 + z)/6]^{-4.5} (N_{HI}/10^{15} \, \mathrm{cm}^{-2})^{0.5}$: it is only on scales greater than this value that the clumping can be averaged over. (Systems with $N_{HI} \ll 10^{14} \, \mathrm{cm}^{-2}$ do not dominate the mass, Miralda-Escudè et al. 1996; if the overdensity is an increasing function of N_{HI}, this implies, a fortiori, that low-N_{HI} absorbers do not contribute significantly to the volume-averaged recombination rate.) On smaller scales underdense regions are ionized first, and only afterwards the UV photons start to gradually penetrate into the higher density gas.

At high redshifts, and for an IGM with $C \gg 1$, one can expand around t to find

$$Q_{HII}(t) \approx \frac{\dot{n}_{ion}}{\bar{n}_H} \bar{t}_{rec}. \qquad (6.22)$$

The porosity of ionized bubbles is then approximately given by the number of ionizing photons emitted per hydrogen atom in one recombination time. In other words, because of hydrogen recombinations, only a fraction $\bar{t}_{rec}/t$ ($\sim$ a few per cent at $z = 5$) of the photons emitted above 1 ryd are actually used to ionize new IGM material. The universe is completely reionized when $Q = 1$, i.e., when

$$\dot{n}_{ion} \bar{t}_{rec} = \bar{n}_H. \qquad (6.23)$$

While this last expression has been derived assuming a constant comoving ionizing emissivity and a time-independent clumping factor, it is also valid in the case where $\dot{n}_{ion}$ and C do not vary rapidly over a timescale $\bar{t}_{rec}$.

6.4.2 Reionization by Massive Stars

In a CDM universe, structure formation is a hierarchical process in which nonlinear, massive structures grow via the merger of smaller initial units. Large numbers of low-mass galaxy halos are expected to form at early times in these popular theories, leading to an era of reionization, reheating, and chemical enrichment. Most models predict that intergalactic hydrogen was reionized by an early generation of stars or accreting black holes at $z = 7 - 15$. One should note, however, that while numerical N-body + hydrodynamical simulations have convincingly shown that the IGM does fragment into structures at early times in CDM cosmogonies (e.g., Cen et al. 1994; Zhang et al. 1995; Hernquist et al. 1996), the same simulations are much less able to predict the efficiency with which the first gravitationally collapsed objects lit up the universe at the end of the dark age.

The scenario that has received the most theoretical studies is one where hydrogen is photoionized by the UV radiation emitted either by quasars or by stars with masses $\gtrsim 10 \, M_\odot$ (e.g., Shapiro & Giroux 1987; Haiman & Loeb 1998; Madau et al. 1999; Chiu & Ostriker 2000; Ciardi et al. 2000), rather than ionized by collisions with electrons heated up by, e.g., supernova-driven winds from early pregalactic objects. In the former case a high degree of ionization requires about $13.6 \times (1 + t/\bar{t}_{\rm rec})\,{\rm eV}$ per hydrogen atom, where $\bar{t}_{\rm rec}$ is the volume-averaged hydrogen recombination timescale, $t/\bar{t}_{\rm rec}$ being much greater than unity already at $z \sim 10$ according to numerical simulations. Collisional ionization to a neutral fraction of only a few parts in 10^5 requires a comparable energy input, i.e., an IGM temperature close to 10^5 K or about 25 eV per atom.

Massive stars will deposit both radiative and mechanical energy into the interstellar medium of protogalaxies. A complex network of "feedback" mechanisms is likely at work in these systems, as the gas in shallow potential is more easily blown away (Dekel & Silk 1986; Tegmark et al. 1993; Mac Low & Ferrara 1999; Mori et al. 2002) thereby quenching star formation. Furthermore, as the blastwaves produced by supernova explosions reheat the surrounding intergalactic gas and enrich it with newly formed heavy elements (see below), they can inhibit the formation of surrounding low-mass galaxies due to "baryonic stripping" (Scannapieco et al. 2002). It is therefore difficult to establish whether an early input of mechanical energy will actually play a major role in determining the thermal and ionization state of the IGM on large scales. What can be easily shown is that, during the evolution of a a "typical" stellar population, more energy is lost in ultraviolet radiation than in mechanical form. This is because in nuclear burning from zero to solar metallicity ($Z_\odot = 0.02$), the energy radiated per baryon is $0.02 \times 0.007 \times m_{\rm H} c^2$, with about one third of it going into H-ionizing photons. The same massive stars that dominate the UV light also explode as supernovae (SNe), returning most of the metals to the interstellar medium and injecting about 10^{51} ergs per event in kinetic energy. For a Salpeter initial mass function (IMF), one has about one SN every 150 $M_\odot$ of baryons that forms stars. The mass fraction in mechanical energy is then approximately 4×10^{-6}, ten times lower than the fraction released in photons above 1 ryd.

The relative importance of photoionization versus shock ionization will depend, however, on the efficiency with which radiation and mechanical energy actually escape into the IGM. Consider, for example, the case of an early generation of subgalactic systems collapsing at redshift 9 from 2–σ fluctuations. At these epochs their dark matter halos would have virial radii $r_{\rm vir} = 0.75h^{-1}$ Kpc and circular velocities $V_c(r_{\rm vir}) = 25$ km s^{-1}, corresponding in top-hat spherical collapse to a virial temperature $T_{\rm vir} = 0.5\mu m_p V_c^2 / k \approx 10^{4.3}$ K and halo mass $M = 0.1 V_c^3 / GH \approx 10^8 h^{-1}\,M_\odot$.[2] Halos in this mass range are characterized by very short dynamical timescales (and even shorter gas cooling times due to atomic hydrogen) and may therefore form stars in a rapid but intense burst before SN feedback quenches

---

[2] This assumes an Einstein–de Sitter universe with Hubble constant $H_0 = 100\,h$ km s^{-1} Mpc^{-1}.

further star formation. For a star formation efficiency of $f = 0.1$, $\Omega_b h^2 = 0.02^3$, $h = 0.5$, and a Salpeter IMF, the explosive output of $10,000$ SNe will inject an energy $E_0 \approx 10^{55}$ ergs. This is roughly a hundred times higher than the gas binding energy: a significant fraction of the halo gas will then be lifted out of the potential well ("blow-away") and shock the intergalactic medium (Madau et al. 2001). If the explosion occurs at cosmic time $t = 4 \times 10^8$ yr (corresponding in the adopted cosmology to $z = 9$), at time $\Delta t = 0.4t$ after the event it is a good approximation to treat the cosmological blast wave as adiabatic, with proper radius given by the standard Sedov–Taylor self–similar solution,

$$R_s \approx \left(\frac{12\pi G \eta E_0}{\Omega_b} \right)^{1/5} t^{2/5} \Delta t^{2/5} \approx 17 \, \text{kpc}. \tag{6.24}$$

Here $\eta \approx 0.3$ is the fraction of the available SN energy that is converted into kinetic energy of the blown–away material (Mori et al. 2002). At this instant the shock velocity relative to the Hubble flow is

$$v_s \approx 2R_s/5\Delta t \approx 40 \, \text{km s}^{-1}, \tag{6.25}$$

lower than the escape velocity from the halo center. The gas temperature just behind the shock front is $T_s = 3\mu m_p v_s^2/16k \gtrsim 10^{4.3}$ K, enough to ionize all incoming intergalactic hydrogen. At these redshifts, it is the onset of Compton cooling off cosmic microwave background photons that ends the adiabatic stage of blastwave propagation; the shell then enters the snow-plough phase and is finally confined by the IGM pressure. According to the Press–Schechter formalism, the comoving abundance of collapsed dark halos with mass $M = 10^8 \, h^{-1} \, M_\odot$ at $z = 9$ is $dn/d\ln M \sim 80 \, h^3 \, \text{Mpc}^{-3}$, corresponding to a mean proper distance between neighboring halos of $\sim 15 \, h^{-1}$ kpc. With the assumed star formation efficiency, only a small fraction, about 4%, of the total stellar mass inferred today (Fukugita et al. 1998) would actually form at these early epochs. Still, our simple analysis shows that the blastwaves from such a population of pregalactic objects could drive vast portions of the IGM to a significantly higher adiabat, $T_{\text{IGM}} \sim 10^5$ K, than expected from photoionization, so as to "choke off" the collapse of further $M \lesssim 10^9 \, h^{-1} \, M_\odot$ systems by raising the cosmological Jeans mass. In this sense the process may be self-regulating.

6.4.3   Reionization by Miniquasar

6.4.3.1   Formation of IMBHs

Since, at zero metallicity, mass loss through radiatively driven stellar winds or nuclear-powered stellar pulsations is expected to be negligible (Kudritzki

---

[3] Here Ω_b is the baryon density parameter, and $f\Omega_b$ is the fraction of halo mass converted into stars.

2002; Baraffe, Heger, & Woosley 2001), Population III stars will likely die losing only a small fraction of their mass. Nonrotating very massive stars in the mass window $140 \lesssim m_* \lesssim 260$ M$_\odot$ will disappear as pair-instability supernovae (Bond, Arnett, & Carr 1984), leaving no compact remnants and polluting the universe with the first heavy elements (e.g., Schneider et al. 2002; Oh et al. 2001; Wasserburg & Qian 2000). Stars with $40 < m_* < 140$ M$_\odot$ and $m_* > 260$ M$_\odot$ are predicted instead to collapse to black holes (BHs) with masses exceeding half of the initial stellar mass (Heger & Woosley 2002). Barring any fine tuning of the IMF of Population III stars, intermediate-mass black holes (IMBHs) – with masses above the 4–18 M$_\odot$ range of known "stellar-mass" BHs (e.g., McClintock & Remillard 2003) – may then be the inevitable endproduct of the first episodes of pregalactic star formation. Since they form in high-σ density fluctuations, "relic" IMBHs and their descendants will tend to cluster in the bulges of present-day galaxies as their host halos aggregate into more massive systems (Madau & Rees 2001, hereafter Paper I; Islam, Taylor, & Silk 2003). The first IMBHs may also have seeded the hierarchical assembly of the supermassive variety (SMBHs) observed at the center of luminous galaxies (Volonteri, Haardt, & Madau 2003; Volonteri, Madau, & Haardt 2003).

6.4.3.2 *Accretion onto IMBHs*

Prior to the epoch of reionization, however, cold material may be efficiently accreted onto IMBHs hosted in minihalos. As mentioned in § 6.1, gas condensation in the first baryonic objects is possible through the formation of H$_2$ molecules, which cool efficiently via roto-vibrational transitions even at virial temperatures of a few hundred kelvin. In the absence of a UV photodissociating flux and of ionizing x-ray radiation, three-dimensional simulations of early structure formation show that the fraction of cold, dense gas available for accretion onto IMBHs or star formation exceeds 20% above $M_h = 10^6$ M$_\odot$ (Machacek, Bryan, & Abel 2003). And while radiative feedback (photodissociation and photoionization) from the progenitor massive star may initially quench BH accretion within the original host minihalo, *new cold material will be readily available through the hierarchical merging of small gaseous subunits.*

Assume now that a fraction $f_{\rm UV}$ of the bolometric power radiated by our Population III miniquasars is emitted as hydrogen-ionizing photons with mean energy $\langle h\nu \rangle$. For a given bolometric emissivity, the number of ionizing photons scales is $f_{\rm UV}/\langle h\nu \rangle$. One of the biggest uncertainties in discussing early reionization by miniquasars is their unknown emission spectrum. If the shape of the emitted spectrum from miniquasars followed the mean spectral energy distribution of the quasar sample in Elvis et al. (1994), then $f_{\rm UV} = 0.3$ and $\langle h\nu \rangle = 3$ ryd; hence $f_{\rm UV}/\langle h\nu \rangle = 0.1$ ryd^{-1}.

Miniquasars powered by IMBHs, however, are likely to be harder emitters than quasars. The hottest blackbody temperature in a Keplerian disk damping material onto a BH at the Eddington rate is $kT_{\max} \sim 1 \, \mathrm{keV} \, m_{\mathrm{BH}}^{-1/4}$ (Shakura & Sunyaev 1973), where the hole mass is measured in solar mass units. The characteristic multicolor (cold) disk spectrum follows a power law with $L_\nu \sim \nu^{1/3}$ at $h\nu < kT_{\max}$. Close to the epoch of reionization at $z \sim 15$, most IMBHs holes will have masses in the range 200–1000 $M_\odot$, hence $kT_{\max} \sim 0.2$–$0.3 \, \mathrm{keV}$.

The spectra of "ultraluminous" x-ray sources (ULXs, Colbert & Mushotzky 1999) in nearby galaxies actually require both a soft and a hard component of comparable luminosities to describe the continuum emission. While the soft components are well fit by cool multicolor disk blackbodies with $kT_{\max} \simeq 0.15 \, \mathrm{keV}$, which may indicate IMBHs (e.g., Miller et al. 2003), the nonthermal power law component has spectral slope $L_\nu \propto \nu^{-\alpha}$, with $\alpha \approx 1$. A power-law with slope $\alpha = 1$ extending from 2 keV down to 13.6 eV has $f_{\mathrm{UV}} = 1$ and $\langle h\nu \rangle = 5 \, \mathrm{ryd}$; hence $f_{\mathrm{UV}}/\langle h\nu \rangle = 0.2 \, \mathrm{ryd}^{-1}$. Figure 6.5 shows the cumulative number of ionizing photons produced per hydrogen atom for different values of $f_{\mathrm{UV}}/\langle h\nu \rangle$. A few ionizing hard photons per atom (less than for a stellar spectrum because of secondary ionizations and larger mean free path) should suffice to reionize the universe and keep the gas in overdense regions and filaments photoionized against radiative recombinations. To illustrate the implications of these results, consider the following estimate for the number of H-ionizing photons emitted by the initial population of progenitor massive stars. In our model the fraction, f_*, of cosmic baryons incorporated into Population III massive stars – progenitors of seed IMBHs – at $z = 24$ is

$$f_* = \frac{\langle m_* \rangle}{\langle m_\bullet \rangle} \frac{\Omega_\bullet}{\Omega_b} \approx 5 \times 10^{-7}, \tag{6.26}$$

where Ω_b is the baryon density parameter. Zero-metallicity stars in the range $40 < m_* < 500 \, M_\odot$ emit about 70,000 photons above 1 ryd per stellar baryon (Schaerer 2002). The total number of ionizing photons produced per baryon by progenitor Population III stars is then

$$\frac{n_{\mathrm{ion}}}{n_b} \approx 70,000 f_* \approx 0.04, \tag{6.27}$$

well below what is needed to reionize the universe. Figure 6.5 shows that, if $f_{\mathrm{UV}}/\langle h\nu \rangle$ is greater than $0.1 \, \mathrm{ryd}^{-1}$ and gas is accreted efficiently onto IMBHs, then miniquasars may be responsible for cosmological reionization at redshift ~ 15.

6.5 Preheating and Galaxy Formation

6.5.1 Preheating

Even if the IMF at early times were known, we still would remain uncertain about the fraction of cold gas that gets retained in protogalaxies after the formation of the first stars (this quantity affects the global efficiency of star formation at these epochs) and whether – in addition to ultraviolet radiation – an early input of mechanical energy may also play a role in determining the thermal and ionization state of the IGM on large scales. The same massive stars that emit ultraviolet light also explode as supernovae (SNe), returning most of the metals to the interstellar medium of pregalactic systems and injecting about 10^{51} ergs per event in kinetic energy. A complex network of feedback mechanisms is likely at work in these systems, as the gas in shallow potential is more easily blown away

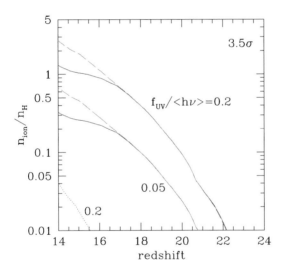

Figure 6.5. Cumulative number of ionizing photons per hydrogen atom produced by miniquasars, for different values of the fraction f_{UV} of the bolomeric power that is emitted as radiation above 1 ryd with mean energy $\langle h\nu \rangle$ (in ryd). Solid curves: in each major merger the BH in the main halo accretes a mass $\Delta m_{acc} = 2m_{BH}$. Dashed curves: same with $\Delta m_{acc} = 10^{-3} M_h$. Dotted curve: same with $\Delta m_{acc} = 10^{-3} M_h$, $f_{UV}/\langle h\nu \rangle = 0.2 \, \text{ryd}^{-1}$, but with gas accretion suppressed in minihalos with virial temperature $T_{vir} \lesssim 10^4$ K.

(Dekel & Silk 1986), thereby quenching star formation. Furthermore, as the blastwaves produced by supernova explosions – and possibly also by winds from miniquasars – sweep the surrounding intergalactic gas, they may inhibit the formation of nearby low-mass galaxies due to "baryonic stripping" (Scannapieco et al. 2002), and drive vast portions of the IGM to a significantly higher temperature than expected from photoionization, so as to choke off the collapse of further galaxy-scale systems. Note that this type of global feedback is fundamentally different from the in situ heat deposition commonly adopted in galaxy formation models, in which hot gas is produced by supernovae within the parent galaxy. We refer here to this global early energy input as "preheating" (Benson & Madau 2003). Note that a large-scale feedback mechanism may also be operating in the intracluster medium: studies of x-ray emitting gas in clusters show evidence for some form of non-gravitational entropy input (Ponman et al. 1999). The energy required there is at a level of $\sim 1\,\mathrm{keV}$ per particle, and must be injected either in a more localized fashion or at late epochs in order not to violate observational constraints on the temperature of the Lyman-α forest at $z \sim 3$. The thermal and ionization history of a preheated universe may be very different from one where hydrogen is photoionized. The gas will be heated up to a higher adiabat, and collisions with hot electrons will be the dominant ionization mechanism. The higher energies associated with preheating may doubly ionize helium at high-z, well before the "quasar epoch" at $z \sim 3$. Galaxy formation and evolution will also be different, as preheating will drive the filtering mass above $10^{10} - 10^{11}\ \mathrm{M_\odot}$ and will tend to flatten the faint-end slope of the present-epoch galaxy luminosity function, in excellent agreement with the data and without the need for SN feedback at late times.

6.5.2   Metal Enrichment

It is interesting to set some general constraints on the early star-formation episode and stellar populations that may be responsible for an early preheating of the IGM at the levels consistent with the temperature of intergalactic gas inferred at $z \approx 3$. Let us characterize the energy input due to preheating by the energy per baryon, E_p, deposited in the IGM at redshift z_p. We examine a homogeneous energy deposition since the filling factor of pregalactic outflows is expected to be large. Let Ω_* be the mass density of stars formed at z_p in units of the critical density, E_{SN} the mechanical energy injected per SN event, and f_w the fraction of that energy that is eventually deposited into the IGM. Denoting with η the number of SN explosions per mass of stars formed, one can write

$$\frac{\Omega_*}{\Omega_b} = \frac{E_p}{f_w \eta E_{\mathrm{SN}} m_p},$$
(6.28)

where m_p is the proton mass. For a Salpeter IMF between 0.1 and 100 $M_\odot$, the number of Type II SN explosions per mass of stars formed is $\eta = 0.0074\ M_\odot^{-1}$, assuming all stars above 8 $M_\odot$ result in SNe II. Numerical simulations of the dynamics of SN-driven bubbles from subgalactic halos have shown that up to 40% of the available SN mechanical luminosity can be converted into kinetic energy of the blown-away material, $f_w \approx 0.4$, the remainder being radiated away (Mori et al. 2002). With $E_{SN} = 1.2 \times 10^{51}$ ergs, Equation (6.28) then implies

$$\left(\frac{\Omega_*}{\Omega_b}\right)_{sp} = 0.05\ (E_p/0.1\ \text{keV}). \qquad (6.29)$$

SN-driven pregalactic outflows efficiently carry metals into intergalactic space (Madau et al. 2001). For a normal IMF, the total amount of metals expelled in winds and final ejecta (in SNe or planetary nebulae) is about 1% of the input mass. Assuming a large fraction, $f_Z = 0.5$, of the metal-rich SN ejecta escape the shallow potential wells of subgalactic systems, the star-formation episode responsible for early preheating will enrich the IGM to a mean level

$$\langle Z \rangle_{sp} = \frac{0.01\ \Omega_*\ f_Z}{\Omega_b} = 0.014\ Z_\odot\ (E_p/0.1\ \text{keV}). \qquad (6.30)$$

The weak C IV absorption lines observed in the Lyman-α forest at $z = 3 - 3.5$ imply a minimum universal metallicity relative to solar in the range $[-3.2]$ to $[-2.5]$ (Songaila 1997). Preheating energies in excess of 0.1 keV appear then to require values of Ω_* and $\langle Z \rangle$ that are too high, comparable to the total mass fraction in stars seen today (Glazebrook et al. 2003), and in excess of the enrichment of the IGM inferred at intermediate redshifts, respectively.

The astrophysics of first light may not be as simple, however. The metal constraint assumes that metals escaping from protogalaxies are evenly mixed into the IGM and the Lyman-α clouds (Voit 1996). Inefficient mixing could instead produce a large variance in intergalactic metallicities. The metal abundances of the Lyman-α clouds may underestimate the average metallicity of the IGM if there existed a significant warm–hot gas phase component with a higher level of enrichment, as detected for example in O VI (Simcoe et al. 2002). Today, the metallicity of the IGM may be closer to $\sim 1/3$ of solar if the metal productivity of galaxies within clusters is to be taken as representative of the universe as a whole (Renzini 1997). Uncertainties in the early IMF make other preheating scenarios possible and perhaps even more likely. Population III stars with main-sequence masses of approximately 140–260 $M_\odot$ will encounter the electron–positron pair instability and be completely disrupted by a giant nuclear-powered explosion (Heger & Woosley 2002). A fiducial 200 $M_\odot$ Population III star will explode with a kinetic energy at infinity of $E_{SN} = 4 \times 10^{52}$ ergs, injecting about 90 $M_\odot$ of metals. For a very "top-heavy" IMF with $\eta = 0.005\ M_\odot^{-1}$, Equation (6.28) now yields

$$\left(\frac{\Omega_*}{\Omega_b}\right)_{III} = 0.001\ (E_p/0.1\,\mathrm{keV}), \tag{6.31}$$

and a mean IGM metallicity

$$\langle Z \rangle_{III} = \frac{0.45\,\Omega_*\,f_Z}{\Omega_b} = 0.02\ Z_\odot\ (E_p/0.1\,\mathrm{keV}) \tag{6.32}$$

(in both expressions above we have assumed $f_w = f_Z = 1$). This scenario can yield large preheating energies by converting only a small fraction of the cosmic baryons into Population III stars, but tends to produce too many metals for $E_p \gtrsim 0.1\,\mathrm{keV}$. The metallicity constraint, of course, does not bound preheating from winds produced by an early, numerous population of faint miniquasars.[4] Accretion onto black holes releases 50 MeV per baryon, and if a fraction f_w of this energy is used to drive an outflow and is ultimately deposited into the IGM, the accretion of a trace amount of the total baryonic mass onto early black holes,

$$\frac{\Omega_{BH}}{\Omega_b} = \frac{E_p}{f_w\,50\,\mathrm{MeV}} = 2 \times 10^{-6}\ f_w^{-1}\ (E_p/0.1\,\mathrm{keV}), \tag{6.33}$$

may then suffice to preheat the whole universe. Note that this value is about $50\,f_w$ times smaller than the density parameter of the supermassive variety found today in the nuclei of most nearby galaxies, $\Omega_{SMBH} \approx 2 \times 10^{-6}\,h^{-1}$ (Merrit & Ferrarese 2001).

6.6 The Intergalactic Medium After Reionization

About half a million years after the Big Bang, the ever-fading cosmic black-body radiation cooled below 3000 K and shifted first into the infrared and then into the radio, and the smooth baryonic plasma that filled the universe became neutral. The universe then entered a "dark age" that persisted until the first cosmic structures collapsed into gravitationally bound systems, and evolved into stars, galaxies, and black holes that lit up the universe again. Some time between redshift of 7 and 15, stars within protogalaxies created the first heavy elements; these systems, together perhaps with an

---

[4] Because the number density of *bright* quasi-stellar objects at $z > 3$ is low (Fan et al. 2001), the thermal and kinetic energy they expel into intergalactic space must be very large to have a global effect, i.e., for their blast waves to fill and preheat the universe as a whole. The energy density needed for rare, luminous quasars to shock-heat the entire IGM would in this case violate the COBE limit on y-distortion.

early population of quasars, generated the ultraviolet radiation that re-heated and reionized the cosmos. The history of the universe during and soon after these crucial formative stages is recorded in the all-pervading intergalactic medium (IGM), which is believed to contain most of the ordinary baryonic material left over from the Big Bang. Throughout the epoch of structure formation, the IGM becomes clumpy and acquires peculiar motions under the influence of gravity, and acts as a source for the gas that gets accreted, cools, and forms stars within galaxies, and as a sink for the metal-enriched material, energy, and radiation which they eject. Observations of absorption lines in quasar spectra at redshifts up to 5 provide invaluable insight into the chemical composition of the IGM and primordial density fluctuation spectrum of some of the earliest formed cosmological structures, as well as of the ultraviolet background radiation that ionizes them.

Because of the overwhelming abundance of hydrogen, the ionization of this element is of great importance for determining the physical state of the IGM. Popular cosmological models predict that most of the intergalactic hydrogen was reionized by the first generation of stars or quasars at $z = 7$–15. The case that has received the most theoretical studies is one where hydrogen is ionized by the absorption of photons, $H + \gamma \rightarrow p + e$ (as opposite to collisional ionization $H + e \rightarrow p + e + e$) shortward of 912 Å, that is, with energies exceeding 13.6 eV, the energy of the Lyman edge. The process of reionization began as individual sources started to generate expanding H II regions in the surrounding IGM; throughout an H II region, H is ionized and He is either singly or doubly ionized. As more and more sources of ultraviolet radiation switched on, the ionized volume grew in size. The reionization ended when the cosmological H II regions overlapped and filled the intergalactic space.

6.6.1 Photoionization Equilibrium

At every point in a optically thin, pure hydrogen medium of neutral density $n_{\rm HI}$, the photoionization rate per unit volume is

$$n_{\rm HI} \int_{\nu_L}^{\infty} \frac{4\pi J_\nu \sigma_{\rm H}(\nu)}{h_P \nu} d\nu, \tag{6.34}$$

where J_ν is the mean intensity of the ionizing radiation (in energy units per unit area, time, solid angle, and frequency interval), and h_P is the Planck constant. The photoionization cross-section for hydrogen in the ground state by photons with energy $h_P \nu$ (above the threshold $h_P \nu_L = 13.6\,{\rm eV}$) can be usefully approximated by

$$\sigma_{\rm H}(\nu) = \sigma_L \left(\nu/\nu_L\right)^{-3}, \qquad\qquad \sigma_L = 6.3 \times 10^{-18}\,{\rm cm}^2. \tag{6.35}$$

At equilibrium, this is balanced by the rate of radiative recombinations $p + e \to H + \gamma$ per unit volume,

$$n_e n_p \alpha_A(T), \tag{6.36}$$

where n_e and n_p are the number densities of electrons and protons, and $\alpha_A = \sum \langle \sigma_n v_e \rangle$ is the radiative recombination coefficient, i.e., the product of the electron capture cross-section σ_n and the electron velocity v_e, averaged over a thermal distribution and summed over all atomic levels n. At the commonly encountered gas temperature of 10^4 K, $\alpha_A = 4.2 \times 10^{-13}$ cm^3 s^{-1}.

Consider, as an illustrative example, a point in an intergalactic H II region at (say) $z = 6$, with density $\bar{n}_H = (1.6 \times 10^{-7}$ cm$^{-3})(1 + z)^3 = 5.5 \times 10^{-5}$ cm^{-3}. The H II region surrounds a putative quasar with specific luminosity $L_\nu = 10^{30} (\nu_L/\nu)^2$ ergs s^{-1} Hz^{-1}, and the point in question is at a distance of $r = 3$ Mpc from the quasar. To a first approximation, the mean intensity is simply the radiation emitted by the quasar reduced by geometrical dilution,

$$4\pi J_\nu = \frac{L_\nu}{4\pi r^2}. \tag{6.37}$$

We then have for the photoionization timescale:

$$t_{\text{ion}} = \left[\int_{\nu_L}^{\infty} \frac{4\pi J_\nu \sigma_H(\nu)}{h_P \nu} d\nu \right]^{-1} = 5 \times 10^{12} \, \text{s}, \tag{6.38}$$

and for the recombination timescale:

$$t_{\text{rec}} = \frac{1}{n_e \alpha_A} = 5 \times 10^{16} \, \text{s} \, \frac{\bar{n}_H}{n_e}. \tag{6.39}$$

As in photoionization equilibrium $n_{HI}/t_{\text{ion}} = n_p/t_{\text{rec}}$, these values imply $n_{HI}/n_p \simeq 10^{-4}$; that is, hydrogen is very nearly completely ionized.

A source radiating ultraviolet photons at a finite rate cannot ionize an infinite region of space, and therefore there must be an outer edge to the ionized volume (this is true unless, of course, there is a population of UV emitters and all individual H II regions have already overlapped). One fundamental characteristic of the problem is the very small value of the mean free path for an ionizing photon if the hydrogen is neutral, $(\sigma_L n_H)^{-1} = 0.9$ kpc at threshold, much smaller than the radius of the ionized region. If the source spectrum is steep enough that little energy is carried out by more penetrating, soft x-ray photons, we have one nearly completely ionized H II region, separated from the outer neutral IGM by a thin transition layer or "ionization-front". The inhomogeneity of the IGM is of primary importance for understanding the ionization history of the universe, as denser gas recombines faster and is therefore ionized at later times than the tenuous gas in underdense regions. An approximate way to study the effect of inhomogeneity is to write the rate of recombinations as

$$\langle n_e n_p \rangle \alpha_A(T) = C \langle n_e \rangle^2 \alpha_A(T) \tag{6.40}$$

(assuming T is constant in space), where the brackets are the space average of the product of the local proton and electron number densities, and the factor $C > 1$ takes into account the degree of clumpiness of the IGM. If ionized gas with electron density n_e uniformly filled a fraction $1/C$ of the available volume, the rest being empty space, the mean square density would be $\langle n_e^2 \rangle = n_e^2/C = \langle n_e \rangle^2 C$.

The IGM is completely reionized when one ionizing photon has been emitted for each H atom by the radiation sources, and when the rate of emission of UV photons per unit (comoving) volume balances the radiative recombination rate, so that hydrogen atoms are photoionized faster than they can recombine. The complete reionization of the universe manifests itself in the absence of an absorption trough in the spectra of galaxies and quasars at high redshifts. If the IGM along the line-of-sight to a distant source were neutral, the resonant scattering at the wavelength of the Lyα $(2p \rightarrow 1s; h_P\nu_\alpha = 10.2\,\text{eV})$ transition of atomic hydrogen would remove all photons blueward of Lyα off the line-of-sight. For any reasonable density of the IGM, the scattering optical depth is so large that detectable absorption will be produced by relatively small column (or surface) densities of intergalactic neutral hydrogen.

6.6.2 Gunn–Peterson Effect

Consider radiation emitted at some frequency ν_e that lies blueward of Lyα by a source at redshift z_e, and observed at Earth at frequency $\nu_o = \nu_e(1 + z_e)^{-1}$. At a redshift z such that $(1+z) = (1+z_e)\nu_\alpha/\nu_e$, the emitted photons pass through the local Lyα resonance as they propagate toward us through a smoothly distributed sea of neutral hydrogen atoms, and are scattered off the line-of-sight with a cross-section (neglecting stimulated emission) of

$$\sigma[\nu_o(1 + z)] = \frac{\pi e^2}{m_e c} f \,\phi[\nu_o(1 + z)], \tag{6.41}$$

where $f = 0.4162$ is the upward oscillator strength for the transition, ϕ is the line profile function [with normalization $\int \phi(\nu)d\nu = 1$], c is the speed of light, and e and m_e are the electron charge and mass, respectively. The total optical depth for resonant scattering at the observed frequency is given by the line integral of this cross-section times the neutral hydrogen proper density $n_{\text{HI}}(z)$,

$$\tau_{\text{GP}} = \int_0^{z_e} \sigma[\nu_o(1 + z)] n_{\text{HI}}(z) \frac{c\,dt}{dz} dz, \tag{6.42}$$

where $cdt/dz = cH_0^{-1}(1+z)^{-1}[\Omega_M(1+z)^3 + \Omega_K(1+z)^2 + \Omega_\Lambda]^{-1/2}$ is the proper line element in a Friedmann–Robertson–Walker metric, and Ω_M, Ω_Λ, and $\Omega_K = 1 - \Omega_M - \Omega_\Lambda$ are the matter, vacuum, and curvature contribution to the present density parameter. As the scattering cross-section is sharply peaked around ν_α, we can write

$$\tau_{\mathrm{GP}}(z) = \left(\frac{\pi e^2 f}{m_e c \nu_\alpha}\right) n_{\mathrm{HI}}(1+z)\frac{cdt}{dz}. \tag{6.43}$$

In an Einstein–de Sitter ($\Omega_M = 1$, $\Omega_\Lambda = 0$) universe, this becomes

$$\tau_{\mathrm{GP}}(z) = \frac{\pi e^2 f}{m_e H_0 \nu_\alpha}\frac{n_{\mathrm{HI}}}{(1+z)^{3/2}} = 6.6 \times 10^3 h^{-1}\left(\frac{\Omega_b h^2}{0.019}\right)\frac{n_{\mathrm{HI}}}{\bar{n}_{\mathrm{H}}}(1+z)^{3/2}. \tag{6.44}$$

The same expression for the opacity is also valid in the case of optically thin (to Lyα scattering) discrete clouds as long as n_{HI} is replaced with the average neutral density of individual clouds times their volume filling factor.

In an expanding universe homogeneously filled with neutral hydrogen, the above equations apply to all parts of the source spectrum to the blue of Lyα. An absorption trough should then be detected in the level of the rest-frame UV continuum of the quasar; this is the so-called Gunn–Peterson effect. Between the discrete absorption lines of the Lyα forest clouds, quasar spectra do not show a pronounced Gunn–Peterson absorption trough. The current upper limit at $z_e \approx 5$ is $\tau_{\mathrm{GP}} < 0.1$ in the region of minimum opacity, implying from Equation (6.44) a neutral fraction of $n_{\mathrm{HI}}/\bar{n}_{\mathrm{H}} < 10^{-6}\,h$. Even if 99% of all the cosmic baryons fragment at these epochs into structures that can be identified with quasar absorption systems, with only 1% remaining in a smoothly distributed component, the implication is a diffuse IGM which is ionized to better than 1 part in 10^4.

In modern interpretations of the IGM, it is difficult to use the Gunn–Peterson effect to quantify the amount of ionizing radiation that is necessary to keep the neutral hydrogen absorption below the detection limits. This is because, in hierarchical clustering scenarios for the formation of cosmic structures (the CDM model being the most studied example), the accumulation of matter in overdense regions under the influence of gravity reduces the optical depth for Lyα scattering considerably below the average in most of the volume of the universe, and regions of minimum opacity occur in the most underdense areas (expanding "cosmic minivoids").

6.6.3   A Clumpy IGM

Owing to the nonlinear collapse of cosmic structures, the IGM is well known to be highly inhomogeneous. The discrete gaseous systems detected in absorption in the spectra of high-redshift quasars blueward of the Lyα

emission line are assigned different names based on the appearance of their absorption features (see Figure 6.1).

The term "Lyα forest" is used to denote the plethora of narrow absorption lines whose measured equivalent widths imply H I column densities ranging from $10^{16}\,\mathrm{cm}^{-2}$ down to $10^{12}\,\mathrm{cm}^{-2}$. These systems, observed to evolve rapidly with redshift between $2 < z < 4$, have traditionally been interpreted as intergalactic gas clouds associated with the era of baryonic infall and galaxy formation, photoionized (to less than a neutral atom in 10^4) and photoheated (to temperatures close to 20,000 K) by a ultraviolet background close to the one inferred from the integrated contribution from quasars. High resolution and high signal-to-noise spectra obtained with the *Keck* telescope have shown that most Lyα forest clouds at $z \sim 3$ down to the detection limit of the data have undergone some chemical enrichment, as evidenced by weak, but measurable C IV lines. The typical inferred metallicities range from 0.3% to 1% of solar values, subject to uncertainties of photoionization models. Clearly, these metals were produced in stars that formed in a denser environment; the metal-enriched gas was then expelled from the regions of star formation into the IGM.

An intervening absorber at redshift z having a neutral hydrogen column density exceeding $2 \times 10^{17}\,\mathrm{cm}^{-2}$ is optically thick to photons having energy greater than 13.6 eV, and produces a discontinuity at the hydrogen Lyman limit, i.e., at an observed wavelength of $912\,(1+z)\,\text{Å}$. These scarcer Lyman-limit systems (LLS) are associated with the extended gaseous halos of bright galaxies near the line-of-sight, and have metallicities that appear to be similar to that of Lyα forest clouds.

In damped Lyα systems the H I column is so large ($N_{\mathrm{HI}} \gtrsim 10^{20}\,\mathrm{cm}^{-2}$, comparable to the interstellar surface density of spiral galaxies today) that the radiation damping wings of the Lyα line profile become detectable. Relatively rare, damped systems account for most of the neutral hydrogen seen at high redshifts. The typical metallicities are about 10% of solar, and do not evolve significantly over a redshift interval $0.5 < z < 4$ during which most of today's stars were actually formed.

Except at the highest column densities, discrete absorbers are inferred to be strongly photoionized. From quasar absorption studies we also know that neutral hydrogen accounts for only a small fraction, $\sim 10\%$, of the nucleosynthetic baryons at early epochs.

6.6.3.1 *Distribution of Column Densities and Evolution*

The bivariate distribution $f(N_{\mathrm{HI}}, z)$ of H I column densities and redshifts is defined by the probability dP that a line-of-sight intersects a cloud with column density N_{HI} in the range dN_{HI}, at redshift z in the range dz,

$$dP = f(N_{\mathrm{HI}}, z)dN_{\mathrm{HI}}dz. \tag{6.45}$$

As a function of the column, a single power law with slope -1.5 appears to provide at high redshift a surprisingly good description over 9 decades in N_{HI}, i.e., from 10^{12} to 10^{21} cm^{-2}. It is a reasonable approximation to use for the distribution of absorbers along the line-of-sight:

$$f(N_{HI}, z) = A\, N_{HI}^{-1.5}(1 + z)^{\gamma}. \tag{6.46}$$

Lyα forest clouds and Lyman-limit systems appear to evolve at slightly different rates, with $\gamma = 1.5 \pm 0.4$ for the LLS and $\gamma = 2.8 \pm 0.7$ for the forest lines. Let us assume, for simplicity, a single redshift exponent, $\gamma = 2$, for the entire range in column densities. In the power-law model (6.46) the number N of absorbers with columns greater than N_{HI} per unit increment of redshift is

$$\frac{dN}{dz} = \int_{N_{HI}}^{\infty} f(N'_{HI}, z)dN'_{HI} = 2AN_{HI}^{-0.5}(1 + z)^2. \tag{6.47}$$

A normalization value of $A = 4.0 \times 10^7$ produces then ~ 3 LLS per unit redshift at $z = 3$, and, at the same epoch, ~ 150 forest lines above $N_{HI} = 10^{13.8}$ cm^{-2}, in reasonable agreement with the observations.

If absorbers at a given surface density are conserved, with fixed comoving space number density $n = n_0(1 + z)^3$ and geometric cross-section Σ, then the intersection probability per unit redshift interval is

$$\frac{dP}{dz} = \Sigma n \frac{cdt}{dz} = \Sigma n_0 (1 + z)^3 \frac{cdt}{dz}. \tag{6.48}$$

If the universe is cosmologically flat, the expansion rate at early epochs is close to the Einstein–de Sitter limit, and the redshift distribution for conserved clouds is predicted to be

$$\frac{dP}{dz} \propto (1 + z)^3 \frac{cdt}{dz} \propto (1 + z)^{1/2}. \tag{6.49}$$

The rate of increase of $f(N_{HI}, z)$ with z in both the Lyα forest and LLS is considerably faster than this, indicating rapid evolution. The mean proper distance between absorbers along the line-of-sight with columns greater than N_{HI} is

$$L = \frac{cdt}{dz}\frac{dz}{dN} \approx \frac{cN_{HI}^{1/2}}{H_0 \Omega_M^{1/2} 2A(1 + z)^{4.5}}. \tag{6.50}$$

For clouds with $N_{HI} > 10^{14}$ cm^{-2}, this amounts to $L \sim 0.7\, h^{-1}\Omega_M^{-1/2}$ Mpc at $z = 3$. At the same epoch, the mean proper distance between LLS is $L \sim 30\, h^{-1}\Omega_M^{-1/2}$ Mpc.

6.6.3.2 *Intergalactic Continuum Opacity*

Even if the bulk of the baryons in the universe are fairly well ionized at all redshifts $z \lesssim 5$, the residual neutral hydrogen still present in the Lyα forest clouds and Lyman-limit systems significantly attenuates the ionizing flux from cosmological distant sources. To quantify the degree of attenuation we have to introduce the concept of an effective continuum optical depth $\tau_{\rm eff}$ along the line-of-sight to redshift z,

$$\langle e^{-\tau} \rangle \equiv e^{-\tau_{\rm eff}}, \tag{6.51}$$

where the average is taken over all lines-of-sight. Neglecting absorption due to helium, if we characterize the Lyα forest clouds and LLS as a random distribution of absorbers in column density and redshift space, then the effective continuum optical depth of a clumpy IGM at the observed frequency ν_o for an observer at redshift z_o is

$$\tau_{\rm eff}(\nu_o, z_o, z) = \int_{z_o}^{z} dz' \int_0^{\infty} dN_{\rm HI}\, f(N_{\rm HI}, z)(1 - e^{-\tau}), \tag{6.52}$$

where $\tau = N_{\rm HI}\sigma_H(\nu)$ is the hydrogen Lyman continuum optical depth through an individual cloud at frequency $\nu = \nu_o(1 + z)/(1 + z_o)$. This formula can be easily understood if we consider a situation in which all absorbers have the same optical depth τ_0 independent of redshift, and the mean number of systems along the path is $\Delta N = \int dz dN/dz$. In this case the Poissonian probability of encountering a total optical depth $k\tau_0$ along the line-of-sight (with k integer) is $p(k\tau_0) = e^{-\Delta N}\Delta N^k/(\tau_0 k!)$, and $\langle e^{-\tau} \rangle = e^{-k\tau_0}p(k\tau_0) = \exp[-\Delta N(1 - e^{-\tau_0})]$.

If we extrapolate the $N_{\rm HI}^{-1.5}$ power law in Equation (6.46) to very small and large columns, the effective optical depth becomes an analytical function of redshift and wavelength,

$$\tau_{\rm eff}(\nu_o, z_o, z) = \frac{4}{3}\sqrt{\pi\sigma_L}\, A \left(\frac{\nu_o}{\nu_L}\right)^{-1.5} (1 + z_o)^{1.5} \left[(1 + z)^{1.5} - (1 + z_o)^{1.5}\right]. \tag{6.53}$$

Due to the rapid increase with lookback time of the number of absorbers, the mean free path of photons at 912 Å becomes so small beyond a redshift of 2 that the radiation field is largely "local". Expanding Equation (6.53) around z, one gets $\tau_{\rm eff}(\nu_L) \approx 0.36(1+z)^2\Delta z$. This means that at $z = 3$, for example, the mean free path for a photon near threshold is only $\Delta z = 0.18$, and sources of ionizing radiation at higher redshifts are severely attenuated.

6.7 Conclusions

The above discussion should make it clear that, despite much recent progress in our understanding of the formation of early cosmic structure and the high-redshift universe, the astrophysics of first light remains one of the missing links in galaxy formation and evolution studies. We are left very uncertain about the whole era from 10^8 to 10^9 yr – the epoch of the first galaxies, stars, supernovae, and massive black holes. Some of the issues discussed above are likely to remain a topic of lively controversy until the launch of the James Webb Space Telescope (JWST), ideally suited to image the earliest generation of stars in the universe. If the first massive black holes form in pregalactic systems at very high redshifts, they will be incorporated through a series of mergers into larger and larger halos, sink to the center owing to dynamical friction, accrete a fraction of the gas in the merger remnant to become supermassive, and form binary systems. Their coalescence would be signaled by the emission of low-frequency gravitational waves detectable by the planned Laser Interferometer Space Antenna (LISA). An alternative way to probe the end of the dark ages and discriminate between different reionization histories is through 21 cm tomography (Madau et al. 1997). Prior to the epoch of full reionization, 21 cm spectral features will display angular structure as well as structure in redshift space due to inhomogeneities in the gas density field, hydrogen ionized fraction, and spin temperature. Radio maps will show a patchwork (both in angle and in frequency) of emission signals from H I zones modulated by H II regions where no signal is detectable against the CMB (Ciardi & Madau 2003). The search at 21 cm for the epoch of first light has become one of the main science drivers of the LOw Frequency ARray (LOFAR). While remaining an extremely challenging project due to foreground contamination from unresolved extragalactic radio sources (Di Matteo et al. 2002) and free-free emission from the same halos that reionize the universe (Oh & Mack 2003), the detection and imaging of large-scale structures prior to reionization breakthrough remains a tantalizing possibility within range of the next generation of radio arrays.

Acknowledgments

We have benefited from many discussions with all our collaborators on the topics described here. Support for this work was provided by NASA grants NAG5-11513 and NNG04GK85G, and by NSF grant AST-0205738 (PM).

References

Barkana R, Haiman Z and Ostriker J P 2001 *ApJ* **558** 482
Becker R H et al. 2001 *AJ* **122** 2850
Benson A J and Madau P 2003 *MNRAS* **344** 835
Bond J R, Arnett W D and Carr B J 1984 *ApJ* **280** 825
Bromm V, Coppi P S and Larson R B 2002 *ApJ* **564** 23
Bromm V, Kudritzki R P and Loeb A 2001 *ApJ* **552** 464
Bruzual A C and Charlot S 1993 *ApJ* **405** 538
Bullock J S, Kravtsov A V and Weinberg D H 2001 *ApJ* **548** 33
Carilli C, Gnedin N Y and Owen F 2002 *ApJ* **577** 22
Cen R 2003 *ApJ* **591** 12
Cen R and Bryan G L 2001 *ApJ* **546** L81
Cen R and Haiman Z 2000 *ApJ* **542** L75
Cen R, Miralda-Escudé J, Ostriker J P and Rauch M 1994 *ApJ* **437** L9
Chiu W A and Ostriker J P 2000 *ApJ* **534** 507
Ciardi B and Madau P 2003 *ApJ* **596** 1
Ciardi B, Ferrara A and White S D M 2003 *MNRAS*
Ciardi B, Ferrara A, Governato F and Jenkins A 2000 *MNRAS* **314** 611
Couchman H M P, Thomas P A and Pearce F R 1995 *ApJ* **452** 797
Dekel A and Silk J 1986 *ApJ* **303** 39
Di Matteo T, Perna R, Abel T and Rees M J 2002 *ApJ* **564** 576
Djorgovski S G, Castro S M, Stern D and Mahabal A A 2001 *ApJ* **560** L5
Ellison S L, Songaila A, Schaye J and Pettini M 2000 *AJ* **120** 1175
Fan X et al. 2001 *AJ* **121** 54
Fan X et al. 2002 *AJ* **123** 1247
Field G B 1958 *Proc I R E* **46** 240
Field G B 1959 *ApJ* **129** 551
Fryer C L, Woosley S E and Heger A 2001 *ApJ* **550** 372
Fukugita M, Hogan C J and Peebles P J E 1998 *ApJ* **503** 518
Furlanetto S and Loeb A 2002 *ApJ* **579** 1
Furlanetto S and Loeb A 2003 *ApJ* **588** 18
Glazebrook K et al. 2003 *ApJ* **587** 55
Gnedin N Y 2000a *ApJ* **535** 530
Gnedin N Y 2000b *ApJ* **542** 535
Gnedin N Y and Hui L 1998 *MNRAS* **296** 44
Haardt F and Madau P 1996 *ApJ* **461** 20
Haiman Z and Loeb A 1998 *ApJ* **503** 505

Haiman Z and Holder G P 2003 *ApJ* submitted (astro-ph/0302403)

Heger A and Woosley S E 2002 *ApJ* **567** 532

Hernquist L, Katz N, Weinberg D and Miralda-Escudé J 1996 *ApJ* **457** L51

Hofmann S, Schwarz D J and Stocker H 2001 *PhRvD* **64** 083507

Iliev I T, Shapiro P R, Ferrara A and Martel H 2002 *ApJ* **572** 123

Klypin A, Kravtsov A V, Valenzuela O and Prada F 1999 *ApJ* **522** 82

Kogut A et al. 2003 *ApJS* **148** 161

Kriss G A et al. 2001 *Science* **293** 1112

Kudritzki R P 2000 in *The First Stars* ed A Weiss T Abel and V Hill (Heidelberg: Springer) 127

Lahav O et al. 1991 *MNRAS* **151** 128

Loeb A and Barkana R 2001 *ARA&A* **39** 19

Mac Low M -M and Ferrara A 1999 *ApJ* **513** 142

Machacek M M, Bryan G L and Abel T 2003 *MNRAS* **338** 273

Madau P 2000 *RSPTA* **358** 2021

Madau P, Ferrara A and Rees M J 2001 *ApJ* **555** 92

Madau P, Haardt F and Rees M J 1999 *ApJ* **514** 648

Madau P, Meiksin A and Rees M J 1997 *ApJ* **475** 429

Merritt D and Ferrarese L 2001 *ApJ* **547** 140

Miralda-Escudé J 1998 *ApJ* **501** 15

Moore B et al. 1999 *ApJ* **524** L19

Mori M, Ferrara A and Madau P 2002 *ApJ* in press (astro-ph/0106346)

Navarro J F, Frenk C S and White S D M 1997 *ApJ* **490** 493

Oh S P and Mack K J 2003 *MNRAS* **346** 871

Peebles P J E 1993 *Principles of Physical Cosmology* (Princeton University Press)

Perlmutter S et al. 1999 *ApJ* **517** 565

Ponman T J, Cannon D B and Navarro J F 1999 *Nature* **397** 135

Purcell E M and Field G B 1956 *ApJ* **124** 542

Rauch M, Sargent W L W and Barlow T A 2001 *ApJ* **554** 823

Rees M J 2000 *RSPTA* **358** 1989

Reiprich T H and Böhringer H 2002 *ApJ* **567** 716

Renzini A 1997 *ApJ* **488** 35

Riess A et al. 1998 *AJ* **116** 1009

Scannapieco E, Ferrara A and Madau P 2002 *ApJ* **574** 590

Schneider R, Ferrara A, Natarajan P and Omukai K 2002 *ApJ* **571** 30

Seager S, Sasselov D D and Scott D 1999 *ApJ* **523** L1

Shapiro P R and Giroux M L 1987 *ApJ* **321** L107

Shaver P, Windhorst R, Madau P and de Bruyn G 1999 *A&A* **345** 380

Shaver P A, Wall J V, Kellerman K I, Jackson C A and Hawkins M R S 1996 *Nature* **384** 439

Simcoe R A, Sargent W L W and Rauch M 2002 *ApJ* **578** 737

Somerville R S, Bullock J S and Livio M 2003 *ApJ*

Songaila A 1997 *ApJ* **490** L1

Songaila A 2001 *ApJ* **561** 153

Spergel D N et al. 2003 *ApJS* **148** 175

Steidel C C, Pettini M and Adelberger K L 2001 *ApJ* **546** 665

Tegmark M, Silk J and Evrard A 1993 *ApJ* **417** 54

Theuns T, Mo H J and Schaye J 2001 *MNRAS* **321** 450

Tozzi P, Madau P, Meiksin A and Rees M J 2000 *ApJ* **528** 59

van den Bosch F C, Robertson B E, Dalcanton J and de Block W J G 2000 *AJ* **119** 1579

Voit G M 1996 *ApJ* **465** 548

Volonteri M, Haardt F and Madau P 2003 *ApJ* **582** 559

Williams R E et al. 1996 *AJ* **112** 1335

Wyithe J S B and Loeb A 2003 *ApJ* **596** 34

Zentner A R and Bullock J S 2002 *PhRvD* **66** 43003

Zhang Y, Anninos P and Norman M L 1995 *ApJ* **453** L57

Chapter 7

Feedback Processes at Cosmic Dawn

Andrea Ferrara & Ruben Salvaterra
International School for Advanced Studies
Trieste, Italy

The word "feedback" is by far one of the most used ones in modern cosmology where it is applied to a vast range of situations and astrophysical objects. However, for the same reason, its meaning in the context is often unclear or fuzzy. Hence a review on feedback should start from setting the definition of feedback on a solid basis. We have found quite useful to this aim to go back to the *Oxford Dictionary* from where we take the following definition.

Feed' back *n. 1. (Electr.) Return of fraction of output signal from one stage of circuit, amplifier, etc. to input of same or preceding stage* (**positive, negative***, tending to increase, decrease the amplification, etc). 2. (Biol., Psych., etc) Modification or control of a process or system by its results or effects, esp. by difference between the desired and actual results.*

In spite of the broad description, we find this definition quite appropriate in many ways. First, the definition outlines the fact that the concept of feedback invokes a back reaction of a process on itself or on the causes that have produced it. Secondly, the character of feedback can be either negative or positive. Finally, and most importantly, the idea of feedback is intimately linked to the possibility that a system can become self-regulated. Although some types of feedback processes are disruptive, the most important ones in astrophysics are probably those that are able to drive the systems towards a steady state of some sort. To exemplify, think of a galaxy that is witnessing a burst of star formation. The occurrence of the first supernovae will evacuate/heat the gas thus suppressing the star formation activity. Such feedback is then acting back on the energy source (star formation); it is of a negative type, and it could regulate the star formation activity in such a way that only a sustainable amount of stars is formed (regulation). However, feedback can fail to produce such regulation either in small galaxies where the gas can be ejected by the first SNe or in cases when the star formation timescale is too short compared to the feedback one. As we will see there are at least three types of feedback, and even the stellar feedback described in the example above is part of a larger class of feedback phenomena related to the energy deposition of massive stars. We then start by briefly describing the key physical ingredients of feedback processes in cosmology.

7.1 Shock Waves

A shock represents a "hydrodynamic surprise", in the sense that such a wave travels faster than signals in the fluid, which are bound to propagate at the gas sound speed, c_s. It is common to define the shock speed, v_s, in terms of c_s by introducing the shock Mach number, M:

$$v_s = M c_s = M(\gamma P/\rho)^{1/2}, \tag{7.1}$$

where P and ρ are the gas pressure and density, respectively. γ is the ratio of specific heats at constant pressure and constant volume. The change of the fluid velocity $\vec{v}$ is governed by the momentum equation

$$\rho \frac{\partial \vec{v}}{\partial t} + \rho \vec{v} \cdot \vec{\nabla} \vec{v} = -\vec{\nabla} P - \frac{1}{8\pi} \vec{\nabla} B^2 + \frac{1}{4\pi} \vec{B} \cdot \vec{\nabla} \vec{B}, \tag{7.2}$$

where $\vec{B}$ is the magnetic field. The density is determined by the continuity equation

$$\frac{\partial \rho}{\partial t} + \vec{v} \cdot \vec{\nabla} \rho = -\rho \vec{\nabla} \vec{v}. \tag{7.3}$$

7.1.1 Hydrodynamics of Shock Waves

The large-scale properties of a shock in a perfect gas, with $B = 0$ are fully described by the three ratios v_2/v_1 ρ_2/ρ_1 and P_2/P_1, determined in terms of the conditions ahead of the shock (i.e., v_1, ρ_1 and P_1) by the Rankine–Hugoniot jump conditions which relate physical quantities on each side of the front and result from matter, momentum and energy conservation:

$$\rho_1 v_1 = \rho_2 v_2, \tag{7.4}$$

$$P_1 + \rho_1 v_1^2 = P_2 + \rho_2 v_2^2, \tag{7.5}$$

$$v_2 \left(\frac{1}{2} \rho_2 v_2^2 + U_2 \right) - v_1 \left(\frac{1}{2} \rho_1 v_1^2 + U_1 \right) = v_1 P_1 - v_2 P_2, \tag{7.6}$$

where U is the internal energy density of the fluid. If the fluid behaves as a perfect gas on each side of the front we have $U = P/(\gamma - 1)$.

In the adiabatic limit it is sufficient to consider the mass and momentum equations only. By solving such a system we obtain:

$$\frac{v_2}{v_1} = \frac{\rho_1}{\rho_2} = \frac{\gamma - 1}{\gamma + 1} + \frac{2}{\gamma + 1}\frac{1}{M^2}, \tag{7.7}$$

where the Mach number is defined as $M^2 = v_1^2/c_1^2$ and c_1 is the sound velocity in the unperturbed medium ahead of the shock. For strong shocks, i.e., $M \gg 1$ and $\gamma = 5/3$, $\rho_2/\rho_1 = 4$ and

$$T_2 = \frac{3\mu}{16k}v_s^2, \tag{7.8}$$

where $v_s = v_1 - v_2$

In the isothermal limit, i.e., when the cooling time of the postshock gas becomes extremely short and the initial temperature is promptly re-established,

$$\frac{\rho_2}{\rho_1} = \frac{v_1^2}{c_s^2} = M^2, \tag{7.9}$$

where c_s is the sound velocity on both sides of the shock. While in the adiabatic case the compression factor of the gas density is limited to four, in an isothermal shock much larger compressions are possible due to the softer equation of state characterizing the fluid.

7.1.2 Hydromagnetic Shock Waves

Assuming B is parallel to the shock front (so that $B \cdot \nabla B = 0$), Equation (7.5) can be replaced by

$$P_1 + \rho_1 v_1^2 + \left(\frac{B_1^2}{8\pi}\right) = P_2 + \rho_2 v_2^2 + \left(\frac{B_2^2}{8\pi}\right). \tag{7.10}$$

Magnetic flux conservation requires $B_1/\rho_1 = B_2/\rho_2$, so that in the adiabatic limit the magnetic pressure $B^2/8\pi$ increases by only a factor 16.

In the isothermal limit the compression depends only linearly on the Alfvénic Mach number M_a

$$\frac{\rho_2}{\rho_1} = \sqrt{2}\frac{v_1}{v_{a,1}} = M_a, \tag{7.11}$$

where $v_a = B/\sqrt{4\pi\rho}$ is the Alfvén velocity. Hence, some of the shock kinetic energy is stored in the magnetic field lines rather than being used to compress the gas as in the field-free case; in other words a magnetized gas is less compressible than an unmagnetized one.

7.1.2.1   Structure of Radiative Shocks

When a shock wave has propagated long enough that radiative losses have cooled the first-shocked material to a small fraction of its initial postshock temperature, then the shock wave is referred to as a *radiative shock*. The structure of a radiative shock can be divided into four regions: radiative precursor, shock front, radiative zone, and thermalization region (McKee & Draine 1991). The radiative precursor is the region upstream of the shock in which radiation emitted by the shock acts as a precursor of the shock arrival. The shock front is the region in which the relative kinetic energy difference of the shocked and unshocked gas is dissipated. If the dissipation is due to collisions among the atoms or molecules of the gas, the shock is collisional. On the other hand, if the density is sufficiently low that collisions are unimportant and the dissipation is due to the collective interactions of the particles with the turbulent electromagnetic fields, the shock is collisionless. Next comes the radiative zone, in which collisional processes cause the gas to radiate. The gas cools and the density increases. Finally, if the shock lasts long enough, a thermalization region is produced, in which radiation from the radiative zone is absorbed and re-radiated.

7.1.3   Supernova Explosions

In the explosion of a supernova three different stages in the expansion may be distinguished (Spitzer 1978). In the initial phase the interstellar material has little effect because of its low moment of inertia; the velocity of expansion of the supernova envelope will then remain nearly constant with time. This phase terminates when the mass of the swept up gas is about equal to the initial mass M_e expelled by the supernova; i.e., $(4\pi/3)r_s^3\rho_1 = M_e$, where ρ_1 is the density of the gas in front of the shock and r_s is the radius of the shock front. For $M_e = 0.25\ M_\odot$ and $\rho_1 = 2\times 10^{-24}\ \text{g/cm}^3$, $r_s \simeq 1$ pc, which will occur about 60 yr after the explosion. During the second phase (the so-called adiabatic or Sedov–Taylor phase) the temperature of the gas is so high that the radiation can be neglected. Thus the total energy of the gas within the shock front is constant, and equal to the initial energy, and the mass behind the shock is determined primarily by the amount of interstellar gas swept up. When radiative cooling becomes important (radiative phase), the temperature of the gas will fall to a relatively low value. The motion of the shock is supported by the momentum of the outward moving gas and the shock may be regarded as isothermal. The velocity of the shell can be computed from the condition of momentum conservation (snowplow model). This phase continues until $v_s = \max\{v_t, c_s\}$, where v_t is the turbulent velocity of the gas, when the shell loses its identity due to gas random motions.

7.1.3.1 Blastwave Evolution

To describe the evolution of a blastwave (assuming spherical symmetry), the so-called thin shell approximation is frequently used. In this approximation, one assumes that most of the mass of the material swept up during the expansion is collected in a thin shell. We start by applying (we neglect the ambient medium density and shell self-gravity) the virial theorem to the shell (see Ostriker & McKee 1988 for all the details):

$$\frac{1}{2}\frac{d^2 I}{dt^2} = 2E = 2(E_k + E_t), \tag{7.12}$$

where $I = \int dm\, r^2$ is the moment of inertia, and E_k (E_t) is the kinetic (thermal) energy of the shell. Let us introduce the (time independent) structure parameter K in the following form,

$$K = \frac{1}{MRv_2}\int_0^R dm\, rv; \tag{7.13}$$

where M is the ejected mass, R the shell radius, and v_2 is the postshock velocity (equal to $3/4v_s$ for an adiabatic shock whose velocity is $v_s = dR/dt$). For a linear velocity field *inside* the ejecta, $v = v_2(r/R)$, then $I = MR^2 K$; substituting into Equation (7.12), we obtain

$$K\frac{d}{dt}(Mrv_s) = 2E(v_s/v_2) = 8E/3. \tag{7.14}$$

To recover the Sedov-Taylor adiabatic phase we integrate up to R and assume $\rho(R) = \rho_0 R^{-m}$, $E = const.$, $M = (4\pi/3)\rho_0 R^{3-m}$; with this assumption we also find that $K = (6 - 2m)/(7 - 2m)$ and by integrating up to R,

$$R(t) = \left[\frac{(5-m)E}{\pi\rho_0 K}\right]^{1/(5-m)} t^{2/(5-m)}. \tag{7.15}$$

The previous expression can be obtained also via a dimensional approach. In fact, in the adiabatic phase the energy is conserved and hence

$$E \propto Mv_s^2 \propto \rho_0 R^{5-m}t^{-2}, \tag{7.16}$$

so that

$$R^{5-m} \propto \frac{E}{\rho_0}t^2; \tag{7.17}$$

the fraction of the total energy transformed in kinetic form is

$$E_k = \frac{1}{2}KM\left(\frac{3}{4}v_s^2\right) = \frac{3E}{2(5-m)}. \tag{7.18}$$

We note that in the limit of a homogeneous ambient medium ($m = 0$) we recover the standard Sedov–Taylor evolution $R \propto t^{2/5}$, with 30% of the explosion energy in kinetic form.

A similar reasoning can be followed to recover the radiative phase behavior. In this phase $v_2 = v_s$ (i.e., the shock wave is almost isothermal) and $K = 1$ (i.e., the shell is thin). Under these assumptions the energy is lost at a rate

$$\dot{E} = -4\pi R^2 \left(\frac{1}{2} \rho_0 v_s^2 \right). \tag{7.19}$$

There are two possible solutions of Equation (7.12) in this case: (a) the pressure-driven snowplow in which

$$R(t) \propto t^{2/(7-m)}, \tag{7.20}$$

or (b) the momentum-conserving solution

$$R(t) \propto t^{1/(4-m)}. \tag{7.21}$$

In both cases, the expansion proceeds at a lower rate with respect to the adiabatic phase.

In order to study the evolution of an explosion in a more realistic way it is necessary to resort to numerical simulations. As an illustrative example, in Figure 7.1 (Cioffi, McKee, & Bertschiger 1988) the structure of a spherical supernova remnant is calculated using a high-resolution numerical hydrodynamical code. In the figure, the inner structure of the remnant is clearly visible: indeed most of the mass is concentrated in a thin shell (located at about 80 pc after 0.5 Myr from the explosion); a reverse shock is also seen which is proceeding towards the center and thermalizing the ejecta up to temperatures of the order of 10^7 K. In between the two shocks is the contact discontinuity between the ejecta and the ambient medium, through which the pressure remains approximately constant.

7.2   Photoionization

Let us consider a source of ionizing photons in a medium of neutral gas with density n. The ionization balance equation around the source is given by

$$4\pi R^2 n \frac{dR}{dt} + \frac{4}{3}\pi R^3 n^2 \alpha = 4\pi S_i(0), \tag{7.22}$$

where $S_i(0)$ is the ionizing photon rate from the source, and α is the recombination coefficient. The solution of the ionization balance equation is

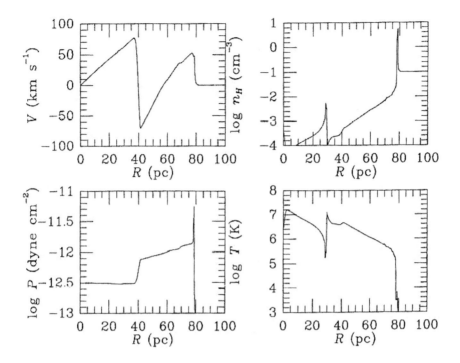

Figure 7.1. The structure of spherical supernova remnant calculated using a high-resolution numerical hydrodynamical code. The remnant age is $t = 5 \times 10^5$ yr. This figure is taken from Cioffi, McKee, & Bertschiger (1988).

$$R(t) = R_S[1 - \exp(-2n\alpha t/3)]^{1/3}. \qquad (7.23)$$

As time goes to infinity the radius tends to the Stromgren radius

$$R_S = \left(\frac{3S_i(0)}{\alpha n^2}\right)^{1/3}. \qquad (7.24)$$

The above equation implies that R_S is reached in

$$t_S = \frac{3}{2}n\alpha \simeq 2 \times 10^5 n \text{ yr}, \qquad (7.25)$$

and the typical ionization front (I-front) velocity is $v_I \propto S_i(0)/R^2 n$.

Depending on the value of v_I we can distinguish three types of I-fronts, where c_i (c_n) is the sound speed in the ionized (neutral) gas:

(i) R(arefied)-TYPE: $v_I > (c_i, c_n)$
(ii) Strong D(ense)-TYPE: $c_n < v_I < c_i$

(iii) Weak D(ense)-TYPE: $v_I < (c_i, c_n)$

In the case of R-TYPE I-front there is no hydrodynamical reaction to the passage of the ionizing front, since its motion is supersonic relative to the gas behind as well as ahead of the front. In the case of D-TYPE fronts the dynamical timescale becomes short enough for the gas to react to photoionization and a shock will form and precede the ionizing front thus increasing the ambient gas density.

7.2.0.2 *Photoionization Equilibrium*

It is convenient to divide the ionizing flux into two parts, a stellar part $S(\nu)$, resulting directly from the input radiation from the star, and a diffuse part $S_d(\nu)$, resulting from the emission of the ionized gas, that in a pure H gas is due to recombinations into 1^2 S state.

$$S_t(\nu) = S(\nu) + S_d(\nu). \qquad (7.26)$$

Both the recombination radiation and photoionization cross-section ($\sigma \propto \nu^3$) are peaked at the threshold, ν_L. We can identify two approximate cases (Osterbrok 1989):

(i) Menzel Case A: optically thin nebula; i.e., $\tau(\nu = \nu_L) = n_H \sigma(\nu_L) R < 1$, and the diffuse photons can escape;
(ii) Menzel Case B: optically thick nebula; i.e., diffuse photons are absorbed very close to the point at which they are generated (on-the-spot approximation).

Menzel case B is usually a good approximation for many astrophysical purposes because the diffuse radiation-field photons have $\nu \simeq \nu_L$, and therefore a short mean free path before absorption. Under the hypothesis of the on-the-spot approximation, the recombination coefficient is given by

$$\alpha_{eff} = \alpha - \alpha_{1S}, \qquad (7.27)$$

resulting roughly in a 60% decrease. The physical meaning is that in optically thick nebulae, the ionizations caused by stellar radiation-field photons are balanced by recombinations to excited levels of H, while recombinations to the ground level generate ionizing photons that are absorbed elsewhere in the nebula but have no net effect on the overall ionization balance.

7.3 Thermal Instability

A final crucial ingredient to understand feedback and self-regulation is represented by the thermal instability. Let us rewrite the usual mass, momentum and energy equations in a slightly different form (Balbus 1995):

$$\frac{d\ln\rho}{dt} + \rho\nabla\cdot v = 0, \tag{7.28}$$

$$\rho\frac{dv}{dt} = -\nabla P, \tag{7.29}$$

$$\frac{3}{2}P\frac{d\ln P}{dt}\rho^{-5/3} = -\rho L(\rho, T), \tag{7.30}$$

where L is the net cooling function (i.e., heating–cooling) and is taken to be a function of density and temperature; in the equilibrium state $v = L = 0$. We perturb the equilibrium by introducing a small disturbance. We write such perturbations as $\delta q \propto e^{(kr+nt)}$; nt is then the growth rate of the instability. The linearized system yields the dispersion relation

$$n^3 + \frac{2}{3}T\Theta_{T,\rho}n^2 + k^2c_s^2n + \frac{2}{5}k^2c_s^2T\Theta_{T,\rho} = 0, \tag{7.31}$$

where $\Theta \equiv (\rho/P)L$ and $\Theta_{A,B} \equiv (\partial\Theta/\partial A)_B$.

For large k (small wavelengths) the instability occurs isobarically; that is, pressure is kept approximately constant as the density increases following the process of condensation and cooling. In this case such equations have three solutions:

$$n_1 = -\frac{2}{5}T\Theta_{T,P}, \tag{7.32}$$

$$n_{2,3} = \pm ikc_s - \frac{2}{15}T\Theta_{T,S}. \tag{7.33}$$

It is straightforward to recognize that $n_{2,3}$ are the modified sound wave modes, whereas n_1 corresponds to the condensation mode. Note also that the sign of the temperature derivative of the cooling function determines whether a mode is unstable (instability occurs if $\Theta_{T,P} < 0$).

In the opposite (isochoric) limit characterized by small k (large wavelengths), the condensation mode has the following expression,

$$n_1 = -\frac{2}{3}T\Theta_{T,\rho}. \tag{7.34}$$

Although formally a solution of the dispersion equation, the thermal instability cannot physically occur under isochoric conditions. The reason is that on such large scales there is not enough time to establish pressure equilibrium as the cooling time is much shorter than the sound crossing time over a wavelength. Thus the density remains approximately constant while the gas cools, and pressure equilibrium is likely to be restored promptly and rather dramatically by a propagating shock front.

To better understand the nature of the thermal instability let us plot the curve of the equilibrium solution $L = 0$ in the P–T plane for a hypothetical cooling function L (Figure 7.2). Above the curve, heating dominates over cooling; below $L = 0$, the gas is cooled. Between P_{min} and P_{max} a constant pressure line will intersect the curve in three points (A, B, and C); A and C are thermally stable. This is readily understood since a small isobaric increase (decrease) in temperature at either of these points takes the system into the cooling (heating) region of the P–T plane from where it is immediately driven back to equilibrium position. On the other hand, a small increase (decrease) in the temperature in the unstable point B takes one into the heating (cooling) region, with the result that the system is driven yet further from its equilibrium. This leads to the coexistence of two (or more) thermodynamic phases and provides the main argument to interpret the multiphase medium commonly observed in galaxies and in cosmic gas. We will see later on how this concept is closely linked with the self-regulation connected to feedback processes.

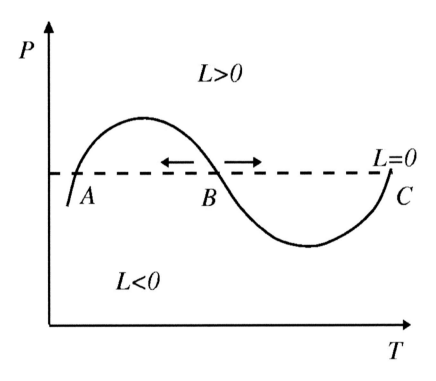

Figure 7.2. Equilibrium solution curve $L = 0$ (solid line) in the P–T plane for a hypothetical cooling function. See text.

7.4   The Need for Feedback in Cosmology

One of the main aims of physical cosmology is to understand in detail galaxy formation starting from the initial density fluctuation field with a given spectrum (typically a CDM one). Such *ab initio* computations require a tremendous amount of physical processes to be included before it becomes possible to compare their predictions with experimental data. In particular, it is crucial to model the interstellar medium of galaxies, which is known to be turbulent, have a multiphase thermal structure, can undergo gravitational instability and form stars. To account for all this complexity, in addition one should treat correctly all relevant cooling processes, radiative and shock heating (let alone magnetic fields!). This has proven to be essentially impossible even with the best present-day supercomputers. Hence one has to resort to heuristic models where simplistic prescriptions for some of these processes must be adopted. Of course such an approach suffers from the fact that a large number of free parameters remain which cannot be fixed from first principles. These are essentially contained in each of the ingredients used to model galaxy formation, that is, the evolution of dark halos, cooling and star formation, chemical enrichment, and stellar populations. Fortunately, there is a large variety of data against which the models can be tested: these data range from the fraction of cooled baryons to cosmic star formation histories, the luminous content of halos, luminosity functions and faint galaxy counts. The feedback processes enter the game as part of such iterative try-and-learn processes to which we are bound by our ignorance in dealing with complex systems such as the galaxies. Still, we are far from a full understanding of galaxies in the framework of structure formation models. The hope is that feedback can help us to solve some "chronic" problems found in cosmological simulations adopting the CDM paradigm. Their (partial) list includes:

(i) **Overcooling**: the predicted cosmic fraction of cooled baryons is larger than observed. Moreover, models predict too many faint, low-mass galaxies;

(ii) **Disk Angular Momentum**: the angular momentum loss is higher and galactic disk scale lengths are smaller than observed;

(iii) **Halo Density Profiles**: profiles are more centrally concentrated than observed;

(iv) **Dark Satellites**: too many satellites predicted around our galaxy.

7.4.1   The Overcooling Problem

Among the various CDM problems, historically overcooling has been the most prominent and yet unsolved one. In its original formulation it has been first spelled out by White & Frenk (1991). Let us assume that, as a halo forms, the gas initially relaxes to an isothermal distribution which exactly parallels that of the dark matter. The virial theorem then relates the gas temperature T_{vir} to the circular velocity of the halo V_c,

$$kT_{vir} = \frac{1}{2}\mu m_p V_c^2 \text{ or } T_{vir} = 36 V_{c,km/s}^2 \text{K}, \qquad (7.35)$$

where μm_p is the mean molecular weight of the gas. At each radius in this distribution we can then define a cooling time as the ratio of the specific energy content to the cooling rate,

$$t_{cool}(r) = \frac{3\rho_g(r)/2\mu m_p}{n_e^2(r)\Lambda(T)}, \qquad (7.36)$$

where $\rho_g(r)$ is the gas density profile and $n_e(r)$ is the electron density. $\Lambda(T)$ is the cooling function. The cooling radius is defined as the point where the cooling time is equal to the age of the universe; i.e. $t_{cool}(r_{cool}) = t_{Hubble} = H(z)^{-1}$.

Considering the virialized part of the halo to be the region encompassing a mean overdensity of 200, its radius and mass are defined by

$$r_{vir} = 0.1 H(z)^{-1} V_c, \qquad (7.37)$$

$$M_{vir} = 0.1 (GH(z))^{-1} V_c^3. \qquad (7.38)$$

Let us distinguish two limiting cases. When $r_{cool} \gg r_{vir}$ (accretion limited case), cooling is so rapid that the infalling gas never comes to hydrostatic equilibrium. The supply of cold gas for star formation is then limited by the infall rate rather than by cooling. The accretion rate is obtained by differentiating Equation (7.38) with respect to time and multiplying by the fraction of the mass of the universe that remains in gaseous form, $f_g\Omega_b$:

$$\dot{M}_{acc} = f_g\Omega_b \frac{d}{dt} 0.1 (GH(z))^{-1} V_c^3 = 0.15 f_g\Omega_b G^{-1} V_c^3. \qquad (7.39)$$

Note that, except for a weak time dependence of the fraction of the initial baryon density that remains in gaseous form, f_g, this infall rate does not depend on redshift.

In the opposite limit, $r_{cool} \ll r_{vir}$ (quasi-static case), the accretion shock radiates only weakly, a quasi-static atmosphere forms, and the supply of cold gas for star formation is regulated by radiative losses near r_{cool}. A simple expression for the inflow rate is

$$\dot{M}_{qst} = 4\pi\rho_g(r_{cool})r_{cool}^2 \frac{d}{dt} r_{cool}. \qquad (7.40)$$

The gas supply rates predicted by Equation (7.39) and Equation (7.40) are illustrated in Figure 7.3. In any particular halo, the rate at which cold

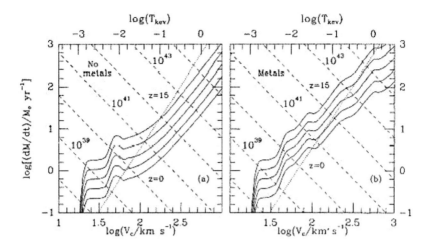

Figure 7.3. Gas infall rate and cooling rates in dark matter halos as a function of circular velocity and redshift. The infall rate (*dotted line*) is essentially independent of redshift; the cooling rates (*solid lines*) are given for redshift $z = 0, 1, 3, 7$ and 15 (*from bottom to top*). Dashed lines give present-day x-ray luminosities in erg s^{-1} produced by gas cooling at the given rate in each halo. The predicted temperature of this emission is given on the upper abscissa. (a) A cooling function for gas of zero metallicity is assumed. (b) A cooling function for metal enriched gas. This figure is taken from White & Frenk (1991).

gas becomes available for star formation is the minimum between $\dot{M}_{acc}$ and $\dot{M}_{qst}$.

According to the models of Thomas et al. (1986), the bolometric x-ray luminosity of the region within the cooling radius of a galactic cooling flow is

$$L_X = 2.5\dot{M}_{cool}V_c^2.$$

The predictions of this formula are superposed on Figure 7.3 (dashed lines). For large circular velocities the mass cooling rates correspond to quite substantial x-ray luminosities.

Integrating the gas supply rates $\dot{M}_{cool}$ over redshift and halo mass distribution, we find that for $\Omega_b = 0.04$ *most of the gas was used before the present time.* This is unacceptable, since the density contributed by the observed stars in galaxies is less than 1% of the critical density. So, the star formation results to be too rapid without a regulating process, i.e., feedback.

To solve this puzzle, Larson (1974) proposed that the energy input from young stars and supernovae could quench star formation in small protogalaxies before more than a small gas fraction has been converted into stars.

Stellar energy input would counteract radiative losses in the cooling gas and tend to reduce the supply of gas for further star formation. We can imagine that the star formation process is self-regulating in the sense that the star formation rate $\dot{M}_\star$ takes the value required for heating to balance dissipation in the material which does not form stars. This produces the following prescription for the star formation rate,

$$\dot{M}_\star(V_c, z) = \epsilon(V_c)\min(\dot{M}_{acc}, \dot{M}_{qst}), \tag{7.41}$$
$$\epsilon(V_c) = [1 + \epsilon_0(V_0/V_c)^2]^{-1}, \tag{7.42}$$

where V_0 and ϵ_0 are some reference values. For large V_c the available gas turns into stars with high efficiency because the energy input is not sufficient to prevent cooling and fragmentation; for smaller objects the star formation efficiency ϵ is proportional to V_c^2. The assumption of self-regulation at small V_c seems plausible because the time interval between star formation and energy injection is much shorter than either the sound crossing time or the cooling time in the gaseous halos. However, other possibilities can be envisaged. For example, Dekel & Silk (1986) suggested that supernovae not only would suppress cooling in the halo gas but would actually expel it altogether. The conditions leading to such an event are discussed next.

7.4.2   Dwarf Galaxies: Feedback Lab

The observation of dwarf galaxies (i.e., galaxies with circular velocity < 100 km/s; see Mateo 1998 for a review) has shown well-defined correlations between their measured properties, and in particular

(i)   Luminosity–Radius $L \propto R^r$ with $r = 4$
(ii) Luminosity–Metallicity $L \propto Z^z$ with $z = 5/2$
(iii) Luminosity–Velocity dispersion $L \propto V^v$ with $v = 4$

A simple model can relate the observed scaling parameters r, z and v to each other (Dekel & Silk 1986). Consider a uniform cloud of initial mass $M_i = M_g + M_\star$ (M_g is the mass of gas driven out of the system, and $M_\star$ is the mass in stars) in a sphere of radius R_i that undergoes star formation. The metallicity for a constant yield in the instantaneous recycling approximation is given by $Z = y\ln(1 + M_\star/M_g)$ where y are the yields.

Let us impose simple scaling relations: $M_i \propto M$ (gas mass proportional to the dark matter mass, M), $R_i \simeq R$ and $V_i \simeq V$ (gas loss has no dynamical effect). From the observed scaling relation we have $L \propto R^r \propto V^v$. Now we should write the analogous relations for structure and velocity that hold before the removal; i.e., $M_i \propto R^{ri}$ and $M_i \propto V^{vi}$.

Consider now the case in which the gas is embedded in a dark halo, and assume that when it forms stars the mass in gas is proportional to the dark matter inside R_i, $M_i \propto M$. If the halo is dominant, the gas loss would have no dynamical effect on the stellar system that is left behind, so $R \simeq R_i$ and $V \simeq V_i$. The relations for structure and velocity that hold before the removal give

$$L \propto R^r \propto V^v, \quad M \propto R^{ri}, \quad V^2 \propto M/R, \tag{7.43}$$

so that we obtain

$$v = 2r/(ri - 1). \tag{7.44}$$

In the limit $M_\star \ll M_g$ and $L \propto M_\star$; from the metallicity relation we have

$$Z \propto \frac{M_\star}{M_g} \propto \frac{L}{M} \propto L^{1/z}, \tag{7.45}$$

so that

$$r/ri = (z - 1)/z. \tag{7.46}$$

The final equation comes from the energy condition. For thermal energy, in the limit of substantial gas loss, $M_g \simeq M_i$, we have $L \propto MV^2$, and so

$$v = 2z. \tag{7.47}$$

For a CDM spectrum $P(k) = Ck^{n_s}$, we obtain $r_i = 6/(5 + n_s)$ and introducing $z = 2r/(r - 1)$ and $n_s = 12/(r + 1) - 5$. For $r = 4$ we find $n_s = -2.6$ consistent with the CDM and the scaling relations are

$$L \propto R^4 \propto Z^{2.7} \propto V^{5.3}. \tag{7.48}$$

7.4.2.1 Conditions for Gas Removal

Let us now investigate the critical conditions, in terms of gas density n and virial velocity V, for a global supernova-driven gas removal from a galaxy while it is forming stars. Here, spherical symmetry and the presence of a central point source are assumed. The basic requirement for gas removal is that the energy that has been pumped into the gas is enough to expel it from the protogalaxy. The energy input in turn depends on the supernova rate, on the efficiency of energy transfer into the gas, and on the time it takes for the SNRs to overlap and hence affect a substantial fraction of the gas. The first is determined by the rate of star formation, the second by the evolution of the individual SNRs and the third by both. When all these are expressed as a function of n and V, the critical condition for removal takes the form

$$E(n, V) \geq \frac{1}{2} M_g V^2. \tag{7.49}$$

This relation defines a locus in the n–V diagram shown in Figure 7.4 within which substantial gas loss is possible. In Figure 7.4, the cooling curve, above which the cooling time is less than the free-fall time, confines the region where the gas can contract and form stars. The almost vertical line V_{crit} divides the permissible region for galaxy formation in two; a protogalaxy with $V > V_{crit}$ would not expel a large fraction of its original gas but rather turn most of its original gas into stars to form a "normal" galaxy. A protogalaxy with $V < V_{crit}$ can produce a supernova-driven wind out of the first burst of star formation, which would drive a substantial fraction of the protogalactic gas out, leaving a diffuse dwarf.

The short-dashed curve marked "1σ" corresponds to density perturbations $\delta M/M$ at their equilibrium configuration after a dissipationless collapse from a CDM spectrum, normalized to $\delta M/M$ at a comoving radius $8h^{-1}$ Mpc. The density n is calculated for a uniformly distributed gas in the CDM halos, with a gas-to-total mass ratio $\chi = 0.1$. The corresponding parallel short-dashed curve corresponds to the protogalactic gas clouds, after a contraction by a factor $\chi^{-1} = 10$ inside isothermal halos, to densities such that star formation is possible. The vertical dashed arrow marks the largest galaxy that can form out of a typical 1σ peak in the initial distribution of density fluctuations. The vast majority of such protogalaxies, when they form stars, have $V < V_{crit}$, so they would turn into dwarfs. The locus where "normal" galaxies are expected to be found is the shaded area. It is evident that most of them must originate from 2σ and 3σ peaks in the CDM perturbations.

So, the theory hence predicts two distinct types of galaxies that occupy two distinct loci in the n–V diagram: the "normal" galaxies are confined to the region of larger virial velocities and higher densities, and they tend to be massive, while the diffuse dwarfs are typically of smaller velocities and lower densities, and their mass in stars is less than 5×10^9 $M_\odot$.

This simple model, in spite of its enlightening power, has clear limitations. The most severe is that it assumes a spherical geometry and a single supernova explosion. These assumptions have been released by a subsequent study based on a large set of numerical simulations (Mac Low & Ferrara 1999). These authors modeled the effects of repeated supernova explosions from starbursts in dwarf galaxies on the interstellar medium of these galaxies, taking into account the gravitational potential of their dom-

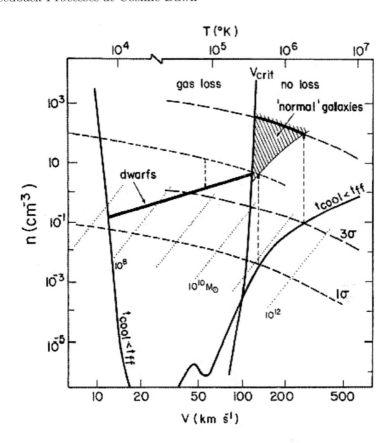

Figure 7.4. Gas number density vs. virial velocity (or viral temperature), the formation of dwarf vs. "normal" galaxies in CDM halos and the origin of biased galaxy formation. This figure is taken from Dekel & Silk (1986).

inant dark matter halos. They explored supernova rates from one every 30,000 yr to one every 3 Myr, equivalent to steady mechanical luminosities of $L = 0.1 - 10 \times 10^{38}$ ergs s^{-1}, occurring in dwarf galaxies with gas masses $M_g = 10^6 - 10^9\,M_\odot$. Surprisingly, Mac Low & Ferrara (1999) found that the mass ejection efficiency is very low for galaxies with mass $M_g \geq 10^7\,M_\odot$. Only galaxies with $M_g \lesssim 10^6\,M_\odot$ have their interstellar gas blown away, and then virtually independently of L (see Table 7.1). On the other hand, metals from the supernova ejecta are accelerated to velocities larger than the escape speed from the galaxy far more easily than the gas. They found[1] that for $L_{38} = 1$, only about 30% of the metals are retained by a $10^9\,M_\odot$

---

[1] Throughout the text we use the standard notation $Y_X = Y/10^X$.

galaxy, and virtually none by smaller galaxies (see Table 7.2).

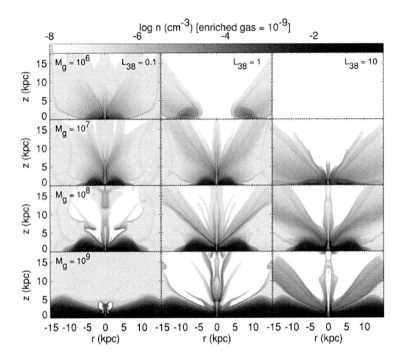

Figure 7.5. Distribution of metal-enriched stellar outflow material and SN ejecta from the starburst energy source at time 200 Myr. Note that in most cases, no enriched gas remains in the disks of the galaxies. This figure is taken from Mac Low & Ferrara (1999).

Table 7.1. Mass ejection efficiency

Visible Mass	Mechanical Luminosity [10^{38} erg s^{-1}]		
[$M_\odot$]	0.1	1.0	10
10^6	0.18	1.0	1.0
10^7	3.5×10^{-3}	8.4×10^{-3}	4.8×10^{-2}
10^8	1.1×10^{-4}	3.4×10^{-4}	1.3×10^{-3}
10^9	0.0	7.6×10^{-6}	1.9×10^{-5}

Table 7.2. Metal ejection efficiency

Visible Mass $[M_\odot]$	Mechanical Luminosity $[10^{38}$ erg s$^{-1}]$		
	0.1	1.0	10
10^6	1.0	1.0	1.0
10^7	1.0	1.0	1.0
10^8	0.8	1.0	1.0
10^9	0.0	0.69	0.97

7.4.3 Blowout, Blowaway and Galactic Fountains

The results of the Mac Low & Ferrara study served to clearly classify the various events induced by starbursts in galaxies. We can distinguish between blowout and blowaway processes, depending on the fraction of the parent galaxy mass involved in the mass loss phenomena. Whereas in the blowout process the fraction of mass involved is the one contained in cavities created by the supernova or superbubble explosions, the blowaway is much more destructive, resulting in the complete expulsion of the gas content of the galaxy. The two processes lie in different regions of the $(1 + \phi) - M_{g,7}$ plane shown in Figure 7.6, where $\phi = M_h/M_g$ is the dark-to-visible mass ratio and $M_{g,7} = M_g/10^7$ $M_\odot$ (Ferrara & Tolstoy 2000). Galaxies with gas mass content larger than 10^9 $M_\odot$ do not suffer mass losses, due to their large gravitational well. Of course, this does not rule out the possible presence of outflows with velocities below the escape velocity (fountain) in which material is temporarily stored in the halo and then returns to the main body of the galaxy. For galaxies with gas mass lower than this value, outflows cannot be prevented. If the mass is reduced further, and for $\phi \lesssim 20$, a blowaway, and therefore a complete stripping of the galactic gas, should occur. To exemplify, the expected value of ϕ as a function of M_g empirically derived by Persic et al. (1996) is also plotted, which should give an idea of a likely location of the various galaxies in the $(1+\phi) - M_{g,7}$ plane.

7.4.3.1 Blowout

The evolution of a point explosion in an exponentially stratified medium can be obtained from dimensional analysis. Suppose that the gas density distribution is horizontally homogeneous and that $\rho(z) = \rho_0 \exp(-z/H)$. The velocity of the shock wave is $v \sim (P/\rho)^{1/2}$, where the pressure P is roughly equal to E/z^3, and E is the total energy of the explosion. Then it follows that

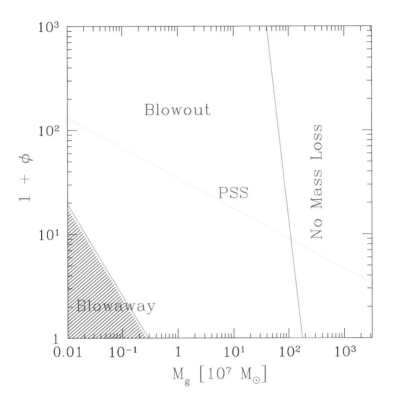

Figure 7.6. Regions in the $(1 + \phi) - M_{g,7}$ plane in which different dynamical phenomena may occur. Also shown is the locus point describing the Persic et al. (PSS) relation (dotted line). This figure is taken from Ferrara & Tolstoy (2000).

$$v(z) \simeq E^{1/2} \rho_0^{-1/2} \exp(z/2H) z^{-3/2}. \tag{7.50}$$

This curve has a minimum at $z = 3H$ and this value defines the height at which the shock wave, initially decelerating, is accelerated to infinity and a blowout takes place. Therefore, $3H$ can be used as the fiducial height where the velocity $v_b = v(3H)$ is evaluated.

There are three different possible fates for SN-shocked gas, depending on the value of v_b. If $v_b < c_{s,eff}$, where c_s is the sound speed in the ISM, then the explosion will be confined in the disk and no mass loss will occur; for $c_{s,eff} < v_b < v_e$ (v_e is the escape velocity) the supershell will break out of the disk into the halo, but the flow will remain bound to the galaxy; finally, $v_e < v_b$ will lead to a true mass loss from the galaxy.

7.4.3.2 Blowaway

The requirement for blowaway is that the momentum of the shell is larger than the momentum necessary to accelerate the gas outside the shell at velocity larger than the escape velocity, $M_s v_c \leq M_0 v_e$. Defining the disk axis ratio as $\epsilon = R/H \geq 1$, the blowaway condition can be rewritten as

$$\frac{v_b}{v_e} \geq (\epsilon - a)^2 a^{-2} e^{3/2}, \tag{7.51}$$

where $a = 2/3$. The above equation is graphically displayed in Figure 7.7. Flatter galaxies (large ϵ values) preferentially undergo blowout, whereas rounder ones are more likely to be blown away; as v_b/v_e is increased the critical value of ϵ increases accordingly. Unless the galaxy is perfectly spherical, blowaway is always preceded by blowout; between the two events the aspect of the galaxy may look extremely perturbed, with one or more huge cavities left after blowout.

7.4.4   Further Model Improvements

All models discussed so far make the so-called SEX BOMB assumption, i.e. a Spherically EXpanding blastwave. This is a good assumption as long as a single burst region exists whose size is small compared to the size of the system and it is located at the center of the galaxy. This however is a rather idealized situation.

A more detailed study about the possibility of mass loss due to distributed SN explosions at high redshift has been carried on by Mori, Ferrara, & Madau (2002). These authors presented results from 3D numerical simulations of the dynamics of SN-driven bubbles as they propagate through and escape the grasp of subgalactic halos with masses $M = 10^8 h^{-1} M_\odot$ at redshift $z = 9$. Halos in this mass range are characterized by very short dynamical timescales (and even shorter gas cooling times) and may therefore form stars in a rapid but intense burst before SN feedback quenches further star formation. These hydrodynamic simulations use a nested grid method to follow the evolution of explosive multi-SN events operating on the characteristic timescale of a few $\times 10^7$ yr, the lifetime of massive stars. The results confirm that, if the star formation efficiency of subgalactic halos is $\approx 10\%$, a significant fraction of the halo gas will be lifted out of the potential well (blow-away), shock the intergalactic medium, and pollute it with metal-enriched material, a scenario recently advocated by Madau, Ferrara, & Rees (2001). Depending on the stellar distribution,

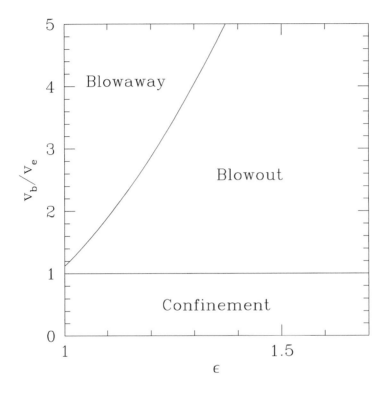

Figure 7.7. Conditions for blowaway, blowout and confinement as a function of the major-to-minor axis ratio ϵ of dwarf galaxies; $\epsilon = 1$ corresponds to spherical bodies. This figure is taken from Ferrara & Tolstoy (2000).

Mori et al. (2002) found that less than 30% of the available SN energy gets converted into kinetic energy of the blown away material, the remainder being radiated away.

However, it appears that realistic models lead to the conclusion that mechanical feedback is less efficient than expected from SEX BOMB simple schemes. The reason is that off-nuclear SN explosions drive inward-propagating shocks that tend to collect and pile up cold gas in the central regions of the host halo. Low-mass galaxies at early epochs then may survive multiple SN events and continue forming stars.

Figures 7.8 and 7.9 show the fraction of the initial halo baryonic mass contained inside (1, 0.5, 0.1) of the virial radius r_{vir} as a function of time for two different runs: an extended stellar distribution (case 1) and a more concentrated one (case 2). The differences are striking: in case 1, the amount of gas at the center is constantly increasing, finally collecting inside $0.1r_{vir}$ about 30% of the total initial mass. On the contrary, in case 2, the central regions remain practically devoid of gas until 60 Myr, when the accretion process starts. The final result is a small core containing a fraction of only 5% of the initial mass. In the former case, 50% of the halo mass is ejected together with the shell, whereas in case 2 this fraction is $\sim 85\%$; i.e., the blowaway is nearly complete.

As a first conclusion, it seems that quenching star formation in galaxies by ejecting a large fraction of their gas seems very difficult and hence unviable as a feedback scheme (although this might be possible in very small galaxies). Heavy elements can instead escape much more easily as they are carried away by the mass-unloaded, hot, SN-shocked gas; energy is carried away efficiently as well. This conclusion raises the issue of whether star formation must be rather governed by more gentle and self-regulating processes, i.e., a "true" feedback.

7.5 The Gentle Feedback

In numerical simulations an astonishingly large number of attempts to implement feedback schemes have been tried. In spite of this wealth, essentially all of them can be classified in one of the two following categories. (i) "Thermal" feedback schemes in which the energy released by a supernova is assumed to simply heat the ISM: the dynamics of the gas is only slightly affected: this is because the dense gas in the star forming region is able to radiate away this heat input very quickly. Moreover it would imply a huge amount of hot gas, that is hardly seen in the local universe. (ii) "Kinetic" feedback schemes, in which the explosion energy is assigned to fluid parcels as pure kinetic energy: this algorithm performs slightly better and it is probably more physical. It would be impossible to review in detail here even only a few of these schemes. We have thus chosen to describe a minimum feedback model that tries to catch the most important feature that is missing in the previous ones, that is, the multiphase structure and the matter exchange among gas phases characterizing real galaxies.

7.5.1 Feedback as ISM Sterilization

Dynamical models of the InterStellar Medium (ISM) describe this complex system in terms of a dense, Cold Neutral Medium (CNM) co-spatial with a Warm Neutral intercloud Medium (WNM) by which it is pressure confined. Such a multiphase medium offers an efficient self-regulatory mechanism for

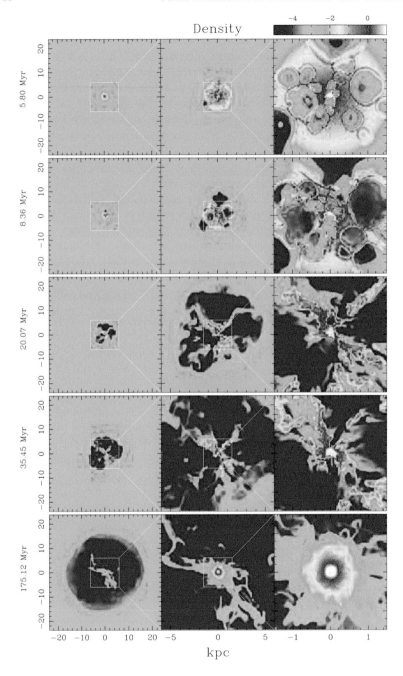

Figure 7.8. Snapshots of the logarithmic number density of the gas at five different elapsed times. The three panels in each row show the spatial density distribution in the x–y plane on the nested grids. The density range is $-5 \leq \log(n/\mathrm{cm}^{-3}) \leq 1$. This figure is taken from Mori et al. (2002).

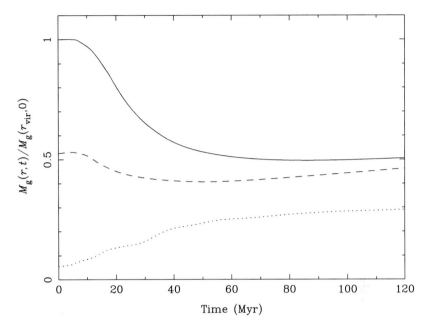

Figure 7.9. The evolution of the gas mass inside the gravitational potential well of the CDM halo for an extended stellar distribution (case 1) as a function of time. Curves correspond to the gas mass inside the virial radius $r_{\rm vir}$ (solid line), $0.5r_{\rm vir}$ (dashed line), and $0.1r_{\rm vir}$ (dotted line). This figure is taken from Mori et al. (2002).

star formation. Schematically, as more stars are formed, the extra heating caused by the enhancement of their UV field tends to transfer mass from the CNM into the WNM, where the conditions are highly unfavorable to star formation, i.e., a *sterile* phase. This acts to decrease the star formation rate and slowly brings back gas from the WNM reservoir in the cold phase where stars can start to form again. Under most conditions this cycle is stable and tends to regulate the amount of gas turned in stars quite effectively. In this scenario starburst can be only produced either by external triggers (mergings and interactions with other galaxies) and/or during the very first phase of galaxy formation, when the gentle feedback has not yet had the time to control the system. The physical basis of the gentle feedback is the thermal instability already discussed early on. Field (1995) discussed the criteria for stability of gas in thermal equilibrium (i.e., $L(n,T) = n\Gamma(n,T) - n^2\Lambda(n,T) = 0$, see also Sect. 7.3). In a plot of the thermal pressure P/k versus the hydrogen density n, the region of thermal stability occurs for $d(\log P)/d(\log n) > 0$. Thus, Figure 7.11 shows that a stable two-phase medium can be maintained in pressure equilibrium for $P_{\rm min} < P <$

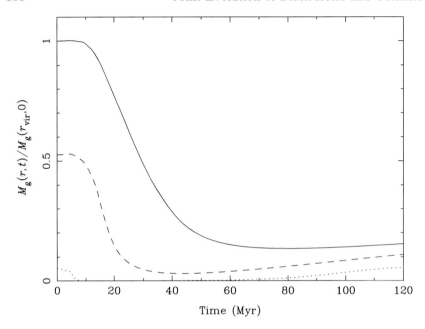

Figure 7.10. The same as Figure 7.9 but for a more concentrated stellar distribution (case 2). This figure is taken from Mori et al. (2002).

P_{max}. For pressure less than P_{min} only the warm phase is possible, while at pressures greater than P_{max} only the cold phase is possible (Wolfire et al. 1995). Moreover, in Figure 7.11 it is shown that when the metallicity of the gas Z, assumed to be equal to the dust-to-gas ratio D/G, is decreased, the pressure range in which a multiphase medium can exist becomes wider; in addition, the mean equilibrium pressure and density of the cold gas increase (Ricotti et al. 1997).

The first advantage of such a feedback scenario is that there is no need for gas loss as required by canonical feedback schemes: typically, the loss rate necessary to regulate the star formation activity is difficult to sustain for galaxies with mass of the order of the Milky Way for interestingly long periods. In addition, the production of large amounts of hot, x-ray emitting gas is expected. Such gas is rarely seen in/around observed galaxies.

An additional advantage of these models is the self-regulatory behavior of the process, that we have seen to be typical of feedback. In fact, there is no need of infall or outflow of gas from the galaxy (close box). As cold gas is available, the galaxy has a star formation burst, leading to the increase of the metallicity and of the dust content. Moreover, extra heating is provided by the SN explosions and from the UV background, so that some gas mass is transferred from the CNM to the WIM. The star formation is quenched since the gas is in the warm, *sterile* phase, and the heating drops down. After a cooling time, the gas again becomes available for star formation.

The disadvantage of these models is that they are physically complex, since they involve the interplay between dynamics and thermodynamics. Hence, a number of physical processes await inclusion in current cosmological simulations.

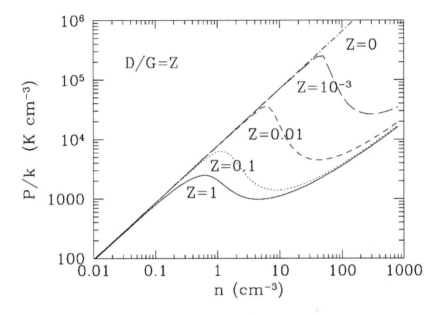

Figure 7.11. Thermal pressure P/k vs. hydrogen density n; the curves refer to different values of the dust-to-gas ratio, D/G, and metallicity Z, with $D/G = Z$. The gas is thermally stable for $(d \log P/d \log n) > 0$ (i.e., positive slope of the curves). Unless $Z = 0$, a stable two-phase medium is supported. This figure is taken from Ricotti et al. (1997).

7.5.2 A Porosity-Regulated Feedback

A similar feedback scheme has been recently proposed by Silk (2003). In such a feedback prescription, the filling factor of the hot, SN-shocked gas plays a central role. The filling factor f_{hot} of hot gas can be expressed in terms of the porosity Q as

$$f_{hot} = 1 - e^{-Q}, \tag{7.52}$$

where Q is regulated by the pressure confinement of the supernova bubbles.

As $f_{hot} \rightarrow 1$, the star formation rate reaches a saturation level and for $f = 1$ a "buoyant" wind is produced with outflow rate approximately equal to the star formation rate.

This feedback has the advantage to be based on the physical treatment of the supernova explosion, that current numerical simulations fail to follow in detail due to the lack of resolution. Moreover the expression of the porosity Q is identical for all the masses: the outflow rate is of the order of the star formation rate. Only the star formation efficiency depends on potential well depth ($\epsilon_{cr} \propto V_c^{2.7}$). The wind efficiency depends only weakly on V_c. In summary, outflows can occur even for massive galaxies since star formation efficiency is larger in these systems.

However, there are some problems. The porosity-driven feedback is not a true multiphase model and has to be generalized to include the time evolution of Q, metallicity, UV background, etc. Moreover, it again implies a large amount of hot gas that is not detected in galaxies.

7.5.3 Advanced Multiphase/Feedback Schemes

The main problem affecting numerical simulations of feedback processes is that, owing to the limited numerical resolution, there is a large disparity between the minimum simulation timescale and the true physical timescale associated with supernova feedback (Thacker & Couchman 2001). Physically, following a supernova event, the gas cooling time very rapidly increases to $O(10^7)$ yr over the time it takes the shock front to propagate. Thus, it is the timescale - $t_{\dot{E}}$ - for the cooling time to increase markedly that is of critical importance. In the SPH simulations, "stars" form and "supernova feedback" occurs in gas cores that are typically less than $h_{\min}/5$ in diameter ($h_{\min}$ being the effective spatial resolution). Any rearrangement of the particles in such a small region leads to a density estimate varying by at most $\sim 7\%$. Since the cooling time in the model is dependent on the local SPH density, any energy deposited into a feedback region only reduces the local cooling rate by increasing the temperature; the SPH density cannot respond on a timescale that in real supernova events would very rapidly increase the cooling time. A successful feedback algorithm in an SPH model must thus overcome the fact that ρ_{SPH} does not change on the same timescale as $t_{\dot{E}}$ and consequently that cooling times for dense gas cores remain short.

Standard implementations of SPH overestimate the density for particles that fall near the boundary of a higher density phase (Pearce et al. 1999). The usual assumption of small density gradients across the smoothing kernel breaks down in this regime, and nearby clusters of high-density particles cause an upward bias in the standard SPH estimate:

$$< \rho_i >= \sum_{j}^{N} m_j W_{ij}, \tag{7.53}$$

where N is the number of neighbors j of particle i and W_{ij} is a symmetric smoothing kernel. In order to avoid this bias, which leads to artificially high cooling rates and to spatial exclusion effects, the neighbor search and the density evaluation of Equation (7.53) have to be modified in a way that leaves the numerical scheme as simple as possible. It is important to recognize that the local density is the gas property responsible for phase segregation (since it determines the local cooling rate). An appropriate density estimator for a particle of any given phase should use only local material which is also part of that same phase.

Marri & White (2003) proposed a new feedback scheme that takes into account these considerations. This works as follows. As a supernova explodes, the energy is distributed in a fraction ϵ_c (ϵ_w) to the cold (warm) gas. The fundamental hypothesis is that at large scale (of kpc order, say) the net effect of all the complex *microscopic* processes is well described by an energy input shared by the *macroscopic* phases in given proportions. Values for ϵ_c and ϵ_w could be fixed from a complete theory of the ISM, describing all the relevant processes, or through numerical simulations. Feedback to the hot phase is implemented by adding thermal energy to the ten nearest hot neighbors. Feedback to the cold phase is instead accumulated in a reservoir within the star-forming particle itself. This continues until the accumulated energy is sufficient to heat the gas component of the particle above $T \sim 50 T_\star \sim 10^6$ K. This is far enough above $T_\star$ for the promoted particle to be considered "hot" in its subsequent evolution. At the same time any "hidden" stellar content is dumped to the cold neighbors.

A simple test of these new schemes (the initial condition is the rotating, centrally concentrated sphere described in Navarro & White (1993) which consists of 90% by mass in dark matter) shows that hot particles do survive near the central disk in the multiphase SPH case. They have an almost spherically symmetric distribution with density peaked at the center (Figure 7.12). The cold disk rotates within this ambient hot medium. In contrast, in the standard SPH model hot particles are excluded from the vicinity of the disk so that the hot phase actually has a density minimum at the center of the galaxy. This is seen most clearly in the gas density profile of Figure 7.13. Such behavior is clearly unphysical.

7.6 Feedback in the Early Universe

At $z = 1000$ the universe has cooled down to 3000 K and the hydrogen becomes neutral (recombination). Then, at $z \lesssim 20$ the first stars (Population or Pop III stars) form and these gradually photoionize the hydrogen

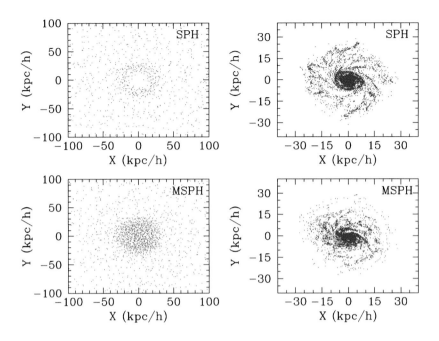

Figure 7.12. Face-on projections at $t \sim 1.2$ Gyr of 8000 gas particle, cooling only simulations of a collapsing, rotating sphere. The standard SPH simulation is in the upper row with the MSPH simulation below it. The left-hand plots show "hot" particles ($T > 10^5$ K) while the right hand ones show "cold" particles ($T < 10^5$ K and $n_H > 0.1$ cm^{-3}). This figure is taken from Marri & White (2003).

in the IGM (reionization). This epoch, witnessing the return of light in the universe after the Big Bang, is usually dubbed the Cosmic Dawn. At $z < 6$, galaxies form most of their stars and grow by merging. Finally at $z < 1$, the massive clusters are assembled.

We can distinguish three main feedback types that are fundamentally shaping the universe at cosmic dawn:

(i) **Stellar Feedback**: the process of star formation produces the conditions (destruction of cold gas) such that astration (or even galaxy formation) comes to a halt or is temporarily blocked;

(ii) **Chemical Feedback**: massive stars by polluting the ambient gas with heavies, shift the prevailing formation mode towards low-mass stars (self-killing population);

(iii) **Radiative Feedback**: the process of star formation erases the con-

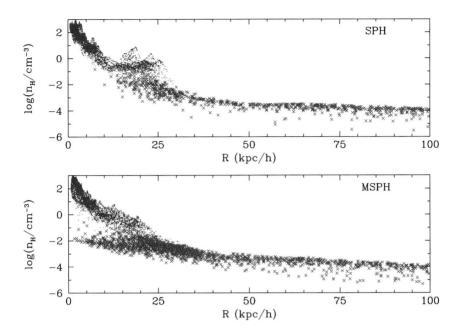

Figure 7.13. Particle densities as a function of galactocentric distance after ~ 1.2 Gyr of evolution in two 8000 gas particles, cooling only simulations of the collapse of a rotating sphere. The upper plot is for an SPH model and the lower for an MSPH model. Gas particles with $T > 10^5$ K are plotted with crosses. This figure is taken from Marri & White (2003).

ditions (molecular formation) necessary to form stars via UV radiation and ionizing photons. Such radiative input heats and ionizes the gas decreasing its ability to produce sources (stars) to further sustain reionization.

In the following we will describe in detail the three feedback types.

7.6.1 Stellar Feedback

7.6.1.1 Oxygen in the Low-Density IGM

One of the most obvious signatures of stellar feedback is the metal enrichment of the intergalactic medium. If powerful winds are driven by supernova explosions, one would expect to see widespread traces of heavy elements away from their production sites, i.e., galaxies. Schaye et al. (2000) have reported the detection of OVI in the low-density IGM at high redshift. They perform a pixel-by-pixel search for OVI absorption in eight high quality quasar spectra spanning the redshift range $z = 2.0$–4.5. At $2 \lesssim z \lesssim 3$ they detect OVI in the form of a positive correlation between the HI Lyα optical depth and the optical depth in the corresponding OVI pixel, down to $\tau_{\mathrm{HI}} \sim 0.1$ (underdense regions), that means that metals are far from galaxies (Figure 7.14). Moreover, the observed narrow widths of metal absorption lines (CIV, SiIV) imply low temperatures $T_{\mathrm{IGM}} \sim$ few $\times 10^4$ K.

A natural hypothesis would be that the Lyα forest has been enriched by metals ejected by Lyman-Break galaxies at moderate redshift. The density of these objects is $n_{\mathrm{LBG}} = 0.013 \ h^3 \ \mathrm{Mpc}^3$. A filling factor of $\sim 1\%$ is obtained for a shock radius $R_s = 140 \ h^{-1}$ kpc, that corresponds at $z = 3$ ($h = 0.5$) to a shock velocity $v_s = 600$ km s^{-1}. In this case, we expect a postshock gas temperature larger than 2×10^6 K, that is around hundred times what we observed. So the metal pollution must have occurred earlier than redshift 3, resulting in a more uniform distribution and thus enriching vast regions of the intergalactic space. This allows the Lyα forest to be hydrodynamically cold at low redshift, as intergalactic baryons have enough time to relax again under the influence of dark matter gravity only (Scannapieco, Ferrara, & Madau 2002).

In Figure 7.15 is shown the thermal history of the IGM as a function of redshift as computed by Madau, Ferrara & Rees (2001). The gas is allowed to interact with the CMB through Compton cooling and either with a time-dependent quasar-ionizing background as computed by Haardt & Madau (1996) or with a time-dependent metagalactic flux of intensity 10^{-22} erg cm^{-2} s^{-1} Hz^{-1} sr^{-1} at 1 Ryd and power–law spectrum with energy slope $\alpha = 1$. The temperature of the medium at $z = 9$ has been either computed self-consistently from photoheating or fixed to be in the range $10^{4.6}$–10^5 K, as expected in SN-driven bubbles with significant filling factors. The various curves show that the temperature of the IGM at $z = 3$–4 will retain little memory of an early era of pregalactic outflows.

The large increase of the IGM temperature at high redshift connected with such an era of pregalactic outflows, causes a larger Jeans mass, thereby preventing gas from accreting efficiently into small dark matter halos (Benson & Madau 2003). For typical preheating energies, the IGM is driven to temperatures just below the virial temperature of halos hosting $L_\star$ galaxies. Thus we may expect preheating to have a strong effect on the galaxy luminosity function at $z = 0$. Moderate preheating scenarios, with $T_{\mathrm{IGM}} \gtrsim 10^5$ K at $z \sim 10$, are able to flatten the faint-end slope of the luminosity function, producing excellent agreement with observations, without the need for any local strong feedback within galaxies (Figure 7.16).

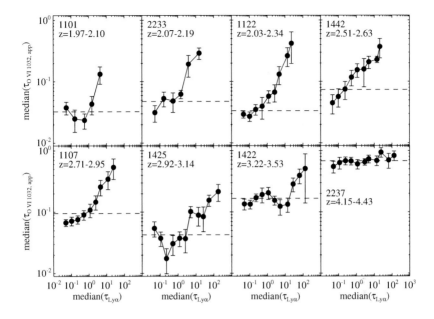

Figure 7.14. IGM Lyα-OVI optical depth correlation from the pixel analysis of the spectra of 8 QSOs. Vertical error bars are 1σ errors, horizontal error bars are smaller than the symbols and are not shown. For $z \lesssim 3$ $\tau_{\rm OVI,app}$ and $\tau_{\rm HI}$ are clearly correlated, down to optical depths as low as $\tau_{\rm HI} \sim 10^{-1}$. A correlation between $\tau_{\rm OVI,app}$ and $\tau_{\rm HI}$ implies that OVI absorption has been detected in the Lyα forest (Schaye et al. 2000).

7.6.1.2 Gas Stripping from Neighbor Galaxy Shocks

The formation of a galaxy can be inhibited also by the outflows from neighboring dwarfs as the result of two different mechanisms (Scannapieco et al. 2000). In the "mechanical evaporation" scenario, the gas associated with an overdense region is heated by a shock above its virial temperature. The thermal pressure of the gas then overcomes the dark matter potential and the gas expands out of the halo, preventing galaxy formation. In this case, the cooling time of the collapsing cloud must be shorter than its sound crossing time, otherwise the gas will cool before it expands out of the gravitational well and will continue to collapse.

Alternatively, the gas may be stripped from a collapsing perturbation by a shock from a nearby source. In this case, the momentum of the shock is sufficient to carry with it the gas associated with the neighbor, thus emptying the halo of its baryons and preventing a galaxy from forming.

In principle, outflows can suppress the formation of nearby galaxies

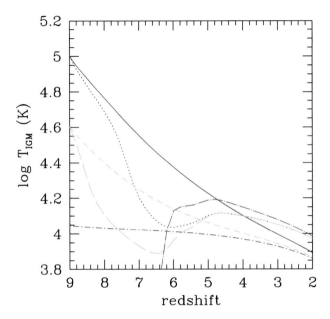

Figure 7.15. Thermal history of intergalactic gas at the mean density in an Einstein–de Sitter universe with $\Omega_b h^2 = 0.019$ and $h = 0.5$. Short dash-dotted line: temperature evolution when the only heating source is a constant ultraviolet (CUV) background of intensity 10^{-22} erg cm^{-2} s^{-1} Hz^{-1} sr^{-1} at 1 Ryd and power-law spectrum with energy slope $\alpha = 1$. Long dash-dotted line: same for the time-dependent quasar ionizing background as computed by Haardt & Madau (1996; HM). Short dashed line: heating due to a CUV background but with an initial temperature of 4×10^4 K at $z = 9$ as expected from an early era of pre-galactic outflows. Long dashed line: same but for a HM background. Solid line: heating due to a CUV background but with an initial temperature of 10^5 K at $z = 9$. Dotted line: same but for a HM background. This figure is taken from Madau, Ferrara, & Rees (2001).

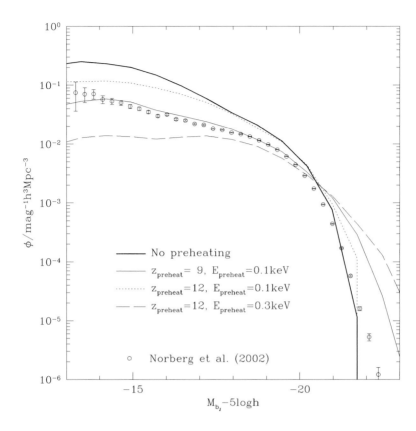

Figure 7.16. B-band luminosity functions of galaxies at $z = 0$, as predicted by the semi-analytic model of Benson et al. (2002). The observational determination of Norberg et al. (2002) is shown as circles. This figure is taken from Benson & Madau (2003).

both by shock heating/evaporation and by stripping of the baryonic matter from collapsing dark matter halos; in practice, the short cooling times for most dwarf-scale collapsing objects suggest that the baryonic stripping scenario is almost always dominant (Figure 7.17). This mechanism has the largest impact in forming dwarfs in the $\lesssim 10^9 M_\odot$ range which is sufficiently large to resist photoevaporation by UV radiation, but too small to avoid being swept up by nearby dwarf outflows.

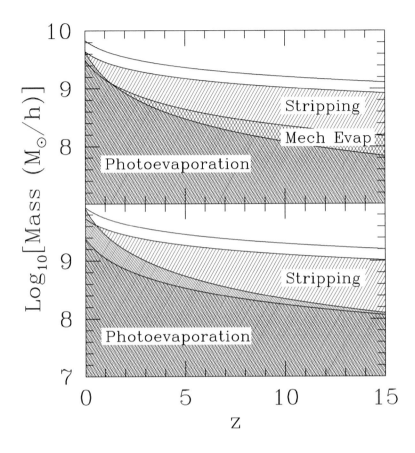

Figure 7.17. Relevant mass scales for suppression of dwarf galaxy formation. The upper lines are the masses below which halos will be heated beyond their virial temperatures, although cooling prevents mechanical evaporation from occurring for halos with masses above the cross-hatched regions. The second highest set of lines, bounding the lightly shaded regions, shows the masses below which baryonic stripping is effective. Finally the heavily shaded regions show objects that are susceptible to photoevaporation. The upper panel is a flat CDM model and the lower panel is a flat ΛCDM model with $\Omega_0 = 0.3$. In all cases $(\epsilon N h) = 5000\Omega_0^{-1}$, the overdensity $\delta = \rho/\rho_0 = 2.0$, $\Omega_b = 0.05$, $h = .65$. Note that photoevaporation affects a larger mass range than mechanical evaporation in the ΛCDM cosmology. This figure is taken from Scannapieco et al. (2000).

7.6.2 Chemical Feedback

Recent studies (Bromm, Coppi, & Larson 1999, 2002; Abel, Bryan, & Norman 2000; Schneider et al. 2002; Omukai & Inutsuka 2002; Omukai & Palla 2003) have shown that in the absence of heavy elements the formation of stars with masses 100 times that of the sun would have been strongly favored, and that low-mass stars could not have formed before a minimum level of metal enrichment had been reached.

Bromm et al. (2001) have studied the effect of the metallicity on the evolution of the gas in a collapsing dark matter mini-halo, simulating an isolated 3σ peak of mass 2×10^6 $M_\odot$ that collapses at $z \sim 30$, using smoothed particle hydrodynamics. The gas has a supposed level of pre-enrichment of either $Z = 10^{-4}$ $Z_\odot$ of 10^{-3} $Z_\odot$. The H_2 is assumed to be radiatively destroyed by the presence of a soft UV background. Moreover Bromm et al. (2001) do not consider the presence of molecules or dust.

The evolution proceeds very differently for the two cases. The gas in the lower metallicity case fails to undergo continued collapse and fragmentation, remaining in the post-virialization, pressure-supported state in a roughly spherical configuration. The final result is likely to be the formation of a single very massive ($\sim 10^3 - 10^4$ $M_\odot$) object (Figure 7.18). On the other hand (Figure 7.19), the gas in the higher metallicity case can cool efficiently and collapse into a disk-like configuration. The disk material is gravitationally unstable and fragments into a large number of high-density clumps, that are the seeds of low-mass stars. So, they conclude that there exists a critical metallicity Z_{crit} at which the transition from the formation of high-mass objects to a low-mass star formation mode, as we observe at present, occurs. In the following we refer to Pop III stars in the mass range 100–600 $M_\odot$ forming out of the collapse of $Z < Z_{crit}$ gas clouds.

More recently, Schneider et al. (2002) have investigated the problem more in detail including the cooling due to H_2, molecular, and dust, using the model of Omukai (2001). The gas within the dark matter halo is given an initial temperature of 100 K, and the subsequent thermal and chemical evolution of the gravitationally collapsing cloud is followed numerically until a central protostellar core forms. The results for different metallicities are summarized in Figure 7.20, showing the temperature and adiabatic index evolution as a function of the hydrogen number density of protostellar clouds for different metallicity ($Z = 0, 10^{-6}, 10^{-4}, 10^{-2}, 1$ $Z_\odot$). The necessary conditions to stop fragmentation and start gravitational contraction within each clump are that cooling becomes inefficient, and the Jeans mass of the fragments does not decrease any further, thus favoring fragmentation into sub-clumps. This condition depends somewhat on the geometry of the fragments, and translates into $\gamma > 4/3$ ($\gamma > 1$) for spherical (filamentary) clumps, where γ is the adiabatic index defined as $T \propto n^{\gamma-1}$.

For a metal-free gas, the only efficient coolant is molecular hydro-

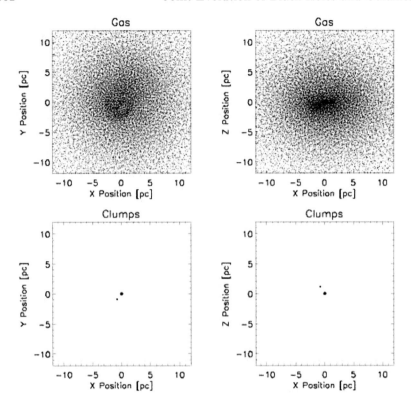

Figure 7.18. Simulation of a collapsing dark matter halo with mean metallicity of $Z = 10^{-4} Z_\odot$. Morphology at $z = 22.7$. Top row: the remaining gas in the diffuse phase. Bottom row: distribution of clumps. Dot sizes are proportional to the mass of the clumps. Left panels: face-on view. Right panels: edge-on view. The box size is 30 pc. It can be seen that no fragmentation occurs. This figure is taken from Bromm et al. (2001).

gen. Cooling due to molecular line emission becomes inefficient at densities above $n > 10^3 \, \mathrm{cm}^{-3}$, and fragmentation stops when the minimum fragment mass is of order 10^3–$10^4 \, M_\odot$. Clouds with mean metallicity $Z = 10^{-6} \, Z_\odot$ follow the same evolution as that of the gas with primordial composition in the (n, T) plane. For metallicities larger than $10^{-4} \, Z_\odot$ the fragmentation proceeds further until the density is $\sim 10^{13} \, \mathrm{cm}^{-3}$ and the corresponding Jeans mass is of the order $10^{-2} \, M_\odot$. Schneider et al. (2002) concluded that the critical metallicity locates in the range 10^{-6}–$10^{-4} \, Z_\odot$.

According to the scenario proposed above, the first stars that form out of gas of primordial composition tend to be very massive, with masses

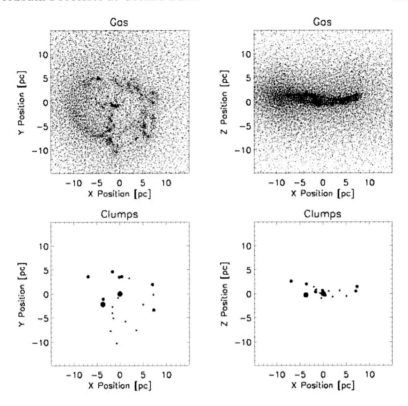

Figure 7.19. Simulation of a collapsing dark matter halo with mean metallicity of $Z = 10^{-3}\ Z_\odot$. Morphology at $z = 28$. The convention of Figure 7.18 is adopted for the row and columns. It can be seen that the gas has undergone vigorous fragmentation. This figure is taken from Bromm et al. (2001)

$\sim 10^2$–10^3 $M_\odot$. It is only when metals change the composition of the gas that further fragmentation occurs, producing stars with significantly lower masses. A series of numerical studies (Heger & Woosley 2001; Fryer et al. 2001; Umeda & Nomoto 2002) has investigated the nucleosynthesis and final state of metal-free massive stars. Heger & Woosley (2001) delineate three mass ranges characterized by distinct evolutionary paths:

(i) $M_\star \gtrsim 260\ M_\odot$. – The nuclear energy released from the collapse of stars in this mass range is insufficient to reverse the implosion. The final result is a very massive black hole (VMBH) locking up all heavy elements produced.

(ii) $140\ M_\odot \lesssim M_\star \lesssim 260\ M_\odot$. – The mass regime of the pair-unstable supernovae ($SN_{\gamma\gamma}$). Precollapse winds and pulsations result in little mass loss; the star implodes to a maximum temperature that depends on its mass and then explodes, leaving no remnant. The explosion expels metals into the surrounding ambient ISM.

(iii) $30 \ M_\odot \lesssim M_\star \lesssim 140 \ M_\odot$. – Black hole formation is the most likely outcome, because either a successful outgoing shock fails to occur or the shock is so weak that the fallback converts the neutron star remnant into a black hole (Fryer 1999).

Stars that form in the mass ranges 1 and 3 above fail to eject most of their heavy elements. If the first stars have masses in excess of 260 $M_\odot$, they invariably end their lives as VMBHs and do not release any of their synthesized heavy elements. However, as long as the gas remains metal-free, the subsequent generations of stars will continue to be top-heavy. This "star formation conundrum" can be solved only if a fraction $f_{\gamma\gamma}$ of the first generation of massive stars are in the $SN_{\gamma\gamma}$ range and enrich the gas with heavy elements up to a mean metallicity of $Z \geq 10^{-5\pm1} \ Z_\odot$.

Schneider et al. (2002) computed the transition redshift z_f at which the mean metallicity is $10^{-4} \ Z_\odot$ for various values of $f_{\gamma\gamma}$. The results are plotted in Figure 7.21 along with the corresponding critical density Ω_{VMBH} contributed by the VMBHs formed. In order to set some limits on $f_{\gamma\gamma}$, we have to compare the predicted critical density of VMBH remnants to present observational data. This depends on the assumptions about the fate of these VMBHs at late times.

Under the hypothesis that all VMBHs have, during galaxy mergers, been used to build up the supermassive black holes (SMBHs) detected today so that $\Omega_{VMBH} = \Omega_{SMBH}$, from the top panel of Figure 7.21 we can derive $f_{\gamma\gamma} \simeq 0.06$ and $z_f \simeq 18.5$.

At the other extreme, wherein the assembled SMBHs in galactic centers have formed primarily via accretion and are unrelated to VMBHs, two possibilities can be distinguished: (1) VMBHs would still be in the process of spiraling into the center of galaxies because of dynamical friction but are unlikely to have reached the center within a Hubble time because of the long dynamical friction timescale (Madau & Rees 2001), or (2) VMBHs contribute the entire baryonic dark matter in galactic halos. In case (1) some fraction of these VMBHs might appear as off-center accreting sources that show up in the hard x-ray wave band. In fact, ROSAT and Chandra have detected such objects. Using the contribution to Ω from X-ray-bright, off-center *ROSAT* sources as a limit for the density of VMBHs, from the lower dashed line in the bottom panel of Figure 7.21 $f_{\gamma\gamma}$ has to be nearly unity and $z_f \gtrsim 22.1$. In case (2), assuming that the baryonic dark matter in galaxy halos is entirely contributed by VMBH (upper dashed line in the bottom panel of Figure 7.21), $f_{\gamma\gamma} \simeq 3.15 \times 10^{-5}$ and $z_f \gtrsim 5.4$. Therefore for the case in which SMBHs are unrelated to VMBHs, only weak limits for $f_{\gamma\gamma}$ and z_f can be obtained. Although the actual data do not allow us at

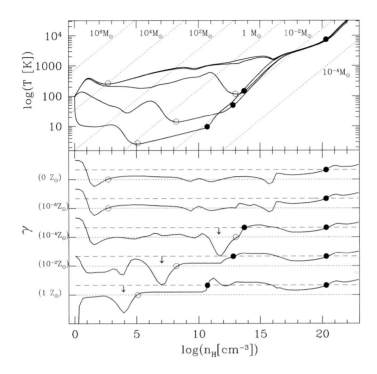

Figure 7.20. Top: Evolution of the temperature as a function of the hydrogen number density of protostellar clouds with the same initial gas temperature but varying the metallicities $Z = (0, 10^{-6}, 10^{-4}, 10^{-2}, 1)\ Z_\odot$ (Z increasing from top to bottom curves). The dashed lines correspond to the constant Jeans mass for spherical clumps; open circles indicate the points where fragmentation stops; filled circles mark the formation of hydrostatic cores. Bottom: The adiabatic index γ as a function of the hydrogen number density for the curves shown in the top panel. Dotted (dashed) lines correspond to $\gamma = 1$ ($\gamma = 4/3$); open and filled circles as above. This figure is taken from Schneider et al. (2002).

present to strongly constrain these two quantities, they provide interesting bounds on the proposed scenario.

We have seen that there exists a critical metallicity $Z_{cr} = 10^{5\pm1}\ Z_\odot$ (Schneider et al. 2002, 2003) at which we expect the transition between a top-heavy and a normal Salpeter-like IMF. One point that might have important observational consequences is the fact that cosmic metal enrichment has proceeded very inhomogeneously (Scannapieco, Ferrara, & Madau

2002; Furlanetto & Loeb 2003; Marri et al. 2003), with regions close to star formation sites rapidly becoming metal-polluted and overshooting Z_{cr}, and others remaining essentially metal-free. Thus, the two modes of star formation, Pop III and normal, must have been active at the same time and possibly down to relatively low redshifts, opening up the possibility of detecting Pop III stars.

Scannapieco, Schneider, & Ferrara (2003) have studied using an analytic model of inhomogeneous structure formation, the evolution of Pop III objects as a function of the star formation efficiency, IMF, and efficiency of outflow generation. They parametrized the chemical feedback through a single quantity, E_g, which represents the kinetic energy input per unit gas mass of outflows from Pop III galaxies. This quantity is related to the number of exploding Pop III stars and therefore encodes the dependence on the assumed IMF. For all values of the feedback parameter, E_g, Scannapieco et al. (2003) found that, while the peak of Pop III star formation occurs at $z \sim 10$, such stars continue to contribute appreciably to the star formation rate density at much lower redshift, even though the mean IGM metallicity has moved well past the critical transition metallicity. This finding has important implications for the development of efficient strategies for the detection of Pop III stars in primeval galaxies. At any given redshift, a fraction of the observed objects have a metal content that is low enough to allow the preferential formation of Pop III stars. As metal-free stars are powerful Lyα emitters (Tumlinson, Giroux, & Shull 2001; Schaerer 2002), it is natural to use this indicator as a first step in any search for primordial objects. Scannapieco et al. (2003) derived the probability that a given high-redshift Lyα detection is due to a cluster of Pop III stars. In Figure 7.22 the isocontours in the Lyα luminosity-redshift plane are shown, indicating this probability for various feedback efficiencies E_g^{III}. We see that Pop III objects populate a well-defined region, whose extent is governed by the feedback strength. Above the typical flux threshold, Lyα emitters are potentially detectable at all redshifts beyond 5. Furthermore, the fraction of Pop III objects increases with redshift, independently of E_g^{III}. For the fiducial case, $E_g^{III} = 10^{-3}$, the fraction is only a few percent at $z = 4$ but increases to approximately 15% by $z = 6$. So, the Lyα emission from already observed high-z sources can indeed be due to Pop III objects, if such stars were biased toward high masses. Hence collecting large data samples to increase the statistical leverage may be crucial for detecting the elusive first stars.

7.6.3　Radiative Feedback and Reionization

Before metals are produced, the primary molecule that acquires sufficient abundance to affect the thermal state of the pristine cosmic gas is molecular hydrogen, H_2. The main reactions able to produce molecular hydrogen are

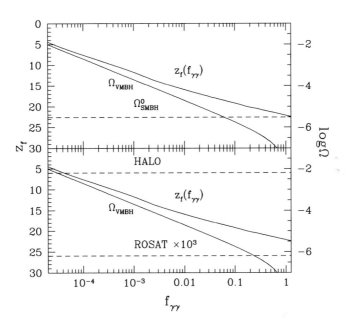

Figure 7.21. Top-heavy to normal IMF transition redshift, z_f, as a function of $SN_{\gamma\gamma}$ progenitor mass fraction and the mass density contributed by VMBHs, Ω_{VMBH}. Top: The computed critical density of VMBH remnants is compared to the observed values for SMBHs (upper dashed line). Bottom: The computed critical density of VMBH remnants is compared to the contribution to Ω from the X-ray-bright, off-center ROSAT sources (lower dashed line) and to the abundance predicted assuming that the baryonic dark matter in galaxy halos is entirely contributed by VMBHs (upper dashed line). The observations on Ω_{VMBH} constrain the value of $f_{\gamma\gamma}$. For a given $f_{\gamma\gamma}$, the corresponding value for the transition redshift can be inferred by the z_f curve. This figure is taken from Schneider et al. (2002).

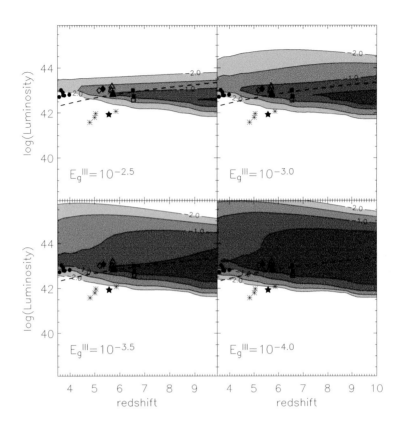

Figure 7.22. Fraction of PopIII objects as a function of Lyα luminosity and redshift. Isocontours of fractions $\geq 10^{-2}, 10^{-1.5}, 10^{-1}$ and $10^{0.5}$ are shown. Burst-mode star formation with a $f_\star^{II} = f_\star^{III} = 0.1$ is assumed for all objects. In the PopIII case, a lower cutoff mass of $50\,\mathrm{M}_\odot$ is assumed. Each panel is labeled by the assumed E_g^{III} value. For reference, the dashed line gives the luminosity corresponding to an observed flux of 1.5×10^{-17} ergs cm^{-2} s^{-1}, and the various points correspond to observed galaxies. The filled diamond is from Dey et al. (1998), the filled triangle is from Hu et al. (1999), the filled star is from Ellis et al. (2001), the open diamond is from Dawson et al. (2002), the open square is from Hu et al. (2002), the asterisks are from Lehnert & Bremer (2003), the open triangles are from Rhoads et al. (2003), the filled circles are from Fujita et al. (2003), and the filled squares are from Kodaira et al. (2003). The curves have been extended slightly past $z = 4$ for comparison with the Fujita et al. (2003) data-set. This figure is taken from Scannapieco, Schneider, & Ferrara (2003).

(Abel et al. 1997)

$$H + e^- \rightarrow H^- + h\nu, \qquad (7.54)$$
$$H^- + H \rightarrow H_2 + e^-, \qquad (7.55)$$

and

$$H^+ + H \rightarrow H_2^+ + h\nu, \qquad (7.56)$$
$$H_2^+ + H \rightarrow H_2 + H^+, \qquad (7.57)$$

In objects with baryonic masses $\gtrsim 3 \times 10^4$ M$_\odot$, gravity dominates and results in the bottom-up hierarchy of structure formation characteristic of CDM cosmologies; at lower masses, gas pressure delays the collapse. The first objects to collapse are those at the mass scale that separates these two regimes. Such objects have a virial temperature of several hundred degrees and can fragment into stars only through molecular cooling (Tegmark et al. 1997). In other words, there are two independent minimum mass thresholds for star formation: the Jeans mass (related to accretion) and the cooling mass. For the very first objects, the cooling threshold is somewhat higher and sets a lower limit on the halo mass of $\sim 5 \times 10^4$ M$_\odot$ at $z \sim 20$.

As the first stars form, their photons in the energy range 11.26–13.6 eV are able to penetrate the gas and photodissociate H$_2$ molecules both in the IGM and in the nearest collapsing structures, if they can propagate that far from their source. Thus, the existence of an UV background below the Lyman limit due to Pop III objects, capable of dissociating the H$_2$, could deeply influence subsequent small structure formation.

Ciardi, Ferrara, & Abel (2000) have shown that the UV flux from these objects results in a soft (Lyman–Werner band) UV background (SUVB), J_{LW}, whose intensity (and hence radiative feedback efficiency) depends on redshift. At high redshift the radiative feedback can be induced also by the direct dissociating flux from a nearby object. In practice, two different situations can occur: (i) the collapsing object is outside the dissociated spheres produced by pre-existing objects: then its formation could be affected only by the SUVB ($J_{LW,b}$), as by construction the direct flux ($J_{LW,d}$) can only dissociate molecular hydrogen inside this region on timescales shorter than the Hubble time; (ii) the collapsing object is located inside the dissociation sphere of a previously collapsed object: the actual dissociating flux in this case is essentially given by $J_{LW,\mathrm{max}} = (J_{LW,b} + J_{LW,d})$. It is thus assumed that, given a forming Pop III, if the incident dissociating flux ($J_{LW,b}$ in the former case, $J_{LW,\mathrm{max}}$ in the latter) is higher than the minimum flux required for negative feedback (J_s), the collapse of the object is halted. This implies the existence of a population of "dark objects" which were not able to produce stars and, hence, light.

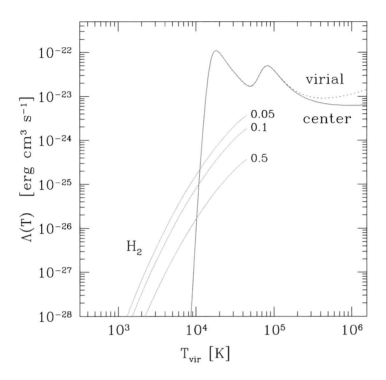

Figure 7.23. Equilibrium cooling curve at the center (solid lines) and virial radius (dashed line) of an isothermal halo at $z = 9$, as a function of virial temperature T_{vir}, and in the absence of a photoionizing background (i.e., prior to the reionization epoch). The halo has an assumed baryonic mass fraction of $\Omega_b = 0.019h^{-2}$. The gas density dependence of the cooling function at high temperatures is due to Compton cooling off comic microwave background photons. The labeled curves extending to low temperatures show the contribution due to H_2 for three assumed values of the metagalactic flux in the Lyman–Werner bands, $4\pi J_{\mathrm{LW}}$ (in units of 10^{-21} erg cm^{-2} s^{-1} Hz^{-1}). This figure is taken from Madau, Ferrara, & Rees (2001).

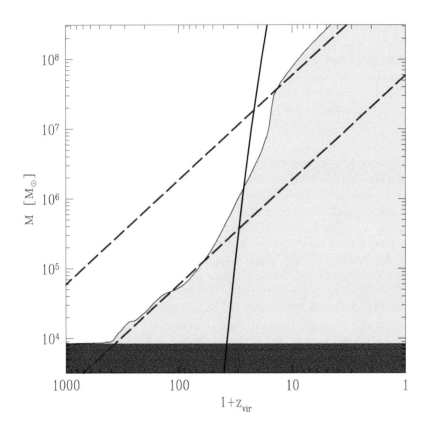

Figure 7.24. The minimum mass needed to collapse. The function $M_c(z_{vir})$ is plotted as a function on virialization redshift for standard CDM ($\Omega = 1$, $\Omega_b = 0.06$, $h = 0.5$). Only lumps whose parameters (z_{vir}, M) lie above the shaded area can collapse and form luminous objects. The dashed straight lines corresponding to $T_{vir} = 10^4$ K and $T_{vir} = 10^3$ K are shown for comparison (dashed). The dark-shaded region is that in which no radiative cooling mechanism whatsoever could help collapse, since T_{vir} would be lower than the CMB temperature. The solid line corresponds to $3-\sigma$ peaks in standard CDM, normalized to $\sigma_8 = 0.7$, so such objects with baryonic mass $\Omega_b \times 2 \times 10^6$ $M_\odot \sim 10^5$ $M_\odot$ can form at $z = 30$. This figure is taken from Tegmark et al. (1997).

To assess the minimum flux required at each redshift to drive the radiative feedback Ciardi et al. (2000) have performed non-equilibrium multifrequency radiative transfer calculations for a stellar spectrum (assuming a metallicity $Z = 10^{-4}$) impinging onto a homogeneous gas layer, and studied the evolution of the following nine species: H, H^-, H^+, He, He^+, He^{++}, H_2, H_2^+ and free electrons for a free fall time. We can then define the minimum total mass required for an object to self-shield from an external flux of intensity $J_{s,0}$ at the Lyman limit, as $M_{sh} = (4/3)\,\pi\langle\rho_h\rangle R_{sh}^3$, where $\langle\rho_h\rangle$ is the mean dark matter density of the halo in which the gas collapses; R_{sh} is the shielding radius beyond which molecular hydrogen is not photodissociated and allowing the collapse to take place on a free-fall time scale.

Values of M_{sh} for different values of $J_{s,0}$ have been obtained at various redshifts (see Figure 7.25). Protogalaxies with masses above M_H for which cooling is predominantly contributed by the Lyα line are not affected by the radiative feedback studied here. The collapse of very small objects with mass $< M_{crit}$ is on the other hand not possible since the cooling time is longer than the Hubble time. Thus radiative feedback is important in the mass range 10^6–$10^8 M_\odot$, depending on redshift. In order for the negative feedback to be effective, fluxes of the order of $10^{-24} - 10^{-23}$erg s^{-1} cm^{-2} Hz^{-1} sr^{-1} are required. These fluxes are typically produced by a Pop III with baryonic mass $10^5 M_{b,5}\,M_\odot$ at distances closer than $\simeq 21 - 7 \times M_{b,5}^{1/2}$ kpc for the two above flux values, respectively, while the SUVB can reach an intensity in the above range only after $z \approx 15$. This suggests that at high z negative feedback is driven primarily by the direct irradiation from neighbor objects in regions of intense clustering, while only for $z \leq 15$ the SUVB becomes dominant.

Figure 7.26 illustrates all possible evolutionary tracks and final fates of primordial objects, together with the mass scales determined by the various physical processes and feedback. There are four critical mass scales in the problem: (i) M_{crit}, the minimum mass for an object to be able to cool in a Hubble time; (ii) M_H, the critical mass for which hydrogen Lyα line cooling is dominant; (iii) M_{sh}, the characteristic mass above which the object is self-shielded, and (iv) M_{by} the characteristic mass for stellar feedback, below which blowaway can not be avoided. Starting from a virialized dark matter halo, condition (i) produces the first branching, and objects failing to satisfy it will not collapse and form only a negligible amount of stars. In the following, we will refer to these objects as *dark objects*. Protogalaxies with masses in the range $M_{crit} < M < M_H$ are then subject to the effect of radiative feedback, which could either impede the collapse of those with mass $M < M_{sh}$, thus contributing to the class of dark objects, or allow the collapse of the remaining ones ($M > M_{sh}$) to join those with $M > M_H$ in the class of *luminous objects*. This is the class of objects that convert

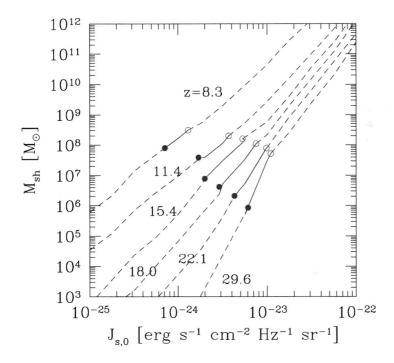

Figure 7.25. Minimum total mass for self-shielding from an external incident flux with intensity $J_{s,0}$ at the Lyman limit. The curves are for different redshift: from the top to the bottom z = 8.3, 11.4, 15.4, 18.0, 22.1, 29.6. Circles show the value of M_H (open) and M_{crit} (filled). Radiative feedback works only in the solid portions of the curves. This figure is taken from Ciardi et al. (2000)

a considerable fraction of their baryons in stars. Stellar feedback causes the final bifurcation by inducing a blowaway of the baryons contained in luminous objects with mass $M < M_{by}$; this separates the class in two subclasses, namely "normal" galaxies (although of masses comparable to present-day dwarfs) that we dub *gaseous galaxies* and tiny stellar aggregates with negligible traces (if any) of gas to which consequently we will refer to as *naked stellar clusters*.

The relative numbers of these objects are shown in Figure 7.27. The straight lines represent, from the top to the bottom, the number of dark matter halos, dark objects, naked stellar clusters and gaseous galaxies, respectively. The dotted curve represents the number of luminous objects with large enough mass ($M > M_H$) to make the H line cooling efficient and become insensitive to the negative feedback. We remind the reader that the naked stellar clusters are the luminous objects with $M < M_{by}$, while the gaseous galaxies are the ones with $M > M_{by}$; thus, the number of luminous objects present at a certain redshift is given by the sum of naked stellar clusters and gaseous galaxies. We first notice that the majority of the luminous objects that are able to form at high redshift will experience blowaway, becoming naked stellar clusters, while only a minor fraction, and only at $z \lesssim 15$, when larger objects start to form, will survive and become gaseous galaxies. An always increasing number of luminous objects is forming with decreasing redshift, until $z \sim 15$, where a flattening is seen. This is due to the fact that the dark matter halo mass function is still dominated by small mass objects, but a large fraction of them cannot form due to the following combined effects: (i) toward lower redshift the critical mass for the collapse (M_{crit}) increases and fewer objects satisfy the condition $M > M_{crit}$; (ii) the radiative feedback due to either the direct dissociating flux or the SUVB increases at low redshift as the SUVB intensity reaches values significant for the negative feedback effect. When the number of luminous objects becomes dominated by objects with $M > M_H$, by $z \sim 10$ the population of luminous objects grows again, basically because their formation is now unaffected by negative radiative feedback. A steadily increasing number of objects is prevented from forming stars and remains dark; this population is about $\sim 99\%$ of the total population of dark matter halos at $z \sim 8$. This is also due to the combined effect of points i) and ii) mentioned above. This population of halos that have failed to produce stars could be identified with the low-mass tail distribution of the dark galaxies that reveal their presence through gravitational lensing of quasars. It has been argued that this population of dark galaxies outnumbers normal galaxies by a substantial amount, and Figure 7.27 supports this view.

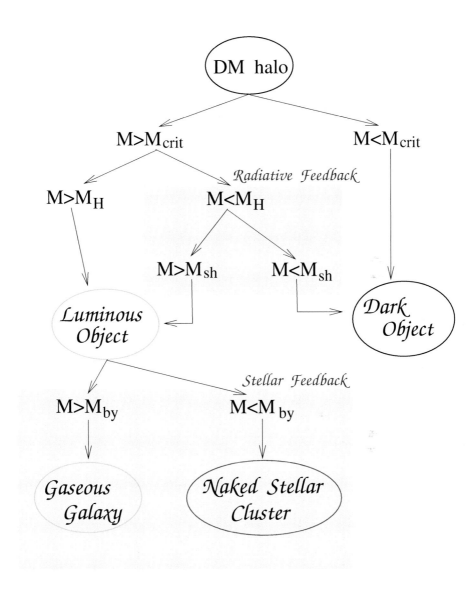

Figure 7.26. Possible evolutionary tracks of objects as determined by the processes and feedback included in the model. See text. This figure is taken from Ciardi et al. (2000).

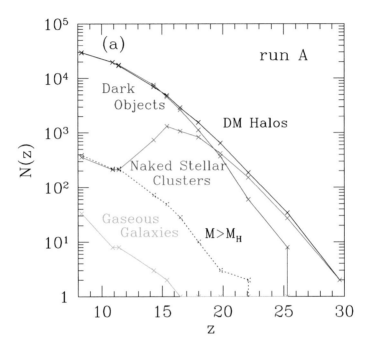

Figure 7.27. Number evolution of different objects in the simulation box. This figure is taken from Ciardi et al. (2000).

7.6.3.1 *Escape of Ionizing Photons*

One crucial point to understand cosmic reionization is to determine the fraction of ionizing photons that can escape from their parent galaxy. By solving the time-dependent radiation transfer problem of stellar radiation through evolving superbubbles within a smoothly varying HI distribution, Dove et al. (2000) have estimated the fraction of ionizing photons emitted by OB associations that escapes the HI disk of our galaxy into the halo and intergalactic medium (IGM). They considered both coeval star formation and a Gaussian star formation history with a time spread $\sigma_t = 2$ Myr;

the calculations are performed with both a uniform HI distribution and a two-phase (cloud/intercloud) model, with a negligible filling factor of hot gas. The shells of the expanding superbubbles quickly trap or attenuate the ionizing flux, so that most of the escaping radiation escapes shortly after the formation of the superbubble. Superbubbles of large associations can blow out of the HI disk and form dynamic chimneys, which allow the ionizing radiation to escape the HI disk directly. However, blowout occurs when the ionizing photon luminosity has dropped well below the association's maximum luminosity. For the coeval star formation history, the total fraction of Lyman Continuum photons that escape both sides of the disk in the solar vicinity is $\langle f_{esc} \rangle \simeq 0.15 \pm 0.05$ (Figure 7.28). For the Gaussian star formation history, $\langle f_{esc} \rangle \simeq 0.06 \pm 0.03$.

Bianchi, Cristiani, & Kim (2001) derived the HI-ionizing background, resulting from the integrated contribution of QSOs and galaxies, taking into account the opacity of the intervening IGM. They have modeled the IGM with pure-absorbing clouds, with a distribution in column density of neutral hydrogen, N_{HI}, and redshift, z, derived from recent observations of the Lyα forest (Kim et al. 2002) and from Lyman Limit systems. The QSOs emissivity has been derived from the recent fits of Boyle et al. (2000), while the stellar population synthesis model of Bruzual & Charlot (2003) and a star-formation history from UV observations for the galaxy emissivity. Due to the present uncertainties in models and observations, three values for the fraction of ionizing photons that can escape a galaxy interstellar medium, f_{esc}= 0.05, 0.1 and 0.4, as suggested by local and high-z UV observations of galaxies, respectively, are used.

In Figure 7.29 is shown the modeled UV background, $J(\nu, z)$, at the Lyman limit as a function of redshift (solid lines) for the flat universe with $\Omega_m = 1$. The total background is shown as the sum of the QSO contribution (the same in each model; dotted line) and the galaxy contribution (scaled with f_{esc}; dashed lines).

At high redshift, the value of the UV background is constrained by the analysis of the proximity effect, i.e., the decrease in the number of intervening absorption lines that is observed in a QSO spectrum when approaching the QSO's redshift (Bajtlik et al. 1988). Using high resolution spectra, Giallongo et al. (1996) derived $J(912\text{Å}) = 5.0^{+2.5}_{-1} \times 10^{-22}$ erg cm^{-2} s^{-1} Hz^{-1} sr^{-1} for $1.7 < z < 4.1$ (see also Giallongo et al. 1999). Larger values are obtained (same units) by Cooke et al. (1997), $J(912\text{Å}) = 1.0^{+0.5}_{-0.3} \times 10^{-21}$ for $2.0 < z < 4.5$ and by a recent re-analysis of moderate resolution spectra by Scott et al. (2000) who found $J(912\text{Å}) = 7.0^{+3.4}_{-4.4} \cdot 10^{-22}$ in the same redshift range. These measurements are shown in Figure 7.29 with a shaded area: the spread of the measurements obtained with different methods and data gives an idea of the uncertainties associated with the study of the proximity effect. Several models can produce a value of $J(912\text{Å})$ compatible with one of the measurements from the prox-

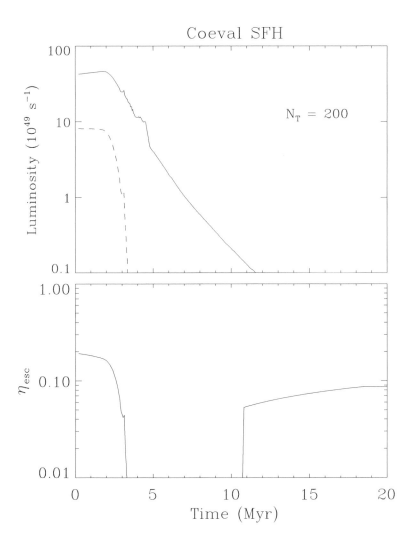

Figure 7.28. Photon luminosity emitted by a single OB association (solid line) and the luminosity of photons escaping each side of the HI disk (dashed line) for coeval star formation. Also shown is the fraction of photons emitted that escape, $\eta_{esc}(N_T, t) = S_{esc}(t)/S(t)$. Here, $Z_{tr} = 2\sigma_h$, the cavity height is $Z_{cy} = 4\sigma_h$ and $\beta = 1$. The I-front is trapped within the shell during the time interval where $\eta_{esc} = 0$. After blowout, η_{esc} suddenly increases even though the photon luminosity of the association is relatively low at these times. Case with $N_\star = 200$. This figure is taken from Dove et al. (2000).

imity effect at $z \sim 3$ shown in Figure 7.29, from a simple QSO-dominated background to models with $f_{esc} \sim 0.2$, leading to a limit of $f_{esc} \lesssim 20\%$.

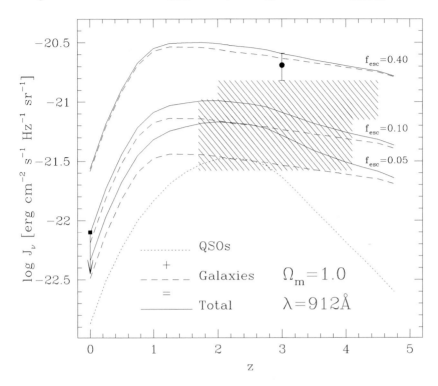

Figure 7.29. UV background at $\lambda = 912$Å for the models with $f_{esc} = 0.05$, 0.1 and 0.4 (solid lines), in a flat $\Omega_m = 1$ universe. Also shown are the separate contribution of the QSOs (dotted line) and of the galaxies (dashed lines, each corresponding to a value of f_{esc}). The shaded area refers to the Lyman limit UV background estimated from the proximity effect (Giallongo et al. 1996; Cooke et al. 1997; Scott et al. 2000). The arrow shows an upper limit for the local ionizing background (Vogel et al. 1995). The datapoint at $z = 3$ is derived from a composite spectrum of Lyman-break galaxies (Steidel et al. 2001). The models and the datapoint of Steidel et al. (2001) have been multiplied by a z-dependent factor, to take into account the cloud emission. This figure is taken from Bianchi et al. (2001).

7.6.3.2 Cosmological Ionization Fronts

The radiative transfer equation in cosmology describes the time evolution of the specific intensity $I_\nu = I(t, x, \omega, \nu)$ of a diffuse radiation field (Ciardi,

Ferrara, & Abel 2000):

$$\frac{1}{c}\frac{\partial I_\nu}{\partial t} + \frac{\hat{n}\cdot\nabla I_\nu}{a_{em}} - \frac{H(t)}{c}\left[v\frac{\partial I_\nu}{\partial \nu} - 3I_\nu\right] = \eta_\nu - \chi I_\nu, \qquad (7.58)$$

where c is the speed of light, χ is the continuum absorption coefficient per unit length along the line of sight and η_ν is the proper space averaged volume emissivity.

If massive stars form in Pop III objects, their photons with $h\nu > 13.6$ eV create a cosmological HII region in the surrounding IGM. Its radius, R_i, can be estimated by solving the following standard equation for the evolution of the ionization front,

$$\frac{dR_i}{dt} - HR_i = \frac{1}{4\pi n_H R_i^2}\left[S_i(0) - \frac{4}{3}\pi R_i^3 n_H^2 \alpha^{(2)}\right]; \qquad (7.59)$$

note that ionization equilibrium is implicitly assumed. H is the Hubble constant, $S_i(0)$ is the ionizing photon rate, n_H is the IGM hydrogen number density and $\alpha^{(2)}$ is the hydrogen recombination rate to levels ≥ 2. For $aR \gg c/H$ the cosmological expansion term HR_i can be safely neglected. In its full form Equation (7.59) must be solved numerically. However, if steady-state is assumed ($dR_i/dt \simeq 0$), then R_i is approximately equal to the Stromgren radius $R_s(z) = [(3dN_\gamma/dt)/(4\pi\alpha_2 n_H^2(z))]^{1/3}$:

$$R_i \simeq R_s = 0.05(\Omega_b h^2)^{-2/3}(1 + z)_{30}^{-2} S_{47}^{1/3}\text{kpc}, \qquad (7.60)$$

where $S_{47} = S_i(0)/10^{47}\text{s}^{-1}$. The approximate expression for $S_i(0)$ is $\simeq 1.2 \times 10^{48} f_b M_6$ s^{-1} when a 20% escape fraction for ionizing photons is assumed and f_b is the fraction of baryon that is able to cool and become available to form stars. In general R_s represents an upper limit for R_i, since the ionization front completely fills the time-varying Stromgren radius only at very high redshift, $z \simeq 100$.

7.6.3.3 Reionization after WMAP

The WMAP satellite has detected an excess in the CMB TE cross-power spectrum on large angular scales ($l < 7$) indicating an optical depth to the CMB last scattering surface of $\tau_e = 0.17 \pm 0.04$ (Kogut et al. 2003). The reionization of the universe must therefore have begun at relatively high redshift. Ciardi, Ferrara, & White (2003) have studied the reionization process using supercomputer simulations of a large and representative region of a universe that has cosmological parameters consistent with the WMAP results. The simulations follow both the radiative transfer of ionizing photons and the formation and evolution of the galaxy population that produces them.

They have shown that the WMAP measured optical depth to electron scattering is easily reproduced by a model in which reionization is caused by the first stars in galaxies with total masses of a few $\times 10^9$ $M_\odot$. Moreover, the first stars are "normal objects", i.e., their mass is in the range 1–50 $M_\odot$, but metal-free. Among the different models explored, the "best" WMAP value for τ_e is matched assuming a moderately top-heavy IMF (Larson IMF with $M_c = 5$ $M_\odot$) and an escape fraction of 20%. A Salpeter IMF with the same escape fraction gives $\tau_e = 0.132$, which is still within the WMAP 68% confidence range. Decreasing f_{esc} to 5% gives $\tau_e = 0.104$, which disagrees with WMAP only at the 1.0 to 1.5σ level. In Figure 7.30 is shown the neutral hydrogen number density at three redshifts ($z = 17.6$, 15.5 and 13.7) and in Figure 7.31 is shown the evolution of the electron optical depth for the different IMFs considered.

In the best-fit model reionization is essentially complete by $z_r \simeq 13$. This is difficult to reconcile with observation of the Gunn–Peterson effect in $z > 6$ quasars (Becker et al. 2001; Fan et al. 2002) which imply a volume-averaged neutral fraction $\gtrsim 10^{-3}$ and a mass-averaged neutral fraction $\sim 1\%$ at $z = 6$. Then a fascinating (although speculative) possibility is that the universe was reionized twice (Cen 2003; Whyte & Loeb 2002) with a relatively short redshift interval in which the IGM became neutral again.

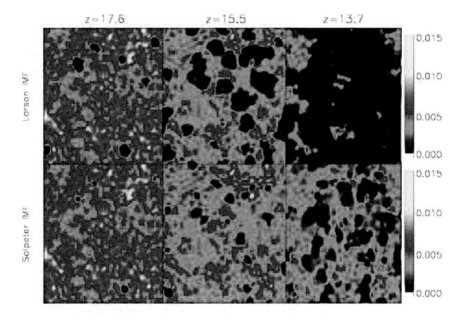

Figure 7.30. Slice through the simulation boxes. The six panels show the neutral hydrogen number density for the case of a Larson IMF with $M_c = 5$ M$_\odot$ (upper panels) and for a Salpeter IMF (lower panels) at redshifts, from left to right, $z = 17.6$, 15.5 and 13.7. The escape fraction is 20%. The box for the radiative transfer simulation has a comoving length of $L = 20h^{-1}$ Mpc. This figure is taken from Ciardi et al. (2003).

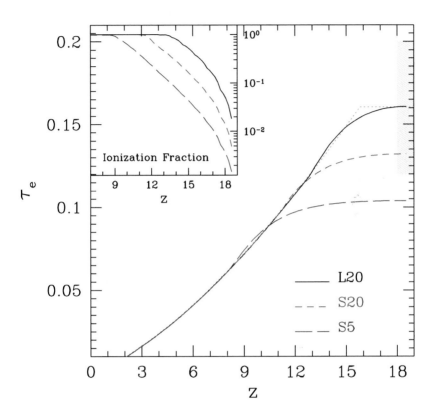

Figure 7.31. Redshift evolution of the electron optical depth, τ_e, for a Salpeter IMF with $f_{esc} = 5\%$ (long-dashed line) and $f_{esc} = 20\%$ (short-dashed line), and for a Larson IMF with $M_c = 5$ $M_\odot$ and $f_{esc} = 20\%$. The dotted line refers to a sudden reionization at $z = 16$. The shaded region indicates the optical depth $\tau_e = 0.16 \pm 0.04$ (68% CL) implied by the Kogut et al. (2003) "model independent" analysis. In the inset the redshift evolution of the volume-averaged ionization fraction, x_v, is shown for the three runs. This figure is taken from Ciardi et al. (2003).

References

Abel T, Anninos P, Zhang Y and Norman M L 1997 *NewA* **2** 181

Abel T, Bryan G L and Norman M L 2000 *ApJ* **540** 39

Bajtlik S, Duncan R C and Ostiker J P 1998 *ApJ* **327** 570

Balbus S A 1995 *The Physics of the Interstellar Medium and Intergalactic Medium* ASP Series Vol 80 (San Francisco: PASP) p 328

Becker R H et al 2001 *AJ* **122** 2850

Benson A J and Madau P 2003 *MNRAS* **344** 835

Benson A J, Lacey C G, Bauch C M, Cole S and Frenk C S 2002 *MNRAS* **333** 156

Bianchi S, Cristiani S and Kim T-S 2001 *A&A* **376** 1

Boyle B J, Shanks T, Croom S M, Smith R J, Miller L, Loaring N and Heymans C 2000 *MNRAS* **317** 1014

Bromm V, Coppi P S and Larson R B 1999 *ApJ* **527** L5

Bromm V, Coppi P S and Larson R B 2002 *ApJ* **564** 23

Bromm V, Ferrara A, Coppi P S and Larson R B 2001 *MNRAS* **328** 969

Bruzual A G and Charlot S 2003 *MNRAS* **344** 1000

Cen R 2003 *ApJ* **591** 12

Ciardi B, Ferrara A and Abel T 2000 *ApJ* **533** 594

Ciardi B, Ferrara A and White S M D 2003 *MNRAS* **344** L7 astro-ph/0302451

Ciardi B, Ferrara A, Governato F and Jenkins A 2000 *MNRAS* **314** 611

Ciardi B, Ferrara A, Marri S and Raimondo G 2001 *MNRAS* **324** 381

Cioffi D, McKee C F and Bertschiger E 1988 *ApJ* **334** 252

Cooke A J, Espey B and Carswell R F 1997 *MNRAS* **284** 552

Dawson S, Spinrad H, Stern D, Dey A, van Breugel W, de Vries W and Reuland M 2002 *ApJ* **570** 92

Dekel A and Silk J 1986 *ApJ* **303** 39

Dey A, Spinrad H, Stern D, Graham J R and Chaffee F H, 1998 *ApJ* **498** 93

Dove J B, Shull J M and Ferrara A 2000 *ApJ* **531** 846

Ellis R, Santos M R, Kneib J-P and Kuijken K 2001 *ApJL* **560** 119

Fan X et al 2002 *AJ* **123** 1247

Ferrara A and Tolstoy E 2000 *MNRAS* **313** 291

Field G B 1995 *ApJ* **142** 531

Fryer G M 1999 *ApJ* **522** 413

Fryer G M, Woosley S E and Weaver T A 2001 *ApJ* **550** 372

Fujita S S et al 2003 *AJ* **125** 13

Furlanetto S R and Loeb A 2003 *ApJ* **588** 18

Giallongo E, Cristiani S, D'Odorico S, Fontana A and Savaglio S 1996 *ApJ* **466** 46

Giallongo E, Fontana A, Cristiani S and D'Odorico S 1999 *ApJ* **510** 605

Haardt F and Madau P 1996 *ApJ* **461** 20

Heger A and Woosley S E 2001 *ApJ* **567** 532

Hu E M, Cowie L L, McMahon R G, Capak P, Iwamuro F, Kneib J-P, Maihara T and Motohara K 2002 *ApJ* **568** L75

Hu E M, McMahon, R G and Cowie, L L 1999 *ApJ* **522** L9

Kim T-S, Cristiani S and D'Odorico S 2002 *A&A* **383** 747

Kodaira K et al 2003 *PASJ* **55** 17

Kogut A et al 2003 *ApJS* **148** 161

Larson R B 1974 *MNRAS* **169** 229

Lehnert M D and Bremer M 2003 *ApJ* **593** 630

Mac Low M-M and Ferrara A 1999 *ApJ* **513** 142

Madau P and Rees M J 2001 *ApJ* **551** L27

Madau P, Ferrara A and Rees M 2001 *ApJ* **555** 92

Marri S and White S D M 2003 *MNRAS* **345** 561

Marri S et al 2003 in preparation

Mateo M L 1998 *AAR&A* **36** 435

McKee C F and Draine B T 1991 *Science* **252** 397

Mori M, Ferrara A and Madau P 2002 *ApJ* **571** 40

Navarro J F and White S D M 1993 *MNRAS* **265** 271

Norberg P et al 2002 *MNRAS* **336** 907

Omukai K 2001 *ApJ* **546** 635

Omukai K and Inutsuka S 2002 *MNRAS* **332** 59

Omukai K and Palla F 2003 *ApJ* **589** 677

Osterbrok D E 1989 *Astrophysics of Gaseous Nebulae and Active Galactic Nuclei* (Mill Valley: University Science Books)

Ostriker J P and McKee C F 1988 *Rev Mod Phys* **60** 1

Pearce F R et al 1999 *ApJ* **521** 99

Persic M, Salucci P and Stel F 1996 *MNRAS* **281** 27

Rhoads J E et al 2003 *AJ* **125** 1006

Ricotti M, Ferrara A and Miniati F 1997 *ApJ* **485** 254

Scannapieco E, Ferrara A and Broadhurst T 2000 *ApJL* **536** 11

Scannapieco E, Ferrara A and Madau P 2002 *ApJ* **574** 590

Scannapieco E, Schneider R and Ferrara A 2003 *ApJ* **589** 35

Schaerer D 2002 *A&A* **382** 28

Schaye J, Rauch M, Sargent W L W and Kim T-S 2000 *ApJL* **541** 1

Schneider R, Ferrara A, Natarayan P and Omukai K 2002 *ApJ* **571** 30

Schneider R, Ferrara A, Salvaterra R, Omukai K and Bromm V, 2003 *Nature* **422** 869

Scott J, Bechtold J, Dobrzycki A and Kulkarni V P 2000 *ApJS* **130** 67

Sedov L 1978 *Similitude et Dimensions en Mecanique* (Moscow: Editions de Moscou)

Silk J 2003 *MNRAS* **343** 249

Spitzer L 1978 *Physical Processes in the Interstellar Medium* (Whiley: New York)

Steidel C C, Pettini M and Adelberger K L 2001 *ApJ* **546** 665

Tegmark M et al 1997 *ApJ* **474** 1

Thacker R J and Couchman H P 2001 *ApJ* **555** L17

Thomas P A, Fabian A C, Arnaud K A, Forman W and Jones C, 1986 *MNRAS* **222** 655

Tumlinson J, Giroux M and Shull J M 2001 *ApJ* **555** 839

Umeda H and Nomoto K 2002 *ApJ* **565** 385

White S D M and Frenk C S 1991 *ApJ* **379** 52

Wolfire M G, Hollenbach D, McKee C F, Tielens A G G M and Bakes E L O 1995 *ApJ* **443** 152

Wyithe S and Loeb A 2003 *ApJ* **588** 69

Chapter 8

The Ecology of Black Holes in Star Clusters

Simon Portegies Zwart
Astronomical Institute Anton Pannekoek
University of Amsterdam
Amsterdam, The Netherlands

8.1   Introduction

In this chapter we investigate the formation and evolution of black holes in star clusters.

The star clusters we investigate are generally rich, containing more than 10^4 stars, and with a density exceeding 10^4 stars/pc^3. Among these are young populous clusters, globular cluster and the nuclei of galaxies.

Under usual circumstances black holes are formed from stars with a zero-age main-sequence (ZAMS) mass of at least 25–30 $M_\odot$ (Maeder 1992; Portegies Zwart, Verbunt, & Ergma 1997; Ergma & van den Heuvel 1998). These stars live less than about 10 Myr, after which a supernova results in a black hole in the range of 5–20 $M_\odot$ (Fryer & Kalogera, 2001). The rest of the mass is lost from the star in the windy phase preceding the supernova or in the explosion itself (Heger 2003). For Scalo (1986) or Kroupa & Weidner (2003) initial mass functions (IMF), variants of Salpeter (1955), one in 2300–3500 stars collapses to a black hole. So black holes are rare, but still, the Milky-Way galaxy is populated by 30 to 40 million black holes. In the remainder we will refer to these relatively common types of black holes as stellar mass black holes or simply as BH.

Supermassive black holes (SMBH) are, with about one per galaxy, among the rarest single objects in the universe. They have masses of about 10^6 to 10^{10} $M_\odot$ which also makes them among the most massive single objects in the universe. The best evidence for the existence of a supermassive black hole comes from the orbit of the star S2 which is in a 15.2 year orbit around an unseen object (Schödel et al. 2002). The mass of this black hole is $3.7 \pm 1.5 \times 10^6 \, M_\odot$, which puts it directly at the bottom of the mass scale for SMBHs (see also Figure 8.17).

The gap in mass between stellar mass black holes and supermassive black holes may be bridged by a third type, often referred to as intermediate mass black holes (IMBH).

In this chapter our main interest is the formation and evolution of these black hole families in star clusters. Also the possible evolutionary link between stellar mass black holes, via intermediate mass black holes to supermassive black holes will be addressed. We mainly focus, however, on the ecology of star clusters. The term star cluster ecology was introduced in 1992 by Douglas Heggie to illustrate the complicated interplay between stellar evolution and stellar dynamics, which by their mutual interactions have similarities with biological systems.

In the here-studied ecology we mainly address the gravitational dynamics, stellar evolution, binary evolution, and external influence and how these seemingly separate effects work together on the star cluster, much in the same way as in an organism.

8.1.1   Setting the Stage

The main objects of our study are clusters of stars. There are a number of distinct families of star clusters, with one common characteristic: a star cluster is a self-gravitating system of stars, all of which are about the same age. We make the distinction between various types, among which are: open cluster like Pleiads (see Figure 8.1), young dense cluster like R 136 in the 30 Doradus region of the large Magellanic cloud or the Arches cluster (see Figure 8.2), and globular cluster like M15 (see Figure 8.3). If galactic nuclei contain a population of stars with similar ages we could include these in the definition. The nucleus of the Andromeda galaxy, depicted in Figure 8.4 is an example, though the stellar population in its nucleus has probably a large spread in ages.

Figure 8.1 shows the $\sim 115 \, \mathrm{Myr}$ old star cluster Pleiads,[1] at a distance of about 135 pc (Pinfield et al. 1998; Raboud & Mermilliod 1998; Bouvier et al 1997). This cluster contains about 2000 stars, half of which are contained in a sphere with a radius of about 8 pc.

Figure 8.2 shows Hubble Space Telescope (HST) images of two well-known young dense star clusters, Arches (to the left) and NGC 2070 (right)

---

[1] According to Greek mythology, one day the great hunter Orion saw the Pleiads as they walked through the Boeotian countryside, and fancied them. He pursued them for seven years, until Zeus answered their prayers for delivery and transformed them into birds (doves or pigeons) placing them among the stars. Later, when Orion was killed (many conflicting stories as to how), he was placed in the heavens behind the Pleiads, immortalizing the chase.

Figure 8.1. The Pleiads open star cluster (image by David Malin) contains a few thousand stars in a volume of several parsecs cubed.

Figure 8.2. Left: Arches star cluster (HST image by Figer et al. 1999). At a projected distance of $\sim 30\,\text{pc}$ from the galactic center this cluster is strongly influenced by external forces. The cluster on the image still looks quite symmetric since we only observe the inner part.

Right: The young dense star cluster R136 (NGC 2070, HST image by N. Walborn) in the 30 Doradus region of the large Magellanic cloud.

in the Large Magellanic Cloud. These clusters are the prototypical examples of young dense star clusters (YoDeCs), which are young ($\lesssim 10\,\mathrm{Myr}$), massive $\sim 10^4\,\mathrm{M_\odot}$ and dense $n \gtrsim 10^5\,\mathrm{stars/pc^3}$. The two clusters show also that the class of young dense star clusters hosts two families of star clusters: the strongly tidally perturbed cluster (such as Arches) and the isolated cluster (such as R 136).

Figure 8.3.   The globular cluster M15 (NGC 7078, HST image by Guhathakurta).

Figure 8.3 gives a Hubble's view of the core collapsed globular cluster M 15 (Guhathakurta 1996). It is one of the old globular star clusters in the Milky Way's halo, contains about a million stars, and the central density is of the order of $n \gtrsim 10^5$ stars/pc^3. No core has been measured in the density profile, and therefore this cluster is identified as a collapsed globular (Harris 1996, http://physun.physics.mcmaster.ca/Globular.html).

In Figure 8.4 we show an image of the Andromeda Galaxy.[2] A total of 693 candidate globular clusters were found in recent 2MASS MIR observations (Galleti 2004). It is still unclear why M31 has more than trice as many globular clusters than the Milky Way galaxy. Note also that there are no YoDeCs observed in M31, which is also somewhat puzzling, as the Milky Way galaxy has at least four.

---

[2] Andromeda was the Ethiopian princess, whom Perseus rescued and married. She became queen of Mycenae and, after her death, a constellation. The galaxy was later named after the constellation.

Figure 8.4. The Andromeda galaxy (M31, image by Jason Ware).

In Figure 8.5 we place a few well-known star clusters in retrospect. All cluster images are on the same scale. It is interesting to see how large the globular cluster M80 is compared to some depicted open clusters (Pleiads) and compared to the known galactic YoDeCs (Arches, Quintuple, NGC3603 and Westerlund 1).

There appears to be a clear relation between the number of globular clusters, such as M15 (see Figure 8.3), and the Hubble type of the host galaxy. This relation is expressed in the specific number of globular clusters S_N. For open clusters and YoDeCs no such relation is known, but young star clusters are particularly abundant in starburst and interacting galaxies such as the Antennae and M82. (See Table 8.1)

Table 8.2 lists the space densities and specific numbers of globular clusters S_N per $M_v = -15$ magnitude (van den Bergh 1984) for various Hubble types of galaxies. The values given for S_N in Table 8.2 are corrected for internal absorption; the absorbed component is estimated from observations in the far infrared.

$$S_N = N_{GC}10^{0.4(M_v+15)}. \tag{8.1}$$

Here N_GC is the total number of globular clusters in the galaxy under

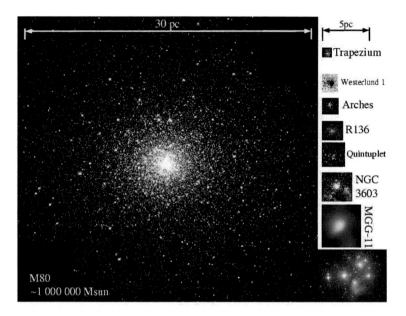

Figure 8.5. Star clusters in the correct physical scale. The galaxy (Figure 8.4) was left out for practical reasons. Images are from: M80 (Ferraro); Trapezium cluster (HST image by Bally, Devine, Sutherland, and Johnson); Westerlund 1 (2MASS image by van Dyk); Arches (HST image by Figer, 1999); R 136 (HST image by Walborn); Quintuplet (HST image by Figer 1999); MGG-11 (VLT image by McCrady et al. 2003); Pleiads (AAT image by Malin).

consideration. The estimated number density of globular clusters in the local universe is

$$\phi_{GC} = 8.4 \, h^3 \, \mathrm{Mpc}^{-3} \tag{8.2}$$

(where $h = H_0/100 \; \mathrm{km \, s^{-1} \, Mpc^{-1}}$), slightly smaller than the result reported by Phinney (1991).

Figure 8.6 illustrates the concentration of clusters, expressed in the structural parameter W_0 (King 1966), which ranges from about 1 for very shallow clusters to about 12 for very concentrated clusters. The figure shows the images of three clusters with quite different concentrations ranging from $W_0 \simeq 6$ (Omega Centauri) to $W_0 \gtrsim 12$ for the core collapsed globular cluster M15. The globular cluster 47 Tuc. is highly concentrated, but not quite in a state of core collapse.

Table 8.1. Overview of selected parameters for young dense clusters (Massey & Hunter 1998), globular clusters (Djorgovski & Meylan, 1994) and galactic nuclei (Schrödel et al. 2002). The first three columns list the cluster type, the total mass (in solar units) and the virial radius (in pc). For globulars the total mass and virial radius are given as distributions with a mean and the standard deviation around the mean. The orbital separation (in solar units) for a $1000 \, kT$ binary consisting of two $10 \, M_\odot$ stars is given in the fourth column. The fifth and sixth columns list the expected number of black hole binaries that are formed by the cluster and the fraction of these binaries which merge within $12 \, Gyr$ (allowing $\sim 3 \, Gyr$ for the formation and ejection of the binaries and assuming a $15 \, Gyr$ old universe (Freedman et al. 2001). The contributions to the total black hole merger rate per star clusters per year (MR) are given in the final column (for details see § 8.3.4). The bottom row contains estimated parameters for the zero-age population of globular clusters in the galaxy.

* Estimate for the parameters at birth for the population of globular clusters.

Cluster Type	$M/M_\odot$ [log]	r_{vir}/pc [log]	$1000 \, kT$ $[R_\odot]$	N_b	f_{merge}	MR $[Myr^{-1}]$
Populous	4.5	-0.4	420	7.9	7.7%	0.0061
Globular	5.5 ± 0.5	0.5 ± 0.3	315	150	51 %	0.0064
Nucleus	~ 7	$\lesssim 0$	$\lesssim 3.3$	2500	100 %	0.21
Globular*	6.0 ± 0.5	0 ± 0.3	33	500	92 %	0.038

Table 8.2. Galaxy morphology class, space densities, average absolute magnitude (Heyl et al. 1997), and the specific frequency of globular clusters S_N (from van den Bergh, 1995 and McLaughlin 1999). The final column gives the contribution to the total number density of globular clusters. The galaxy morphologies are identified as: E (elliptical), S0-Scd (spiral galaxies), Blue E (young blue elliptical galaxies), Sdm (blue spiral galaxies) and StarB (star burst galaxies).

Galaxy Type	ϕ_{GN} $[10^{-3} \, h_0 \, Mpc^{-3}]$	M_v	$S_N h^2$	GC Space Density $[h_0^3 \, Mpc^{-3}]$
E-S0	3.49	-20.7	10	6.65
Sab	2.19	-20.0	7	1.53
Sbc	2.80	-19.4	1	0.16
Scd	3.01	-19.2	0.2	0.03
Blue E	1.87	-19.6	14	1.81
Sdm/StarB	0.50	-19.0	0.5	0.01

Figure 8.6. Three globular clusters to illustrate the effect of core collapse. Left: the rather shallow globular clusters Omega Cen. (NGC 5139, image by P. Seitzer) with a concentration $W_0 \simeq 6$. Middle: the concentrated cluster 47 Tuc. (NGC 104, image by W. Keel) with concentration $W_0 \simeq 9$. Right: the core collapsed cluster M15 (NGC 7078, HST image R. Guhathakurta) with concentration $W_0 \simeq 12$.

8.1.2   Fundamental Timescales

The evolution of a star cluster is dominated by two main effects: the mutual gravitational attraction between the stars and by the evolution of the individual stars. Before we discuss these two ingredients in more detail it is important to understand how they are coupled in the ecological network.

8.1.2.1   Stellar Evolution

The fundamental timescale for stellar evolution is the nuclear burning timescale. This timescale only depends on characteristics of the stars themselves and is unrelated to any of the dynamical cluster parameters.[3] The mass is the most fundamental parameter in the stellar evolution process; more massive stars burn up more quickly than lower-mass stars. The main sequence lifetime of a star with mass m and luminosity l can, to first order, be approximated with

$$t_{\mathrm{ms}} \simeq 1.1 \times 10^{10} \text{ year} \left(\frac{m}{M_\odot} \right) \left(\frac{L_\odot}{l} \right). \qquad (8.3)$$

After the main sequence the star generally grows to giant dimensions, but remains large only for approximately 10% of its main-sequence lifetime.[4] Massive stars continue their evolution after central hydrogen is exhausted by burning carbon, neon, oxygen and silicon until an iron core

---

[3] At least as long as the stars do not strongly influence each other's evolution due to dynamical interactions.

[4] As a rule of thumb, one can adopt that each subsequent burning stage for a star takes about 10% of the previous burning stage; carbon burning lasts for about 10% of the helium burning stage, etc.

forms, which ultimately collapse catastrophically. The result is a supernova and the formation of a compact stars. Lower-mass stars cannot process all material and shed their envelope in a planetary nebula phase. This results in a white dwarf. Rather detailed stellar evolutionary tracks are published in the form of comprehensive look-up tables (Schaller et al 1993; Meynet et al. 1994) or fitting formulae (Eggleton, Fitchet, & Tout 1989; Hurley et al. 2000) using a variety of stellar evolution codes.

8.1.2.2 Dynamical Timescales

The most fundamental dynamical timescale t_d for a star cluster is the crossing time or dynamical timescale, which for a cluster with half-mass radius R and dispersion velocity $\langle v \rangle$ can be written simply as

$$t_d = R/\langle v \rangle. \tag{8.4}$$

We can generalize this by using the cluster half-mass radius and the dispersion velocity to calculate the global crossing time t_{hm} of the star cluster. As long as stellar evolution remains relatively unimportant, the cluster's dynamical evolution is dominated by two-body relaxation, with proceeds via the characteristic half-mass relaxation timescale (Spitzer 1987)

$$t_{rt} = \left(\frac{R^3}{GM}\right)^{1/2} \frac{N}{8\ln\Lambda}. \tag{8.5}$$

Here G is the gravitational constant, M is the total mass of the cluster, $N \equiv M/\langle m \rangle$ is the number of stars and r is the characteristic (half-mass) radius of the cluster. The Coulomb logarithm $\ln\Lambda \simeq \ln(0.1N) = O(10)$ typically. In convenient units the two-body relaxation time becomes

$$t_{rt} \simeq 1.9\,\text{Myr} \left(\frac{R}{1\,\text{pc}}\right)^{3/2} \left(\frac{M}{1\,M_\odot}\right)^{1/2} \left(\frac{1\,M_\odot}{\langle m \rangle}\right) (\ln\Lambda)^{-1}. \tag{8.6}$$

Table 8.3 summarizes the relevant fundamental characteristics of various types of star clusters. The parameters we selected are the stellar evolution timescale, cluster mass, size and velocity dispersion. With these we can compute the relaxation time and crossing time for the various clusters. Near the bottom of the table we present the collision time: $t_{coll} \equiv 1/\Gamma_{coll}$. Here Γ_{coll} is the collision rate for a cluster with stellar density n, and can be written as $\Gamma_{coll} = n\sigma v$, where σ is the cross-section for physical collisions:

$$\sigma = \pi d^2 \left(1 + \frac{v^2}{v_\infty^2}\right). \tag{8.7}$$

Here v_∞ is the relative velocity of two stars with mass m_1 and m_2 at infinity and v is the relative velocity at closest distance d in a parabolic encounter,

Table 8.3. Time scales.

Timescale	Symbol	Bulge	Globular	YoDeC	Open Cluster
Star	t_{ms}	10 Gyr	10 Gyr	10 Myr	10 Myr
Size	R	100 pc	10 pc	$\lesssim 1$ pc	10 pc
Mass	M	$10^9 M_\odot$	$10^6 M_\odot$	$10^5 M_\odot$	1000 $M_\odot$
Velocity	$\langle v \rangle$	100 km s^{-1}	10 km s^{-1}	10 km s^{-1}	1 km s^{-1}
Relaxation	t_{rt}	10^{15} yr	3 Gyr	50 Myr	100 Myr
Crossing	t_{hm}	100 Myr	10 Myr	100 Kyr	1 Myr
Collision	t_{coll}	10 Gyr	100 Myr	10 Kyr	100 Myr
t_{rt}/t_{ms}		10^5	3	5	10
t_{hm}/t_{ms}		0.01	1	10^{-4}	0.1
t_{coll}/t_{ms}		1	0.1	10^{-5}	10

i.e. $v^2 = 2G(m_1 + m_2)/d$. The second term results from the gravitational attraction between the two stars, and is referred to as gravitational focusing.

We can now estimate the timescale for a collision between two stars as $t_{coll} = 1/\Gamma_{coll}$, which for a cluster with low velocity dispersions ($v_\infty \ll v$, can be written as

$$t_{coll} = 7 \times 10^{10} \text{yr} \left(\frac{10^5 \text{pc}^{-3}}{n} \right) \left(\frac{v_\infty}{10\text{km/s}} \right) \left(\frac{R_\odot}{d} \right) \left(\frac{M_\odot}{m} \right) \text{ for } v \gg v_\infty. \tag{8.8}$$

For a collision between two single stars we can adopt $d \sim 3r_\star$ (Davies, Benz, & Hills 1991).

8.1.3   The Effect of Two-Body Relaxation: Dynamical Friction

A star cluster in orbit around the galactic center is subject to dynamical friction, in much the same way as dynamical friction drives massive stars toward the cluster center. This causes clusters to spiral into the galactic center and stars to the cluster center. A star cluster is generally destroyed by the tidal field when it approaches the galactic center (see Gerhard 2001). We derive here the dynamical friction timescale for a mass point in the potential of the galactic center. The derivation of the dynamical friction timescale of a star as it spirals to the cluster center is very similar, with the major exception that the cluster potential is much more complicated than the potential of the galaxy. We therefore opted for showing the derivation for a galaxy instead.

We assume the inspiraling object to have constant mass M, deferring the more realistic case of a time-dependent mass (for example in the case of a star cluster sinking to the galactic center, see Equation (8.37)) to McMillan & Portegies Zwart (2003).

The drag acceleration due to dynamical friction in an infinite homogeneous medium with isotropic velocity distribution that is not self-gravitating is (Equations [7–18] in Binney & Tremaine, 1987)

$$a = -\frac{4\pi \ln \Lambda G^2 M \rho_G(R_{gc})}{v_c^2} \left[\text{erf}(X) - \frac{2X}{\sqrt{\pi}} e^{-X^2} \right]. \tag{8.9}$$

Here $\ln \Lambda$ is the Coulomb logarithm for the galactic central region, for which we adopt $\ln \Lambda \sim R_{gc}/R$, erf is the error function and $X \equiv v_c/\sqrt{2} v_{disp}$, where v_{disp} is the one-dimensional velocity dispersion of the stars at distance R_{gc} from the galactic center, and v_c is the circular speed of the cluster around the galactic center.

The mass of the galaxy lying within the cluster's orbit at distance R_{gc} ($\lesssim 500\,\text{pc}$) from the galactic center is (Sanders & Lowinger 1972; Mezger et al. 1996)

$$M_{\text{Gal}}(R_{gc}) = 4.25 \times 10^6 \left(\frac{R_{gc}}{1\,\text{pc}} \right)^{1.2} M_\odot. \tag{8.10}$$

Its derivative, the local galactic density (see Portegies Zwart et al. 2001a) is

$$\rho_G(R_{gc}) \simeq 4.06 \times 10^5 \left(\frac{R_{gc}}{1\,\text{pc}} \right)^{-1.8} M_\odot\,\text{pc}^{-3}. \tag{8.11}$$

For inspiral through a sequence of nearly circular orbits, the function $\text{erf}(X) - \frac{2X}{\sqrt{\pi}} \exp(-X^2)$ appearing in Equation (8.9) may be determined as follows.

Following Binney & Tremaine (p. 226), we write the equation of dynamical equilibrium for stars near the galactic center as

$$\frac{dP}{dR_{gc}} = -\rho_G \frac{GM_{\text{Gal}}(R_{gc})}{R_{gc}^2}, \tag{8.12}$$

where $P = kT\rho/\langle m \rangle$, $\frac{3}{2}kT = \frac{1}{2}\langle m \rangle \langle v^2 \rangle$. Since $u^2 = \frac{1}{3}\langle v^2 \rangle$, it follows that $P = u^2 \rho$, and Equation 8.12 becomes

$$\frac{d}{dr}(u^2 \rho) = -\frac{\rho}{r} v_c^2, \tag{8.13}$$

where v_c is the circular orbital velocity at R: $v_{disp}^2 = GM_{\text{Gal}}(R_{gc})/R_{gc}$. For $M_{\text{Gal}} \propto R_{gc}^x$ (see Equation (8.10)), and assuming that $v_{disp}^2 \propto v_c^2 \sim R_{gc}^{x-1}$, we find $v_{disp}^2 \rho \sim R_{gc}^{2x-4}$, so

$$r\frac{d}{dr}(v_{disp}^2 \rho) = (2x - 4)v_{disp}^2 \rho = -\rho v_c^2, \tag{8.14}$$

and hence $X = \sqrt{2 - x}$. Equation (8.9) then becomes

$$a = -1.2 \ln \Lambda \frac{GM}{R_{gc}^2} \left[\text{erf}(X) - \frac{2X}{\sqrt{\pi}} \exp(-X^2) \right]. \tag{8.15}$$

For $x = 1.2$, $X = 0.89$ and

$$a = -0.41 \ln \Lambda \frac{GM}{R_{\text{gc}}^2} .$$ (8.16)

Again following Binney & Tremaine, defining $L = R_{\text{gc}} v_{\text{c}}$ and setting $dL/dt = a R_{\text{gc}}$, we can integrate Equation (8.16) with respect to time to find an inspiral time from initial radius R_i of

$$T_{\text{f}} \simeq \frac{1.28}{\ln \Lambda} \frac{M_{\text{Gal}}(R_i)}{M} \left[\frac{G M_{\text{Gal}}(R_i)}{R_i^3} \right]^{-1/2}$$ (8.17)

$$\simeq 1.4 \left(\frac{R_i}{10 \, \text{pc}} \right)^{2.1} \left(\frac{10^6 \, \text{M}_\odot}{M} \right) \text{Myr}.$$ (8.18)

For definiteness, we have assumed $\ln \Lambda \sim 4$ ($\Lambda \sim R_{\text{gc}}/R \sim 100$) in Equation (8.18), corresponding to a distance of about 10–$30 \, \text{pc}$ from the galactic center.

8.1.4 Simulating Star Clusters

Stars move around due to their mutual gravity. This principle was first accurately described by Sir Isaac Newton in 1687 in his *Philosophiae Naturalis Principia Mathematica* (an excellent short biography can be found at http://www-gap.dcs.st-and.ac.uk).

Newton's equation describes the gravitational interaction between two stars with masses m_1 and m_2 with relative positional vector $\mathbf{r} = \mathbf{r}_2 - \mathbf{r}_1$, which can be written as

$$\frac{d\mathbf{r}^2}{dt^2} = -G \frac{m_1 + m_2}{r^3} \mathbf{r}.$$ (8.19)

The minus sign in the right-hand side indicates that the interaction force $(d\mathbf{r}^2/dt^2)$ is attractive.

This second-order differential equation can be integrated in many ways. At this moment Hut and Makino are in the process of writing a series of 10 books about integrating this equation using the so-called direct N-body technique. The first three volumes of this series are available at http://www.ArtCompSci.org. Other recent excellent work has been published by Heggie & Hut (2003) and Aarseth (2003). Therefore instead of worrying about the intricacies of N-body techniques we continue directly with the core problem.

First, however, it may be useful to explain a bit about the methods under consideration and some of the alternatives; it is not my intention to give a thorough overview of all the ways in which you can solve an N-body system, but it is good to have some overview. In a direct N-body solver you integrate the equations of motion of all stars in the system by computing the forces from each star directly. This means that the amount of work for the computer scales roughly with the square of the number of particles N, or in units of CPU time:

$$t_{\mathrm{CPU}} = \binom{N}{2}, \qquad (8.20)$$

which for large N becomes $t_{\mathrm{CPU}} = \lim_{N \to \infty} N^2$. This scaling becomes even worse if one imagines that the dynamical time unit in a star cluster is inversely proportional to N (see Equation (8.4)) resulting in a time complexity that approaches $t_{\mathrm{CPU}} \propto N^3$. Due to this high computational cost it is at this moment not possible to integrate the equations of motion of 10^6 stars for a Hubble time. The largest simulations that so far have been done are $N \simeq 10^5$ stars for a Hubble time (Baumgardt et al. 2003) and of $N = 585.000$ stars for the first 12 Myr of the cluster (Portegies Zwart et al. 2004).

One way to escape this conundrum is by integrating the equations of motion less accurately, or by not integrating them at all but by approximating the time evolution of the (grand-) canonical ensemble of stars that forms the cluster. The (semi)approximate methods are generally harder to code and often impossible to assess, which makes fine-tuning to direct N-body simulations inevitable. The core problem here is that a star cluster is in a delicate way not in perfect virial equilibrium. Methods that assume a vitalized state therefore have a distinct disadvantage over methods that do not explicitly require equilibrium. Often the more complex numerical coding of approximate solvers is well spent for large systems, such as galaxies or cosmological simulations, in which low precision methods are preferred because of the sheer number of "stars" that have to be modeled.

For simulating dense star clusters the best way is probably still the direct integration of the N-body system, though competitive approaches have been taken (see e.g., § 8.1.4.1 and § 8.1.4.2).

An extensive comparison between various types of N-body codes has been performed in Heggie's (1998) collaborative experiment, the results of which can be inspected at
http://www.maths.ed.ac.uk/~douglas/experiment.html.

Recently Spinnato et al. (2003) carried out a comprehensive comparison between three very different N-body techniques. The methods they adopted are a direct integration approach, that is, though accurate, strongly limited in the number of particles which can be integrated. For larger particle numbers they used a tree-code, and for the same system but with up to several million particles they adopted a particle-mesh technique.

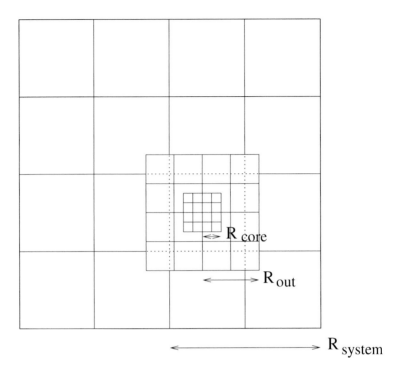

Figure 8.7. The different grids of SUPERBOX (Fellhauer et al. 2000) for 4 cells per dimension. The finest and intermediate grids are focused on the object of interest. Figure from Spinnato et al. (2003).

8.1.4.1   *Particle mesh*

To perform calculations for close to homogeneous particle distributions a particle-mesh code is quite suitable. Grainier systems such as dense star clusters are less suited, though some advances have been made by increasing resolution in substructure regions. However, caution has to be taken to make sure that the studied stellar system does not relax, as relaxation is generally not treated correctly in a particle-mesh technique. A nice example is SUPERBOX (Fellhauer et al. 2000) in which accuracy is sacrificed for speed.

In the particle-mesh technique densities are derived on Cartesian grids. Using a fast Fourier transform algorithm these densities are converted into a grid-based potential. Forces acting on the particles are calculated using these grid-based potentials, making the code nearly collisionless. To achieve high resolution at the places of interest several techniques to improve a better local accuracy are used; SUPERBOX for example, incorporates two levels of subgrids that stay focused on the objects of interest while they are moving through the simulated area (see Figure 8.7), providing higher resolution where required.

Particle-mesh codes, however, will always suffer from discrete effects due to the projection of the system on a grid. The size of a grid cell how-ever, appears to be related directly to the softening parameter ϵ used in tree codes and in some direct N-body codes. The concept of softening as introduced by (Aarseth 1963) is a technique to prevent large angle gravi-tational scattering by increasing the distance between two particles with a small parameter ϵ. As long as $r \gg \epsilon$ the effect of softening is not notice-able, but at close distances it has a profound effect on the behavior of the system.

In order to understand how the cell length of the particle-mesh code and the softening parameter ϵ of direct N-body and tree codes relate to each other, we compare in Figure 8.8 the results from the particle-mesh code SUPERBOX with the GADGET (Spingel et al. 2001) treecode simulations for 80,000 particles. In this example Spinnato et al. (2003) use a black hole of mass 0.00053 (the mass of the inner part of the galaxy is unity in these units) to sink to the center of the galaxy from a normalized distance. We can scale these numbers to astrophysically relevant units, in which case the black hole is $\sim 65,000\,\mathrm{M_\odot}$ and born at a distance of $\sim 8\,\mathrm{pc}$ from the galactic center. The initial orbit of the black hole was circular, but it still sinks slowly to the center of the galaxy, due to dynamical friction (see §8.1.3). Figure 8.8 shows the distance of the black hole to the galactic center as a function of time for the two computer codes. The results are presented for two values of the softening parameter ϵ in the tree code and compared with two values of the cell size in the particle-mesh code.

The infall of the black hole, as shown in Figure 8.8, depends on the values of l and ϵ in a remarkably similar way; l and ϵ seem to play the same qualitative role, but also quantitatively the results are quite similar.

8.1.4.2 Tree Codes

The concept of a hierarchical tree code was introduced by Apple (1985) and Barnes & Hut (1986), and is now widely used for the simulation of (near) collisionless systems. Figure 8.9 illustrates the concept of hierar-chically dividing the spatial coordinates in the tree code. The force on a given particle in a tree code is computed by considering particle groups of ever larger size as their distance from the particle of interest increases. Force contributions from such groups are evaluated by truncated multipole expansions. The grouping is based on a hierarchical tree data structure, which is realized by inserting the particles one by one into initially empty simulation cubes. Each time two particles are in the same cube, it is split

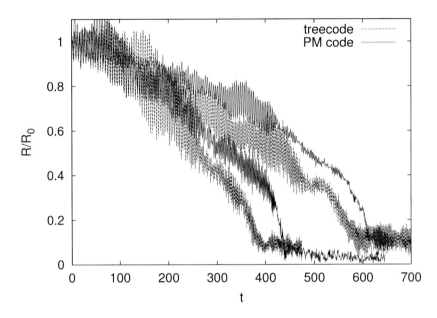

Figure 8.8. Spiral-in of a black hole (of mass $M_{BH} \simeq 0.0005$) to the center of the galaxy (with mass $M_{\mathrm{Gal}} \equiv 1$). The simulations were performed with 80,000 stars, using the same initial realization for both the tree code and the particle-mesh code. The particle-mesh simulations are run with cell size $\simeq 0.037$ (left) and 0.086 (right); softening parameters in the tree code runs are resp. 0.030 (left) and 0.060 (right).

into eight "child" cubes, whose linear size is one half of its parent's. This procedure is repeated until each particle is in a different cube. Hierarchically connecting such cubic cells according to their parental relation leads to the hierarchical tree data structure. The force on the particle of interest is then computed on neighboring cubes, which increase in size as they are farther away. One of the interesting characteristics of tree codes is the relatively simple parallelization by domain decomposition of the spatial coordinates (Olson & Dorband 1994); a disadvantage is the lack of support for special-purpose hardware (see however Kawai et al. 2004).

In the same way we compared a particle-mesh code with a tree code in § 8.1.4.1, we can also compare the tree code with a direct N-body code (see next section). In Figure 8.10 such comparison is illustrated on the same simulation of the black hole that spirals to the galactic center. Both codes are run with $N = 80,000$ and for a softening of $\epsilon \simeq 0.0037$, 0.030 and $\epsilon \simeq 0.060$. In this comparison we adopted `kira` from the `Starlab` package as the direct N-body code. The latter code was also run with zero

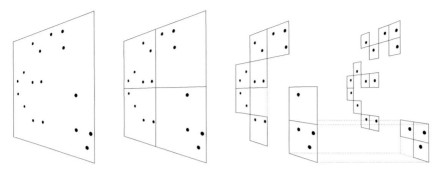

Figure 8.9. Schematic illustration of tree-building in a Barnes & Hut (1986) algorithm in two dimensions (from Springel, Yoshida and White, 2001). The particles are first enclosed in a single square, which is iteratively subdivided in squares of only half the size until a single particle remains per square.

softening ($\epsilon = 0$) but for the tree code this did not produce reliable results.

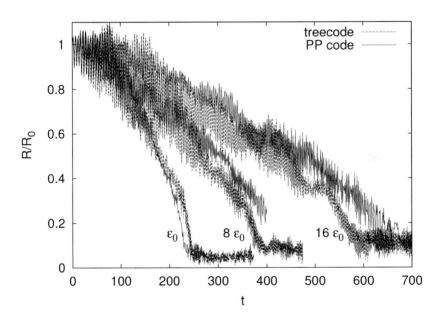

Figure 8.10. Comparison of results from the PP code with results from the tree code, at different values of ϵ. For all cases shown here $N = 80,000$ and $M_{BH} = 0.000528$. The PP simulation with $\epsilon = 8\epsilon_0$ has been already shown in Figure 8.8. The runs were performed with $\epsilon \simeq 0.0037$ (left), 0.030 (middle) and $\epsilon \simeq 0.060$ (right).

Interestingly, from the comparison among the particle-mesh code, the tree code and the direct N-body code in Figures 8.8 and 8.10 it is evident that in this practical case all three codes produce qualitatively and quantitatively the same results. One can wonder why then it is so important to perform an accurate calculation, as low resolution simulations produce results that are consistent with high-precision simulations?[5]

8.1.4.3   Direct N-Body

The most accurate way to simulate the dynamical evolution of a star cluster is by solving Equation (8.19) using direct integration. Direct N-body codes seem simple to write, as it is just a matter of solving Equation (8.19) in small steps, but in fact a computer program that does the job accurately and quickly is very hard to write correctly. This technique was pioneered by von Hoerner (1960, 1963), Aarseth (1963), Aarseth & Hoyle (1964) and van Albada 1968). Excellent reviews were published in the earlier mentioned books by Heggie & Hut (2003), Aarseth (2003) and Hut & Makino (2003), but see also Aarseth & Lecar (1975). A further reference to Heggie & Mathieu (1986) cannot be omitted as this valuable paper discusses the dimensionless units used in N-body techniques, in which $M = R = G = 1$.

In a direct N-body code the forces between all stars are calculated with numerical precision. The time complexity in this calculation is, as discussed earlier $O(N^3)$. In such a code, the particle motion is followed using a high-order integrator often with a predictor–corrector scheme (Makino and Aarseth 1992). These codes can generally not work with shared time steps;[6] it saves time and gains accuracy to allow stars in a strong encounter to be updated by individual time steps.[7] For simpler parallelization one generally adopts block time steps so that groups of strongly interacting particles are integrated more frequently than weakly interacting stars (McMillan 1986a; 1986b; Makino 1991). Still special treatment for binaries and higher-order hierarchies are required to prevent the code from coming to a grinding halt during strong encounters.

During a time step, particle positions and velocities are first predicted to fourth order using the acceleration and "jerk" (time derivative of the acceleration), which are known from the previous step. The new acceleration and jerk are then computed, and the motion is corrected using the additional derivative information.

One of the great advantages of using a direct N-body solver is the simplicity with which extra effects can be incorporated. Since each star in the

---

[5] It is probably worth revisiting this problem by performing a detailed side-by-side analysis of two or more N-body simulation methods (see Heggie et al. 1998 and his collaborative experiment at http://www.maths.ed.ac.uk/~douglas/experiment.html).

[6] All particles share the same time step.

[7] Each particle is integrated with its own special time step; particles in a strong interaction can then be integrated more accurately than weakly interacting particles.

cluster is represented by a particle in the code, individual characteristics, such as stellar properties, can be accounted for relatively easily and without loss of generality. This makes the direct N-body method preferable for simulating star clusters where these effects are important. In our case we are interested in the evolution of black holes, which are relatively rare objects. It would therefore be best to utilize a technique in which we can also treat the black holes individually. The drawback here is that black holes are so rare that large clusters have to be simulated in order to obtain enough statistics on the black hole population.

On the other hand the gravitational N-body problem has many applications over a wide range of research fields, including informatics, computational science, geology and astronomy.

8.1.4.4 GRAPE Family of Computers

The enormous computational requirement for solving the N-body problem with the direct method has been effectively addressed by a small team of researchers, who developed the GRAPE family of special-purpose computers. GRAPE (short for GRAvity PipE) hardware was designed and built by a group of astrophysicists at the University of Tokyo (to name only the most relevant publications: Sugimoto et al. 1990; Fukushige et al. 1991; Ito et al. 1991; Okumura et al. 1993; Taiji et al. 1996; Makino et al. 1997; Kawai et al. 2000; Makino et al. 2003). It may be clear that GRAPE is a very successful endeavor. The history and computational science of the GRAPE project is published by Makino & Taiji (1998).

The GRAPE family of computers is like a graphics accelerator speeding up graphics calculations on a workstation; without changing the software running on that workstation, the GRAPE acts as a Newtonian force accelerator, in the form of an attached piece of hardware. In a large-scale gravitational N-body calculation, where N is the number of particles, almost all instructions of the corresponding computer program are thus performed on a standard workstation, while only the gravitational force calculations, the innermost loop, are replaced by a function call to the special-purpose hardware.

Figure 8.11 shows the fully configured GRAPE-6 at Tokyo University.

8.1.5 Performing a Simulation

Before starting to simulate one may want to consider what technique is most suitable. In our further discussion that will be the direct N-body integrator. Several such computer programs are readily available. The `starlab` environment provides a entire library of functions and routines built around the main N-body integrator. The package can be downloaded from http://www/manybody.org/starlab.html. But also NBODY1–6 are

Figure 8.11. The large 64 Tflops GRAPE-6 configuration in Tokyo, in the summer of 2003.

available by ftp via ftp://ftp.ast.cam.ac.uk/sverre/. Both codes are large and very complicated as they have been evolving to their current sophistication for over more than a decade. But simpler alternatives are available from a variety of sources (from example via: http://www.ids.ias.edu/~piet/ act/comp/algorithms/starter/. A fully operational parallel N-body code based on the above mentioned starter code can be obtained from http:// carol.science.uva.nl/~spz/act/modesta/Software/index.html.

Whatever computer code you select or even if you write one from scratch, make sure that you test it. Test the code against other similar codes, test it with calculations by hand, regardless of how painstaking this often is, and test it against a simple problem for which the solution is known. You should develop a feeling of the regimes where the code can be trusted and in which cases extra care must be taken in interpreting the results.

Let us assume that we have found an interesting problem, which we think we can solve with the code available. The main problem of starting a simulation then is the selection of initial conditions. Starting with the wrong initial conditions is a complete waste of time. It is better to spend enough time thinking about the initial conditions, until you are convinced that they are the best choice. Possibly you want to perform several test calculations to converge to a better understanding of the question asked and the initial conditions required to give the most reliable answer.

For simulating a star cluster the primary initial conditions are:

- What are the basic cluster properties: mass, size, density profile?
- How many stars did the cluster have at birth?
- How are the stellar masses distributed?
- What is the fraction of primordial binaries?
- And what are the binary parameters: semi-major axis, eccentricity, inclination, etc.?
- Do you want to include triples and higher-order systems?
- And what about the shape and strength of the external tidal field of the galaxy?
- Are there tidal shocks, spiral arms or other external potentials to worry about?
- Is there anything else to add such as passing molecular clouds or black holes, etc.?

Many of the above effects have to some extent been incorporated in various calculations. And there is a rich scientific literature about the relevance and effect of many of these ingredients.

8.2 Theory of Star Cluster Evolution

Now that we have set the stage and discussed the tools and the techniques we can continue by discussing the global evolution of star clusters, which is characterized by three quite distinct phases; these are subsequently: A the early relaxation dominated phase, followed by phase B in which the $\sim 1\%$ (by number) most massive stars quickly evolve and lose an appreciable fraction of their mass. Finally, phase C starts when stellar evolution slows down even more and relaxation takes over until the cluster dissolves. To complete the list we can define phase D which is associated with the final dissolution of the cluster due to tidal stripping, but we will not discuss this phase in detail.

8.2.1 Phase A: $t \lesssim 10$ Myr

In an early stage, when stellar evolution is not yet important the star cluster is dominated by its own dynamical evolution; we call this phase A. This stage in the evolution of the cluster is only relevant if $t_{rt} \lesssim 100$ Myr. For most open star clusters and for all globular clusters this stage is probably[8] not very important, but for YoDeCs it is crucial, as we explain below.

[8] Regretfully we know very little about the early stages of globular clusters, and it is therefore hard to say whether phase A was important.

In the following discussion we assume, for clarity, that the location in the cluster where the stars are born is unrelated to the stellar mass; i.e., there is no primordial mass segregation. In that case, the early evolution of the star cluster is dominated by two-body relaxation, or to be more precise by dynamical friction.

In young dense clusters dynamical friction implies a characteristic time scale t_{df} for a massive star in a roughly circular orbit to sink from the half-mass radius R, to the cluster center (Spitzer 1971, 1987):

$$t_{df} \simeq \frac{\langle m \rangle}{100\,\mathrm{M_\odot}} \frac{0.138N}{\ln(0.11M/100\,\mathrm{M_\odot})} \left(\frac{R^3}{GM} \right)^{1/2}. \qquad (8.21)$$

Here $\langle m \rangle$ and M are the mean stellar mass and the total mass of the cluster, respectively, N is the number of stars, and G is the gravitational constant. For definiteness, we have evaluated t_{df} for a $100\,\mathrm{M_\odot}$ star. Less massive stars undergo weaker dynamical friction, and thus must start at smaller radii in order to reach the cluster center on a similar time scale.

Dynamical friction will have two very distinct effects on the cluster: (1) it tends to produce cores in hitherto coreless clusters, and (2) it initiates core collapse in other clusters. These two statements seem to contradict each other but as we will see below, this is not the case.

8.2.1.1   *Dynamical Friction Induced Core Development*

Consider a gravitationally bound stellar system in which most of the mass is in the form of stars of mass $m \simeq \langle m \rangle$, but which also contains a subpopulation of more massive objects with masses m_h. The orbits of the more massive objects decay due to dynamical friction. Assume that the stellar density profile is initially a power law in radius, $\rho(r) \propto (r/a)^{-\gamma} M/a^3$ with density scale length a (Dehnen 1993; Merritt et al. 1994). The orbits of the massive stars decay at a rate that can be computed by equating the torque from dynamical friction with the rate of change of the orbital angular momentum. We adopt the usual approximation (Spitzer 1987) in which the frictional force is produced by stars with velocities less than the orbital velocity of the massive object. The rate at which the orbit decays, assuming a fixed and isotropic stellar background, is

$$\frac{dr}{dt} = -2 \frac{(3-\gamma)}{4-\gamma} \sqrt{\frac{GM}{a}} \frac{m_h}{M} \ln \Lambda \left(\frac{r}{a} \right)^{\gamma/2-2} F(\gamma), \qquad (8.22)$$

with

$$F(\gamma) = \frac{2^\beta}{\sqrt{2\pi}} \frac{\Gamma(\beta)}{\Gamma(\beta - 3/2)} (2-\gamma)^{-\gamma/(2-\gamma)} \int_0^1 dy\, y^{1/2} \left(y + \frac{2}{2-\gamma} \right)^{-\beta}. \qquad (8.23)$$

Here $\beta = (6-\gamma)/2(2-\gamma)$ and $\ln \Lambda$ is the Coulomb logarithm, roughly equal to 6.6 (Spinnato et al. 2003). For $\gamma = 1.0$ (2.0), $F = 0.19$ (0.43).

If we approximate the cluster structure with an isothermal sphere, we find (Binney & Tremaine 1987, Equations 7-25) that a star of mass m_h at distance R from the cluster center drifts inward at a rate given by

$$R\frac{dR}{dt} \simeq -0.43\frac{Gm_h}{\langle v \rangle}\ln \Lambda. \qquad (8.24)$$

Here $\langle v \rangle$ is the cluster's velocity dispersion.

Equation (8.22) implies that the massive object comes to rest at the center of the stellar system in a time

$$t \approx 0.2\sqrt{\frac{a^3}{GM}}\frac{M}{m_h}\left(\frac{R_i}{a}\right)^{(6-\gamma)/2} \qquad (8.25)$$

with R_i the initial orbital radius.

Or, we can express the dynamical friction time in terms of the half-mass relaxation time by substituting Equation (8.6) in Equation (8.24) and integrate with respect to time.

$$t_{\rm df} \simeq 3.3\frac{\langle m \rangle}{m_h}t_{\rm rt}. \qquad (8.26)$$

To estimate the effect on the stellar density profile, consider the evolution of an ensemble of massive particles in a stellar system with initial density profile $\rho \sim R^{-2}$. The energy released as one particle spirals in from radius R_i to R_f is $2m_h\sigma^2\ln(R_i/R_f)$, with σ the 1D stellar velocity dispersion. Decay will halt when the massive particles form a self-gravitating system of radius $\sim GM_h/\sigma^2$ with $M_h = \sum m_h$. Equating the energy released during infall with the energy of the stellar matter initially within r_c, the "core radius," gives

$$R_c \approx \frac{2GM_h}{\sigma^2}\ln\left(\frac{R_i\sigma^2}{GM_h}\right). \qquad (8.27)$$

Most of the massive particles that deposit their energy within R_c will come from radii $R_i \approx$ a few $\times R_c$, implying $R_c \approx$ several $\times GM_h/\sigma^2$ and a displaced stellar mass of $\sim$ several $\times M_h$ (see also Watters et al. 2000). If $M_h \approx 10^{-2}M$ (Portegies Zwart & McMillan, 2000) then $R_c/a \approx$ several $\times 2M_h/M$ and the core radius is roughly 10% of the effective radius. Merritt et al. (2004) discuss this process in more detail and apply it successfully to the evolution of the core radii of large Magellanic cloud star clusters.

Evolution will continue as the massive particles form binaries and begin to engage in three-body interactions with other massive particles. These superelastic encounters will eventually lead to the ejection of most or all of the massive particles. Assume that this ejection occurs via many small "kicks," such that almost all of the binding energy so released can find its way into the stellar system as the particle sinks back into the core after each ejection. The energy released by a single binary in shrinking to a separation such that its orbital velocity equals the escape velocity from the core is $\sim m_h \sigma^2 \ln(4M_h/M)$ (see also § 8.3.4). If all of the massive particles find themselves in such binaries before their final ejection and if most of their energy is deposited near the center of the stellar system, the additional core mass will be

$$M_c \approx M_h \ln\left(\frac{M}{M_h}\right) \tag{8.28}$$

e.g., $\sim 5M_h$ for $M_h/M = 0.01$, similar to the mass displaced by the initial infall. The additional mass displacement takes place over a much longer timescale however and additional processes (e.g. core collapse) may compete with it.

8.2.1.2 Dynamical Friction Induced Core Collapse

The dynamical evolution of the star cluster drives it toward core collapse (Antonov 1962; Spitzer & Hart, 1971a; 1971b) in which the central density runs away to a formally infinite value in a finite time. In an isolated cluster in which all stars have the same mass, core collapse occurs in a time $t_{cc} \simeq 15\, t_{rt}$ (Cohn 1980; Makino 1996; Joshi et al. 2001).

If the dynamical friction timescale of a star cluster is shorter than the lifetime of the most massive stars, the cluster may experience an early phase of core collapse before the first supernova occurs. This can only happen if the initial half-mass relaxation time of the cluster is small $t_{rt} \lesssim 100\,\mathrm{Myr}$ otherwise the most massive stars burn up before they reach the cluster center (Portegies Zwart et al. 1999; Gürkan et al. 2004). The early core collapse in a cluster with small relaxation time is illustrated in Figure 8.13, where we plot the core radius as a function of time for a number of young star clusters with a relaxation time of about 100 Myr.

The details of what exactly happens in the cluster core, and whether the cluster will experience gravothermal oscillations probably depends quite critically on the initial density profile, as we will discuss in more detail in § 8.3.3.

When after core collapse the most massive stars explode the core radius remains highly variable, but small on average. What exactly happens at this stage is still ill understood. Naively one would expect the collapsed core to expand, as stellar mass loss drives an adiabatic expansion of the cluster. This can be been seen in the post-core collapse evolution of the bottom solid curve in Figure 8.13. Stage B starts when dynamical friction and relaxation cannot further drive the core collapse of the cluster, but

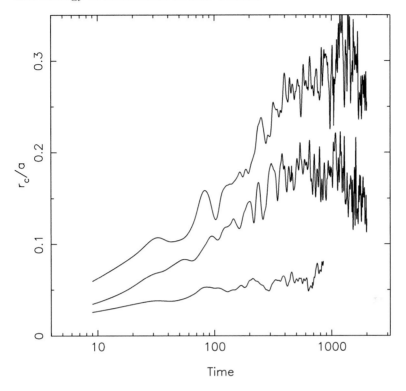

Figure 8.12. Evolution of the core radius, defined as the radius at which the projected density falls to one-half of its central value. Each curve is the average of the various runs at the specified value of γ, $\gamma = 1$ for the top curve, 1.5 for the middle curve and $\gamma = 2$ for the bottom curve (see § 8.2.1.1). The figure is taken from Merritt et al. (2004).

expansion by stellar mass loss starts to dominate. Typically this happens around $\sim 10\,\mathrm{Myr}$.

The cluster simulated for Figure 8.13 was MGG11, in the starburst galaxy M82. It contained 131,072 single stars from a Salpeter initial mass function between $1\,\mathrm{M}_\odot$ and $100\,\mathrm{M}_\odot$ distributed in a King (1966) model density profile with dimension less dept $W_0 = 3$ (shallow) to $W_0 = 12$ (very concentrated). The half mass radius of these simulated clusters was 1.2 pc.

We discuss each curve in Figure 8.13 in turn, starting at the top. The core radius of the shallow model, $c \simeq 0.67$ ($W_0 = 3$), hardly changes with time. The intermediate model $c \simeq 1.8$ ($W_0 = 8$) almost experiences core collapse near $t = 3\,\mathrm{Myr}$ but as stellar mass loss starts to drive the expansion of the core it never really experiences collapse. This is the moment where phase B sets in. Core collapse occurs in the $c \simeq 2.1$ ($W_0 = 9$) model near $t = 0.8\,\mathrm{Myr}$. The $c \simeq 2.7$ ($W_0 = 12$) simulation is so concentrated that it starts virtually in core collapse, and the entire cluster evolution is dominated by a post-collapse phase. At this point it is not a priori clear why the core radius for models $W_0 = 9$ and $W_0 = 12$ fluctuates so much more violently than in models with a smaller concentration.

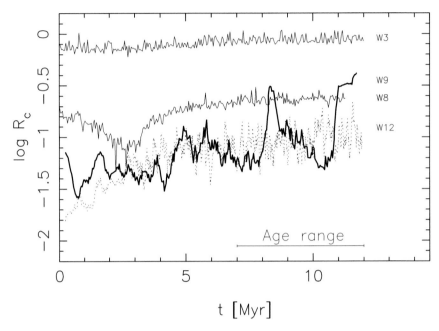

Figure 8.13. Evolution of the core radius for four simulations of the star cluster MGG-11. These calculations are performed with `Starlab` with $c \simeq 0.67$ ($W_0 = 3$), $c \simeq 1.8$ ($W_0 = 8$), $c \simeq 2.1$ ($W_0 = 9$, bold curve) and $c \simeq 2.7$ ($W_0 = 12$, dotted curve) indicated along the right edge of the figure as $W3$, $W8$, $W9$ and $W12$ respectively. The $W9$ curve is plotted with a heavy line to distinguish it from the curves for $W8$ and $W12$. The age range of the observed clusters MGG11 is indicated near the bottom of the figure. In Figure 8.21 we present the evolution of the average core density for the simulations with $W_0 = 3$, 8 and 9.

8.2.2   Phase B: 10 Myr $\gtrsim t \lesssim 100$ Myr

After the first few million years and until the most massive stars have turned into compact remnants, the cluster will be dominated by stellar mass loss. Since the most massive stars evolve first and sink most quickly to the cluster center, mass is lost from deep inside the potential well. A collapsed cluster may recover from its earlier core collapse due to stellar mass loss (see §8.2.1.2). The dotted curve in Figure 8.13 (for the simulations with $W_0 =$

12) illustrates the growth of the core as a result of heating by dynamical friction and stellar evolution mass loss is quite clearly visible; it is hard to separate the two effects as both are taken into account in the calculation self-consistently. For the slightly shallower initial density profile ($W_0 = 9$) the steady growth of the core radius after collapse is less clear, but the rate is similar as for $W_0 = 12$. This indicates that it is indeed mainly stellar mass loss which drives the expansion. At this stage we do not know why the core radius in the $W_0 = 9$ models fluctuates more wildly than in the $W_0 = 12$ simulations. Details probably depend quite sensitively on the presence of an intermediate mass black hole, which could form in the preceding phase A. Such a massive compact object can effectively heat the cluster core as it forms tight multiple systems with other (massive) stars (Baumgardt et al. 2003).

Few studies have been carried out for clusters with short relaxation time to understand this particular evolutionary stage. For a long relaxation time $t_{rt} \gtrsim 100\,\text{Myr}$, it is quite clear that the cluster expands substantially during the first Gyr of its lifetime (Chernoff & Weinberg 1990; Fukushige & Heggie 1995; Takahashi & Portegies Zwart 2000; Baumgardt & Makino 2003).

The evolution of the cluster mass is given in Figure 8.14 (taken from Takahashi & Portegies Zwart 2000) for a variety of particle numbers, ranging from 1024 to 32,768. The first few 100 Myr of all clusters is very similar, but at later time large differences appear. In the first epoch, during the first few hundred million years, about 20% of the cluster mass is lost. This mass loss is the result of stellar evolution and, to a lesser extent by tidal stripping. Tidal stripping and relaxation become important at later time.

8.2.3 Phase C: $t \gtrsim 100$ Myr

After the most massive stars have turned into remnants, stellar evolution slows down, and relaxation processes can take over again. In Figure 8.14 this phase starts at an age of about a Gyr. The reason that stellar mass loss slows down is twofold, (1) low-mass stars remain on the main sequence longer than high mass stars, and (2) once the star turns into a remnant, the mass lost in the process is relatively small; imagine a $12\,\text{M}_\odot$ star loses about $10.6\,\text{M}_\odot$ by stellar wind and in the supernova explosion, whereas a $2\,\text{M}_\odot$ turns into a $0.64\,\text{M}_\odot$ white dwarf, losing only $1.36\,\text{M}_\odot$ in the process.

The effects of mass lost from the evolving stars and mass lost from the dynamical evolution of the cluster are coupled. If stellar mass loss slows down, the cluster responds to this by a slower expansion, which again makes it less prone to tidal stripping. While the importance of stellar evolution diminishes, relaxation gradually takes over until it becomes the dominant mechanism that drives the evolution of the cluster.

The later stage of the low N (1 k and 2 k) clusters in Figure 8.14 are

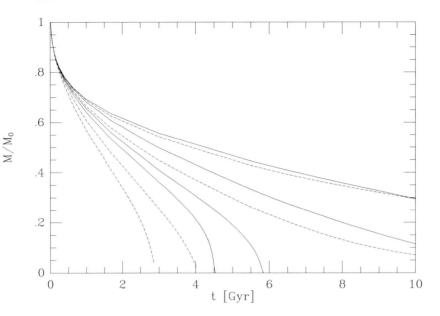

Figure 8.14. Mass as a function of time for a number of Fokker–Planck models. The four solid lines represent the results of model Aa with 32 k, 16 k, 4 k and 1 k particles from left to right, respectively. Dotted curves present model *Ie* for the same numbers of particles as for model *Aa* (model names refer to Takahashi & Portegies Zwart, 2000). The difference between the models *Ie* and *Aa* hides in the way escaping stars are identified.

much more strongly affected by relaxation than the high N (16 k and 32 k) clusters. This effect was named the *ski-jump* problem in Portegies Zwart et al. (1998). The transition between ski-clusters (low N in Figure 8.14) and *jump*-clusters (high N) is a result of the non-linear interaction between the external tidal field of the parent galaxy and relaxation.

At a later time during phase C dynamical friction once again becomes important. This time not driven by massive stars, as these have all gone supernova by now, but by the compact remnants formed in supernovae; black holes and neutron stars, but also heavy white dwarfs, blue stragglers and giants. All these stellar species are generally more massive than the mean mass, and therefore subject to dynamical friction. A similar process as in phase A starts again and the cluster may experience core collapse for the second time. It may therefore be possible that a cluster experiences two very distinct phases of core collapse, one during phase A, and again at a later time, during phase C (see also Deiters & Spurzem 2000, 2001).

In a realistic cluster, however, there are a number of additional complications that are particularly important at this stage, in part because it may take rather long to reach a state of core collapse again because the stellar mass function is rather flat now, with a relatively small difference between the least and the most massive stars. The consequence is that external influences, such as disc shocks, passing molecular clouds and the presence of an external tidal field, may become particularly important at this stage, simply because they have a lot of time to accumulate their effect.

The slowdown of stellar evolution has a second important consequence, which is the termination of active binary evolution. Only relatively low-mass stars are able to evolve from the main sequence, and no supernova will occur after $\sim 10^8$ years. It becomes therefore almost impossible to ionize hard binaries, which may effectively arrest the collapse of the cluster core (Fregeau et al. 2003). The binaries may therefore once more[9] become dynamically important, and heat the cluster by interacting with single stars or other binaries (Heggie 1975).

We illustrate phase C in Figure 8.15 with a simulation of 10,000 identical point masses initially distributed in a Plummer sphere. Figure 8.15 shows the evolution of the core radius of this model. Core collapse occurs at $t_{cc} \simeq 15.2 \pm 0.1\, t_{rt}$ (for these initial conditions $t_{rt} \simeq 150$ N-body time units). This result is consistent with earlier calculations of e.g., Cohn (1980) and Makino (1996). Doubling the mass of 20% of the stars reduces the core collapse time to $t_{cc} \simeq 7.2\, t_{rt}$. Making 20% of the stars 10 or 100 times more massive reduced the time of core collapse further, to $t_{cc} \simeq 1.4\, t_{rt}$ and $t_{cc} \simeq 0.16\, t_{rt}$, respectively. Clearly, the more massive stars drive the core collapse of the cluster by dynamical friction, as $t_{df} \simeq \frac{\langle m \rangle}{m} t_{rt}$ (see Equation (8.21), and also Watters et al. 2000).

The presence of a population of relatively massive stars, such as black holes will therefore shorten the timescale for core collapse in phase C of the cluster evolution, but even in the absence of black holes or other heavy remnants core collapse cannot be prevented. In phase A this role was played by the most massive main-sequence stars.

Note that if it is the population of dark remnants, black holes, neutron star and white dwarfs, which experience the core collapse, the lighter stars will not necessarily follow. So, the cluster may physically be in a state of core collapse, where the optical observer would measure a King profile (see Baumgardt et al. 2003a, 2003b)

8.2.4 The Consistent Picture

In this section we have seen that a star cluster experiences three very distinct evolutionary phases, each of which is dominated either by relaxation or by stellar mass loss.

---

[9] The first time binary interactions were relevant was during phase A.

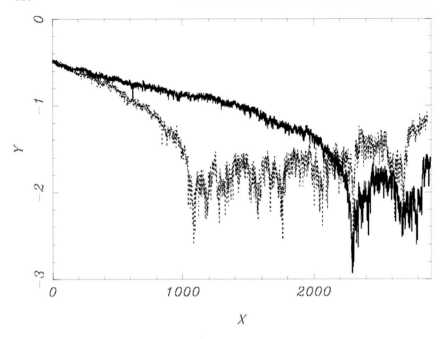

Figure 8.15. Evolution of the core radius for two star clusters well beyond the moment of core collapse. Time (in N-body units) is along the X-axis and radius (log r, also in N-body) units along the Y-axis. Initially each cluster contained 10,000 single stars distributed in a Plummer sphere with the same realization of positions and velocities. The solid curve was computed with equal masses for all stars, the dashed curve is computed with 20% of the stars being twice as massive. These calculations were performed without stellar evolution and the units along the axis are in dimensionless N-body units.

Figure. 8.16 summarized these phases in a single simulation that contains all relevant physics. The simulation presented here is carried out to illustrate the formation mechanism for intermediate mass black holes in dense young star clusters. It was performed with 131,072 stars from a Salpeter initial mass function; 10% (13,107) of the stars form a hard binary system with a close companion. The initial conditions for this simulation are explained in more detail in § 8.3.3. Here we only use the result of the simulation to illustrate the three distinct evolutionary phases in star cluster evolution; as with the adopted parameters each phase is clearly present.

The three phases, A, B and C, are identified with the horizontal bars in Figure 8.16. It is not a priori clear when one phase stops and the next starts, and some gray area has to be allowed in which both, stellar evolution and relaxation, may temporarily have a similar effect on the cluster. These runs were continued till 100 Myr.[10] Since this simulation was performed in isolation, the dissolution of the cluster will take quite a while (Baumgardt, Hut, & Heggie 2002)

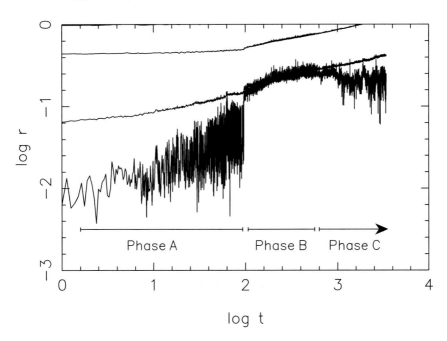

Figure 8.16. Evolution of the core radius (lower line) 10% and the 25% Lagrangian radii for a 128 k star cluster (axis are in N-body units). The various stages of cluster evolution are indicated near the bottom of the figure. In the early evolution of the cluster expansion of the core is driven by mass segregation, followed by phase B in which the cluster expands by stellar evolution. Near the end of the simulation the cluster tends to experience core collapse (Phase C).

8.3 Black Holes in Star Clusters

In the previous sections we have set the stage for the evolution of star clusters, and we have introduced the fundamental physics. We will now continue with the evolution of black holes and their progenitors in star clusters.

---

[10] There was no particular reason to terminate this calculation at about 100 Myr, other than that I got a bit tired of babysitting the run after three months straight on the GRAPE-6 at the University of Tokyo. In time, this cluster will experience core collapse again, to dissolve eventually.

8.3.1 The Formation of Intermediate Mass Black Holes in Phase A Clusters

Young star clusters, with a half-mass relaxation time $t_{rt} \lesssim 100\,\mathrm{Myr}$ are, as we discussed in § 8.2.1, prone to dynamical friction, and therefore are likely to experience core collapse before stellar mass loss drives the expansion of the cluster. Realistic clusters have a broad range in initial stellar masses, generally from $m_{min} \lesssim 0.1\,\mathrm{M_\odot}$ to $m_{max} \gtrsim 100\,\mathrm{M_\odot}$. Adopting such a mass function as a condition at birth, the mean mass $\langle m \rangle$ ranges then from $\langle m \rangle \sim 0.39\,\mathrm{M_\odot}$ (Salpeter 1955) to about $0.65\,\mathrm{M_\odot}$ (Scalo 1986), depending on the specific mass function adopted. Here I like to stress that there is a large variety of initial mass function available; apart from the two above mentioned there are the Miller & Scalo (1979), Kroupa, Tout, & Gilmore (1990) with some adjustments for high mass stars and Kroupa & Weidner (2003, see also Kroupa, 2001). These mass functions seem to differ quite substantially, but for the dynamical evolution of the cluster they do not make a big difference, as for this process it is the ratio of most massive star to the mean mass that counts (see Equation (8.21)).

In a multimass system, core collapse is driven by the accumulation of the most massive stars in the cluster center. This process takes place on a dynamical friction timescale (Equation (8.21)). Empirically, we find, for initial mass functions of interest here, that core collapse (actually, the appearance of the first persistent dynamically formed binary systems) occurs at about (Portegies Zwart & McMillan 2000; Fregeau et al 2002)

$$t_{cc} \simeq 0.2\,t_{rt}\,. \tag{8.29}$$

This core collapse time is taken in the limit where stellar evolution is unimportant, i.e. where stellar mass loss is negligible and the most massive stars survive until they reach the cluster center, i.e., what we called phase A in § 8.2.1.

Experimentally we find that starting with a Scalo (1986) initial mass function and a King (1966) $W_0 = 6$ density distribution the time to reach the first core collapse is $t_{cc} \simeq 0.19 \pm 0.08\,t_{rt}$. Some of our simulations were performed with high $W_0 \gtrsim 9$ concentration. Core collapse in these models occurred within about a million years, which for these simulations corresponded to $t_{cc} \simeq 0.01\,t_{rt}$, or about 7.5% of the core relaxation time. These findings are consistent with the Monte Carlo simulations performed by Joshi, Nave, & Rasio (2001).

The collapse of the cluster core may initiate physical collisions between stars. The product of the first collision is likely to be among the most massive stars in the system, and to be in the core. This star is therefore likely to experience subsequent collisions, resulting in a collision runaway (see Portegies Zwart et al. 1999). The maximum mass that can be grown in a dense star cluster if all collisions involve the same star is $M_{\rm run}$, where

$$\frac{dM_{\rm run}}{dt} = \mathcal{N}_{\rm coll}\delta m_{\rm coll}. \tag{8.30}$$

Here $\mathcal{N}_{\rm coll}$ and $\delta m_{\rm coll}$ are the average collision rate and the average mass increase per collision (assumed independent). We now discuss these quantities in more detail. Interestingly enough, Gürkan et al. (2004) performed comparable calculations with a Monte Carlo N-body code, which produces qualitatively the same results. Also the calculations of Portegies Zwart et al. (2004) who used two independently developed N-body codes (NBODY4 and Starlab), obtain similar results.

8.3.1.1 *The Collision Rate $\mathcal{N}_{\rm coll}$*

A key result from the simulations of Portegies Zwart et al. (1999) was the fact that collisions between stars generally occur in dynamically formed (three-body) binaries. The collision rate is therefore closely related to the binary formation rate, which we can estimate from first principles.

The flux of energy through the half-mass radius of a cluster during one half-mass relaxation time is on the order of 10% of the cluster potential energy, largely independent of the total number of stars or the details of the cluster's internal structure (Goodman 1987). For a system without primordial binaries this flux is produced by heating due to dynamically formed binaries (Makino & Hut 1990). It is released partly in the form of scattering products that remain bound to the system, and partly in the form of potential energy removed from the system by escapers recoiling out of the cluster (Hut & Inagaki 1985). Makino & Hut argue, for an equal-mass system, that a binary generates an amount of energy on the order of $10^2 kT$ via binary–single-star scattering (where the total kinetic energy of the stellar system is $\frac{3}{2}NkT$). This quantity originates from the minimum binding energy of a binary that can eject itself following a strong encounter. Assuming that the large-scale energy flux in the cluster is ultimately powered by binary heating in the core. It follows that the required formation rate of binaries via three-body encounters is

$$\Gamma_{\rm bf} \simeq 10^{-3}\frac{N}{t_{\rm rt}}. \tag{8.31}$$

For systems containing significant numbers of primordial binaries, which segregate to the cluster core, equivalent energetic arguments (Goodman & Hut 1989) lead to a similar scaling for the net rate at which binary encounters occur in the core.

The above arguments apply to star clusters comprising identical point-masses. In a cluster with a range of stellar masses, three-body binaries generally form from stars that are more massive than average. After repeated exchange interactions, the binary will consist of two of the most massive stars in the cluster. Conservation of linear momentum during encounters with lower-mass stars means that the binary receives a smaller recoil velocity, making it less likely to be ejected from the cluster. The binary must therefore be considerably harder— $\gtrsim 10^3 kT$ —before it is ejected following a encounter with another star (see Portegies Zwart & McMillan 2000).

However, taking the finite sizes of real stars into account, it is quite likely that such a hard binary experiences a collision rather than being ejected. A strong encounter between a single star and a hard binary generally results in a resonant interaction. Three stars remain in resonance until at least one of them escapes, or a collision reduces the three-body system to a stable binary. For harder binaries it becomes increasingly likely that a collision occurs instead of ejection (McMillan 1986; Gualandris et al. 2004; Fregeau et al. 2004). In the calculations of Portegies Zwart et al. (1999) most binaries experience a collision at a binding energy of order $10^2 kT$, considerably smaller than the binding energy required for ejection. Accordingly, we retain the above estimate of the binary formation rate (Equation 8.31) and conclude that the collision rate per half-mass relaxation time is

$$\mathcal{N}_{\mathrm{coll}} \sim 10^{-3} f_{\mathrm{c}} \frac{N}{t_{\mathrm{rt}}}. \tag{8.32}$$

Here we introduce $f_{\mathrm{c}} \leq 1$, the effective fraction of dynamically formed binaries that produce a collision. Note again that Equation (8.32) is valid only in the limit where stellar evolution is unimportant.

The most massive star in the cluster is typically a member of the interacting binary and therefore dominates the collision rate. Subsequent collisions cause the runaway to grow in mass, making it progressively less likely to escape from the cluster. The star that experiences the first collision is therefore likely to participate in subsequent collisions. The majority of collisions thus involve one particular object—the runaway merger—generally selected by its high initial mass and proximity to the cluster center.

For systems containing many primordial binaries the above argument must be modified. Since dynamically formed binaries tend to be fairly soft—a few kT— the fraction of interactions with primordial binaries leading to collision is comparable to the value f_{c} above. However, a critical difference is that, in systems containing many binaries, the collisions involve many different pairs of stars, not just the binary containing the massive runaway. The total collision rate is therefore much higher, but most collisions do not contribute to the growth of the runaway merger. The presence of primordial binaries has little influence on the collision runaway.[11]

---

[11] The author realizes that the simple mention of primordial binaries opens-up an entire discussion that would require several pages, which I try to prevent.

8.3.1.2 Average Mass Increase per Collision

Once begun, the collision runaway dominates the collision cross-section. The average mass increase per collision depends on the characteristics of the mass function in the cluster core. A lower limit for stars that participate in collisions can be derived from the degree of segregation in the cluster. Inverting Equation (8.21) results in an estimate (still assuming an isothermal sphere) of the minimum mass of a star that can reach the cluster core in time t due to dynamical friction:

$$m_{\rm df} = 1.9 \, {\rm M}_\odot \left(\frac{1 \, {\rm Myr}}{t} \right) \left(\frac{R}{1 \, {\rm pc}} \right)^{3/2} \left(\frac{m}{1 \, {\rm M}_\odot} \right)^{1/2} (\ln \Lambda)^{-1} \, . \tag{8.33}$$

Thus, at time t and for a given mass m, there is a maximum radius R inside of which stars of mass m will have segregated to the core. The stars contributing to the growth of the runaway are likely to be among those more massive than $m_{\rm df}$; because their number density in the core is enhanced by mass segregation, their collision cross-sections are larger, and they contribute more to $\delta m_{\rm coll}$ when they do collide.

The shape of the central mass function of a segregated cluster is not trivial to derive.[12] In thermal equilibrium, the central number densities of stars of different masses would be expected to scale as

$$n_0(m) \sim m^{3/2} \frac{dN}{dm} \, , \tag{8.34}$$

where dN/dm is the global initial mass function, which scales roughly as $m^{-2.7}$ at the high-mass end ($m \gtrsim 10 {\rm M}_\odot$). The distribution of secondary masses (i.e. the masses of the lighter stars participating in collisions) does not follow the above simple relation. Rather, we find that stars in the core do not reach thermal equilibrium (a result generally consistent with earlier findings by Chernoff and Weinberg 1990 and Joshi, Nave & Rasio 2001), and that the dynamical nature of the collisional processes involved means that more massive stars tend to be consumed before lower-mass stars arrive in the core. In addition, most collisions involve three-body binaries and interactions with higher-order multiples in a multimass environment.

Empirically, we find that the secondary mass distribution is quite well fit by a power law, $dN/dm \propto m^{-2.3}$ (coincidentally very close to a Salpeter distribution). Integrating this expression from a minimum mass of $m_{\rm df}$ (and ignoring the upper limit) results in a mean mass increase per collision of

$$\delta m_{\rm coll} \simeq 4 m_{\rm df} \, . \tag{8.35}$$

[12] In that case, the mass function in the core takes on a rather curious form: the mass functions for stars with masses $m \lesssim m_{\rm df}$ and $m \gtrsim m_{\rm df}$ have roughly the same slopes as the initial mass function, but the more massive stars are overabundant because they have accumulated in the cluster center.

If we neglect stars with masses less than m_{df} and substitute Equation (8.5) into Equation (8.33) and Equation (8.35) then the mass increase per collision can be written as

$$\delta m_{coll} \simeq 4\frac{t_{rt}}{t}\langle m\rangle \ln \Lambda .\qquad (8.36)$$

This quantity remains rather constant over the entire collisional lifetime, e.g., about $3\,\text{Myr}$ (see also §8.3.3.2).

8.3.1.3 Lifetime of a Cluster in a Static Tidal Field

With simple expressions for $\mathcal{N}_{coll}$ and δm_{coll} now in hand, we return to the determination of the runaway growth rate (Equation (8.30)). The evaporation of a star cluster that fills its Jacobi surface in an external potential is driven by tidal stripping. Portegies Zwart et al. (2001a) have studied the evolution of young compact star clusters within $\sim 200\,\text{pc}$ of the galactic center. Their calculations employed direct N-body integration, including the effects of both stellar and binary evolution and the (static) external influence of the galaxy, and made extensive use of the GRAPE-4. They found that the mass of a typical model cluster decreased almost linearly with time:

$$M = M_0 \left(1 - \frac{t}{t_{disr}}\right).\qquad (8.37)$$

Here M_0 is the mass of the cluster at birth and t_{disr} is the cluster's disruption time. Portegies Zwart et al. (2001a) found that their model clusters dissolved within about 30% of the two-body relaxation time at the tidal radius (defined by substituting the tidal radius instead of the virial radius in Equation (8.5)). In terms of the half-mass relaxation time, this translates to $t_{disr} = 1.6$–$5.4\, t_{rt}$, depending on the initial density profile (the range corresponds to King (1966) dimensionless depths $W_0 = 3$–7; more centrally condensed clusters live longer).

Substituting Equations (8.32) and (8.36) into Equation (8.30), and defining $M_0 = N\langle m\rangle$ to rewrite Equation 8.37 in terms of the number of stars in the cluster, we find

$$\frac{dM_{run}}{dt} = 4 \times 10^{-3} f_c \frac{N\langle m\rangle \ln \Lambda}{t}$$
$$= 4 \times 10^{-3} f_c M_0 \ln \Lambda \left(\frac{1}{t} - \frac{1}{t_{disr}}\right).\qquad (8.38)$$

Integrating from $t = t_{cc}$ to $t = t_{disr}$ results in

$$M_{run} = m_{seed} + 4 \times 10^{-3} f_c M_0 \ln \Lambda \left[\ln\left(\frac{t_{disr}}{t_{cc}}\right) + \frac{t_{cc}}{t_{disr}} - 1\right].\qquad (8.39)$$

Here $m_{\rm seed}$ is the seed mass of the star that initiates the runaway growth, most likely one of the most massive stars initially in the cluster. With $t_{\rm cc} \simeq 0.2 t_{\rm rt}$, Eq (8.39) reduces to

$$M_{\rm run} = m_{\rm seed} + 4 \times 10^{-3} f_{\rm c} M_0 \kappa \ln \Lambda \,, \qquad (8.40)$$

where $\kappa \simeq \ln(t_{\rm disr}/t_{\rm cc}) + t_{\rm cc}/t_{\rm disr} - 1 \sim 1$.

In Figure 8.17 we present this relation in the form of a solid curve. We comment further on the left side of this figure where we extrapolate the relation in Equation (8.40) to galactic nuclei masses.

The maximum mass of the runaway merger for clusters that are disrupted by inspiral (which of course always destroys the cluster before it reaches the center) may be calculated by replacing $t_{\rm disr}$ in Equation (8.39) by $T_{\rm f}$. The right-hand side of that equation then becomes a function of

$$\frac{T_{\rm f}}{t_{\rm cc}} \simeq 9.0 \left(\frac{R_i}{10 {\rm pc}} \right)^{2.1} \left(\frac{0.25 {\rm pc}}{R} \right)^{3/2} \left(\frac{10^5 {\rm M}_\odot}{M} \right)^{3/2} \qquad (8.41)$$

We can also estimate the maximum initial distance from the Galactic center for which core collapse occurs (and hence runaway merging may begin) before the cluster disrupts by setting $T_{\rm f} = t_{\rm cc}$. The result is $R_i \gtrsim 0.0025 \, {\rm pc} \, (RM/[{\rm pc}\,{\rm M}_\odot])^{0.71}$. For $R = 0.25 \, {\rm pc}$ and $M = 10^5 \, {\rm M}_\odot$, we find $R_i \gtrsim 3.3 \, {\rm pc}$.

8.3.2 Calibration with N-Body Simulations

The development of the GRAPE (see §8.1.4.4) family of special-purpose computers makes it relatively straightforward to test and tune the above theory using direct N-body calculations. The N-body calculations performed by Portegies Zwart & McMillan (2002) span a broad range of initial conditions in the relevant part of parameter space. The number of stars varied from 1 k (1024) to 64 k (65,536). The initial conditions explored by Portegies Zwart et al. (2004) range from 131,072 to 585,000 stars. Gürkan et al. (2004) performed similar calculations using a Monte Carlo N-body code; they adopted a considerably larger number of particles.

Initial density profiles and velocity dispersion for the models were taken from Heggie–Ramamani models (Heggie & Ramamani 1995) with W_0 ranging from 1 to 7, and from King (1966) models with $W_0 = 3$ to 15. At birth, the Heggie-Ramamani clusters were assumed to fill their zero-velocity (Jacobi) surfaces in the galactic tidal field, while the classical King models were assumed to be isolated. In most cases we adopted an initial mass function between $0.1 \, {\rm M}_\odot$ and $100 \, {\rm M}_\odot$ suggested for the solar neighborhood by Scalo (1986) and Kroupa, Tout, & Gilmore (1990). However, several calculations were performed using power-law initial mass functions with exponents of –2 or –2.35 (Salpeter) and lower mass limits of $1 \, {\rm M}_\odot$.

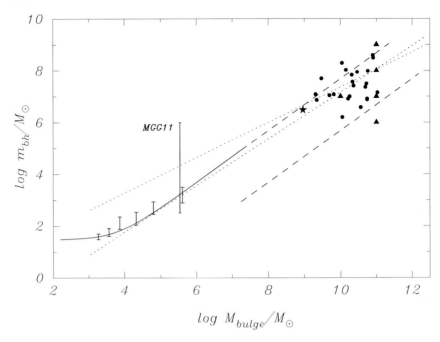

Figure 8.17. The mass after a period of runaway growth as a function of the mass of the star cluster. The solid line is $M_{\rm run} = 30 + 8 \times 10^{-4} M_0 \ln \Lambda$ (see Equation (8.40) with $f_c = 0.2$, $\gamma = 1$ and $\ln \Lambda = \ln M_0/M_\odot$, where M_0 is the initial mass of the cluster or $10^6\,M_\odot$, whatever is smaller). This relation may remain valid for larger systems built up from many clusters having masses $\lesssim 10^6\,M_\odot$. For clusters with $M_0 \gtrsim 10^7\,M_\odot$ we therefore extend the relation as a dashed line. The logarithmic factor, however, remains constant, as it refers to the clusters out of which the bulge formed, not the bulge itself (Ebisuzaki et al. 2001). The bottom dashed line shows $0.01 M_{\rm run}$. The five error bars to the left give a summary of the results presented by Portegies Zwart & McMillan (2002), the two rightmost error bars are taken from Portegies Zwart et al. (2004). The error bar indicating *MGG11* gives the result of estimates for the mass of the intermediate mass black hole in the M82 star cluster MGG11 (Matsumoto & Tsuru 1999; McCrady et al. 2003). The Milky Way is represented by the asterisk using the bulge mass from Dwek (et al. 1995) and the black hole mass from Eckart & Genzel (1997) and Ghez (2000). Bullets and triangles (upper right) represent the bulge masses and measured black hole mass of Seyfert galaxies and quasars, respectively (both from Wandel 1999, 2001). The dotted lines gives the range in solutions to a least squares fit to the bullets and triangles (Wandel 2001).

8.3.2.1 *Collision Rate During Phase A*

In all calculations, the first collision occurred shortly after the formation of the first $\gtrsim 10kT$ binary by a three-body encounter, i.e., close to the time of core collapse. When stars were given unrealistically large radii (100 times larger than normal), the first collisions occurred only slightly (about 5%) earlier.

As discussed earlier, the first star to experience a collision was generally one of the most massive stars in the cluster; this star then became the target for further collisions. In models where the core collapse time exceeds about 3 Myr the target star explodes in a supernova before experiencing runaway growth. The collision rates in these clusters were considerably smaller than for clusters with smaller relaxation times (see Figure 8.18). As discussed in more detail in §8.4; the onset of stellar evolution terminates the collision process; premature disruption of the cluster also ends the period of runaway growth.

The first physical collisions occur at the moment of core collapse. The cluster then enters a phase which is dominated by stellar collisions. In particular one single object experiences many repeated collisions, giving rise to a collision runaway. We identify this particular object as the *designated target*. In our models collisions tend to increase the mass of the collision product; only a small fraction of the mass of the incoming star is lost; the designated target therefore tends to increase in mass.

The number of collisions in the direct N-body simulations ranged from 0 to 100. Figure 8.18 shows the mean collision rate $\mathcal{N}_{\mathrm{coll}}$ per star per million years as a function of the initial half-mass relaxation time. The solid line in Figure 8.18 is a fit to the simulation data, and has

$$\mathcal{N}_{\mathrm{coll}} = 2.2 \times 10^{-4} \frac{N}{t_{\mathrm{rt}}} , \tag{8.42}$$

for $t_{\mathrm{rt}} \lesssim 20 - 30$ Myr, consistent with our earlier estimate (Equation (8.32)) if $f_{\mathrm{c}} = 0.2$. The quality of the fit in Figure 8.18 is quite striking, especially when one bears in mind the rather large spread in initial conditions for the various models. See, however, the prominent square to the right, which is about a factor ~ 3 above the fitted curve. This discrepancy is mainly caused by the high average stellar mass in these models and by the use of much more concentrated King models.

The collision runaway phase lasts until about 3.3 Myr, at which time the designated target tends to collapse to a black hole.

In Figure 8.19 we present the cumulative distribution of the number of mergers of some of the calculations by Portegies Zwart et al. (2004).

Figure 8.20 shows the cumulative mass distributions of the primary (more massive) and secondary (less massive) stars participating in collisions. We include only events in which the secondary experienced its first collision (that is, we omit secondaries which were themselves collision products). In addition, we distinguish between collisions early in the evolution of the cluster and those that happened later by subdividing our data based

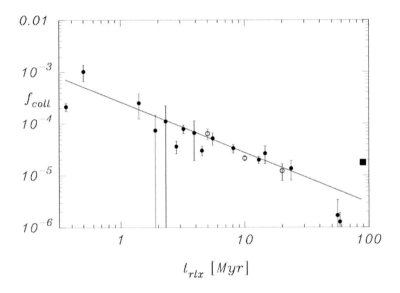

Figure 8.18. Mean collision rate $f_{coll} = N_{coll}/N t_{last}$ as a function of initial relaxation time for the simulations performed by Portegies Zwart et al. (1999, 2004) and Portegies Zwart & McMillan (2002). Here t_{last} is the time of the last collision in the cluster. The open circles give the results of systems which are isolated from the galactic potential (see Portegies Zwart et al. 1999). Vertical bars represent Poissonian 1-σ errors. The solid line is a least squares fit to the data (see Equation (8.42)). The strong reduction in the collision rate for cluster with an initial relaxation time $t_{rt} \gtrsim 30\,\text{Myr}$ is probably real. Note however the filled square to the right from the calculations of Portegies Zwart et al. (2004), which gives a higher collision rate due to the high initial concentration of the models and the top-heavy mass function.

on the ratio $\tau = t_{coll}/t_{df}$, where t_{coll} is the time at which a collision occurred and t_{df} is the dynamical friction timescale of the secondary star (see Equation (8.21)). The solid lines in Figure 8.20 show cuts in the secondary masses at $\tau \lesssim 1$, $\tau \lesssim 5$ and $\tau < \infty$ (rightmost line). The mean secondary masses are $\langle m \rangle = 4.0 \pm 4.8\,\text{M}_\odot$, 8.2 ± 6.5 and $\langle m \rangle = 13.5 \pm 8.8\,\text{M}_\odot$ for $\tau \lesssim 1, 5$ and ∞, respectively.

The distribution of primary masses in Figure 8.20 (dashed line) hardly changes as we vary the selection on τ. We therefore show only the full ($\tau \lesssim \infty$) data set for the primaries. In contrast, the distribution of secondary masses changes considerably with increasing τ. For small τ, sec-

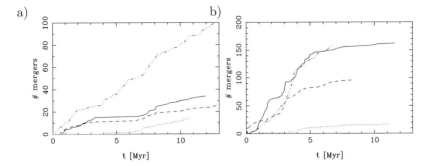

Figure 8.19. Cumulative distributions of the number of mergers that occurred within time t. Panel (a) gives the results of the calculations performed with `starlab`; panel (b) is from `NBODY4`. The dotted, solid and dashed curves are for the models with $W_0 = 8$ (in the right panel we show the results for $W_0 = 7$), $W_0 = 9$ and $W_0 = 12$. The dash-3-dotted curve in panel (a) is for a model with $W_0 = 12$ with 10% primordial binaries. The dash-3-dotted curve in panel (b) is for a model with $W_0 = 9$ using 585,000 stars, distributed according to a Kroupa (2001) initial mass function between $0.1\,M_\odot$ and $100\,M_\odot$.

ondaries are drawn primarily from low-mass stars. As τ increases, the secondary distribution shifts to higher masses while the low-mass part of the distribution remains largely unchanged. The shift from low-mass ($\lesssim 8\,M_\odot$) to high-mass collision secondaries ($\gtrsim 8\,M_\odot$) occurs between $\tau = 1$ and $\tau = 5$. This is consistent with the theoretical arguments presented in § 8.3.1.2. During the early evolution of the cluster ($\tau \lesssim 1$), collision partners are selected more or less randomly from the available (initial) population in the cluster core; at later times, most secondaries are drawn from the mass-segregated population.

Interestingly, although hard to see in Figure 8.20, all the curves are well fit by power laws between $\sim 8\,M_\odot$ and $\sim 80\,M_\odot$ ($0.8\,M_\odot$ and $30\,M_\odot$ for the leftmost curve). The power-law exponents are -0.4, -0.5 and -2.3 for $\tau \lesssim 1$, 5, and ∞, and -0.3 for the primary (dashed) curve. (Note that the Salpeter mass function has exponent -2.35.)

Figure 8.17 shows the maximum mass reached by the runaway collision product as a function of the initial mass of the star cluster. Only the left side ($\log M/M_\odot \lesssim 7$) of the figure is relevant here; we discuss the extrapolation to larger masses in §8.4.2. The N-body results are consistent with the theoretical model presented in Equation (8.39).

8.3.3 Simulating the Star Cluster MGG11

In a recent publication (Portegies Zwart et al. 2004) we simulate a well-observed star cluster in the starburst galaxy M82. Interestingly enough,

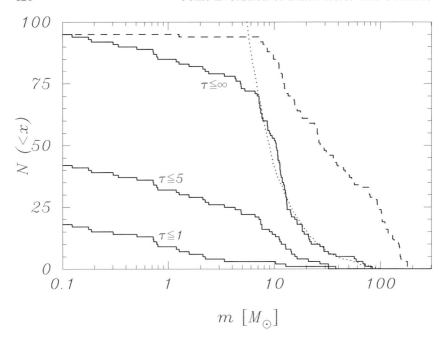

Figure 8.20. Cumulative mass distributions of primary (dashed line) and secondary (solid lines) stars involved in collisions. Only those secondaries experiencing their first collision are included. From left to right, the solid lines represent secondary stars for which $\tau \equiv t_{\mathrm{coll}}/t_{\mathrm{df}} \lesssim 1$, 5, and ∞. The numbers of collisions included in each curve are 18 (for $t_{\mathrm{coll}}/t_{\mathrm{df}} \lesssim 1$), 42 and 95 (rightmost two curves). The dotted line gives a power-law fit with the Salpeter exponent (between $5\,\mathrm{M}_\odot$ and $100\,\mathrm{M}_\odot$) to the rightmost solid curve ($\tau \lesssim \infty$).

this cluster has experienced a prominent phase A evolution, and is currently in a phase B. In this section, I will report some of the interesting results about these simulations. Note that we have been quoting these results in earlier occasions in the text, but here we review the global results.

The observed mass function, as reported by McCrady et al. (2003) for the M82 star cluster MGG11 is consistent with a Salpeter power law (with $x = -2.35$) between $1\,\mathrm{M}_\odot$ and an upper limit that corresponds to the age of the cluster of 7 to 12 Myr. By that time all stars more massive than 17–$25\,\mathrm{M}_\odot$ have experienced a supernova. For the IMF we adopt the same Salpeter slope and lower mass limit of $1\,\mathrm{M}_\odot$, but extend the upper limit to $100\,\mathrm{M}_\odot$. This IMF has an average mass of $\langle m \rangle \simeq 3.1\,\mathrm{M}_\odot$. If we assume that at an age of 12–7 Myr all stars between 17–$25\,\mathrm{M}_\odot$ and $100\,\mathrm{M}_\odot$ are lost from the cluster by supernova explosions, the mean stellar mass drops to $\langle m \rangle \simeq 2.4 - 2.6\,\mathrm{M}_\odot$. With a current total mass of $3.5 \times 10^5\,\mathrm{M}_\odot$ the cluster would contain 130,000 to 140,000 stars. For clarity we decided to select 128 k (131,072) single stars, resulting in an initial mass of about 406,000 $M_\odot$.

McCrady et al. (2003) measured the projected half-light radius of the cluster, $r_{hp} = 1.2\,$pc. De-projection of the half-light radius depends on the density profile. For King (1966) models in the range $W_0 = 3 - 12$ it turns out that $r_h = 0.75 - 0.88 r_{hp}$ (which is somewhat larger than Spitzer's (1987) $r_h \sim \frac{3}{4} r_{hp}$). A number of initial test simulations indicate that over a time span of 7–12 Myr the projected half-light radius of the selected IMF and number of stars, the cluster expands by about a factor 1.3. We then adopt an initial half-mass radius for our model cluster of $r_{hm} = 1.2\,$pc.

We ignore the effect of the tidal field of M82. The star cluster is located at a distance of about 160 pc from the dynamical center of the galaxy, assuming a distance of 3.6 Mpc (Freedman et al. 1994).[13] With the relatively small mass of M82 of $\sim 10^9\,M_\odot$ the gravitational force of the galaxy is negligible compared to the self-gravity of the stars within the cluster.

We performed several calculations starting from King models with different central concentrations W_0. We also performed one run with 10% primordial binaries and one run starting with 585,000 stars and a Salpeter IMF between $0.1\,M_\odot$ and $100\,M_\odot$.

To qualify the results we make the distinction between clusters with a high central concentration and clusters with low concentration. Since the initial half-mass radius is the same for all models, we varied the density profile. For the density profile we adopted King (1966) models. We draw the empirical distinction between high concentration models having $W_0 \gtrsim 9$; low concentration models have $W_0 \lesssim 8$. With the adopted half-mass radius and total mass the high concentration cluster $W_0 \gtrsim 9$ models have core density $\log \rho_c > 6\,M_\odot\,{\rm pc}^{-3}$.

In Figure 8.21 we present the evolution of the central density of various `starlab` models with $W_0 = 7$, 8 and 9. As discussed, in the low-concentration models core collapse is arrested by the copious mass loss from the evolving massive stars. In the high-concentration clusters core collapse occurs early enough that the process is little affected by stellar mass loss.

8.3.3.1 *Clusters with* $\log \rho_c > 6\,M_\odot\,{\rm pc}^{-3}$

According to Portegies Zwart et al. (1999), who studied similar clusters in Portegies Zwart & McMillan (2002), the high collision rate is mainly the result of binaries created in three-body encounters during core collapse. In our latest models, however, only about half (0.52 ± 0.1) the collisions are preceded by the formation of a binary; the other half are direct hits in that

---

[13] Freedman et al. 1994 measured a distance of 3.63 ± 0.34 Mpc to M81, the neighboring galaxy of M82, using Cepheids.

Table 8.4. Resulting collisions for our model clusters as a function of the initial density profile expressed in the King parameter W_0 (identified on the first column) and the average core density is presented in the second column. The third column gives the number of collisions that occurred in 12 Myr. The fourth and fifth columns give the number of collisions in which one particular star participates and the average mass of the star with which it collides. The last three columns give the number of collisions and the mean mass of the more massive and lower mass stars participating in these collisions.

W_0	ρ_c $M_\odot\,pc^{-3}$	Designated Target n_{coll}	Other Stars n_{coll}	$\langle m \rangle$	n_{coll}	$\langle m \rangle$	
			Results from `Starlab`				
3	4.55	2	0	NA	2	17.3	5.0
7	5.51	5	0	NA	5	13.7	8.4
8	5.81	17	0	NA	17	19.0	9.7
9	6.58	36	21	48.1	15	16.4	6.2
12	8.12	27	14	47.9	12	33.2	9.0
12*		101	25	41.7	76	16.6	14.0
			Results from `NBODY4`				
7		19	0	NA	19	24.7	4.6
9		164	99	30.0	65	34.8	8.1
12		98	70	38.7	28	50.1	7.8
9°		161	98	20.9	63	31.5	3.1

* Run performed with 10% hard primordial binaries.
° Run performed with 585,000 stars and a Kroupa (2001) IMF between $0.1\,M_\odot$ and $100\,M_\odot$ (for details see Portegies Zwart et al. 2004).

no third star was bound to the two stars which ultimately collided. Though seemingly a detail, it has far-reaching consequences for the interpretation of the collisional growth which plays an important role in the evolution of these systems.

After the first epoch of runaway growth and the collapse of the designated target, an epoch of about 3 Myr starts in which the collision rate drops dramatically. This phase is visible in Figure 8.19 between ~ 3.3 Myr and ~ 6 Myr.

This quiet phase lasts until the first M5I hypergiants appear, which, in our stellar evolution model, happens at a turn-off mass of $\sim 25\,M_\odot$ (at ~ 6 Myr). (Note that in our stellar evolution model `SeBa` stars below $\sim 25\,M_\odot$ turn into neutron stars where more massive stars become black holes; see Portegies Zwart & Verbunt 1996.) By this time the collision rate picks up again to 2–3 per Myr. The spectral type M0I–M5I star dominates the collision rate; the designated target, now an intermediate mass black hole, participates in only about one-third of the collisions. This phase lasts until 9–12 Myr, after which the rate drops below 0.3 collisions per Myr.

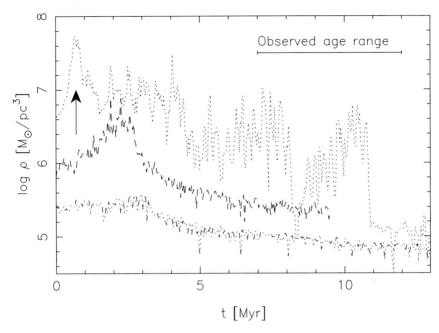

Figure 8.21. Evolution of the central density for star clusters with $W_0 = 7$ (dash-3-dotted line), $W_0 = 8$ (dashes), $W_0 = 9$ (dotted curve). The $W_0 = 9$ model shows a clear core collapse at about 0.7 Myr, indicated by the arrow. The less concentrated models ($W_0 = 8$ and $W_0 = 7$) show a very shallow start of a core collapse near $t \sim 3$ Myr, but in these cases the collapse of the core is arrested by stellar mass loss. In Figure 8.13 we present the evolution of the core radius for the same calculations.

8.3.3.2   Clusters with $\log \rho_c < 6\,M_\odot\,\mathrm{pc}^{-3}$

In low-density cluster $W_0 \lesssim 8$ the initial phase of runaway growth is absent. The subsequent collisional phase at an age $\gtrsim 6$ Myr, however, occurs at a comparable rate as in the high-density simulations (see Figure 8.20). As discussed in the previous Section there is no designated target in this phase; all collisions tend to happen between massive stars (or stellar remnants) of which at least one component is evolved (spectral type M0I to M5I), at a rate that becomes gradually smaller for less-concentrated clusters: 3.4 collisions per Myr for $W_0 = 9$, 3.2 for $W_0 = 8$, 1.6 for $W_0 = 7$, 1.5 for $W_0 = 5$ and 0.5 for $W_0 = 3$. Curiously enough the $W_0 = 12$ models have a collision rate of only 2.3 per Myr, which is lower than the less concentrated $W_0 = 9$ models.

   The average star with which a collision occurs (counting the least massive of the two stars) depends on the initial cluster density profile, ranging from $5.3\,M_\odot$ for $W_0 = 5$, $6.2\,M_\odot$ for $W_0 = 7$, $9.7\,M_\odot$ for $W_0 = 8$

and ~ 30 for both $W_0 = 9$ and $W_0 = 12$ (consistent with the expectations of Portegies Zwart & McMillan 2002).

8.3.4 Black Hole Ejection in Phase B and C Cluster with $t_{rt} \gtrsim 100\,\mathrm{Myr}$

Here we discuss the evolution of star clusters in which the early phase is dominated by stellar mass loss and the subsequent evolution by the stellar interaction: $\tau_{SE} < \tau_{SD}$. In this regime or parameter space, the most massive stars evolve before the structure of the star cluster has appreciably changed; i.e., no intermediate mass black hole can form via the scenario discussed in §8.3.1. The consequence is that the most massive stars turn into black holes and neutron stars before they have a chance to sink to the cluster center by dynamical friction. This regime is valid for most globular clusters, and possibly many open star clusters.

Upon the birth of a cluster we assume that the stars populate the initial mass function from the hydrogen burning limit ($\sim 0.1\,\mathrm{M_\odot}$) all the way to the most massive stars currently observed in the galaxy, which is about $100\,\mathrm{M_\odot}$. Stars with solar abundance between $50\,\mathrm{M_\odot}$ and $100\,\mathrm{M_\odot}$ leave the main sequence at an age of about 3.7 Myr to 3.3 Myr, to explode in a supernova a few hundred thousand years later. The total mass in this range is about 1%, and the cluster therefore loses approximately 1% of its total mass in less than half a million years.[14]

The first black holes are produced at about the same time. Black hole formation proceeds to about 7–9 Myr, until all stars with initial masses exceeding $20\text{–}25\,\mathrm{M_\odot}$ (Maeder 1992; Portegies Zwart et al. 1997) have collapsed to black holes. Assuming a Scalo (1986) mass distribution with a lower mass limit of $0.1\,\mathrm{M_\odot}$ and an upper limit of $100\,\mathrm{M_\odot}$ about 0.071% of the stars are more massive than $20\,\mathrm{M_\odot}$, and 0.045% are more massive than $25\,\mathrm{M_\odot}$. A star cluster containing N stars thus produces $\sim 6 \times 10^{-4}N$ black holes. Known galactic black holes have masses m_{bh} between $6\,\mathrm{M_\odot}$ and $18\,\mathrm{M_\odot}$ (Timmes et al. 1996). For definiteness, we adopt $m_{bh} = 10\,\mathrm{M_\odot}$.

A black hole is formed in a supernova explosion. If the progenitor is a single star (i.e., not a member of a binary), the black hole experiences little or no recoil and remains a member of the parent cluster (White & van Paradijs 1996).[15] If the progenitor is a member of a binary, mass loss during the supernova may eject the binary from the cluster potential via the Blaauw mechanism (Blaauw 1961), where conservation of momentum causes recoil in a binary as it loses mass impulsively from one component.

---

[14] For an entire star cluster a 1% mass loss is not very dramatic, and simply causes the cluster to expand by about the same fraction. For a cluster core, which contains less than 5% of the total cluster mass (for $W_0 \gtrsim 8$), a 1% mass loss may drive a substantial expansion of the core.

[15] There are some arguments that indicate that a black hole may received small "impulse" kicks, relative to neutron stars.

The Blaauw velocity kick can be as large as the relative orbital velocity of the pre-supernova binary (see e.g., Nelemans et al. (2001). The escape speed is $\sim 40\,\mathrm{km\,s^{-1}}$ for a young globular cluster, and somewhat smaller for YoDeCs (see Table 8.1). Such high recoil velocities are generally achieved only if the binary loses $\lesssim 50\%$ of its total mass in the supernova, and if its orbital period is initially quite small ($\lesssim 2\,\mathrm{yr}$); (Portegies Zwart & Verbunt 1996). The binary frequency in globular clusters is between 5 and 40% (Rubenstein 1997), and less than 30% of binaries have orbital periods smaller than 2 years (Rubenstein & Bailyn 1997; 1999); we assume the same distributions of orbital parameters for binaries in YoDeCs. We therefore estimate that no more than $\sim 10\%$ of black holes are ejected from the cluster immediately following their formation (i.e., the black hole retention fraction is $\gtrsim 90\%$). Note the remarkable difference here with the retention fraction of neutron stars, which is less than about 10% (Drukier 1996).

After $\sim 40\,\mathrm{Myr}$ the last supernova has occurred, the mean mass of the cluster stars is $\langle m \rangle \sim 0.56\,\mathrm{M_\odot}$ (Scalo 1986), and black holes are by far the most massive objects in the system. Mass segregation causes the black holes to sink to the cluster core in a fraction $\sim \langle m \rangle / m_{\mathrm{bh}}$ of the half-mass relaxation time scale (see Equation (8.21)). For a young populous cluster, the relaxation time is on the order of 10 Myr; for a globular cluster it is about 1 Gyr (see Table 8.3).

By the time of the last supernova, stellar mass loss has also significantly diminished and the cluster core starts to contract, enhancing the formation of binaries by three-body interactions. Single black holes form binaries preferentially with other black holes (Kulkarni et al. 1993), while black holes born in binaries with a lower-mass stellar companion rapidly exchange the companion for another black hole. The result in all cases is a growing black-hole binary population in the cluster core. Once formed, the black-hole binaries become more tightly bound through superelastic encounters with other cluster members (Heggie 1975; Kulkarni et al. 1993; Sigurdsson & Hernquist 1993). On average, following each close binary–single black hole encounter, the binding energy of the binary increases by about 20% (Hut et al. 1992); roughly one third of this energy goes into binary recoil, assuming equal mass stars. The minimum binding energy of an escaping black hole binary may then be estimated as

$$E_{b,\mathrm{min}} \sim 36\,W_0\,\frac{m_{\mathrm{bh}}}{\langle m \rangle}\,kT\,, \tag{8.43}$$

where $\frac{3}{2}kT$ is the mean stellar kinetic energy and $W_0 = \langle m \rangle |\phi_0|/kT$ is the dimensionless central potential of the cluster (King 1966). By the time the black holes are ejected, $\langle m \rangle \sim 0.4\,\mathrm{M_\odot}$. Taking $W_0 \sim 10$ we find $E_{b,\mathrm{min}} \sim 10000\,kT$.

Portegies Zwart & McMillan (2000) have tested and refined the above estimates by performing a series of N-body simulations within the `Starlab` software environment using the special-purpose computer GRAPE-4 to speed up the calculations (see §8.1.4.4). For most (seven) of these calculations we used 2048 equal-mass stars with 1% of them ten times more massive than the average; two calculations were performed with 4096 stars. One of the 4096-particle runs contained 0.5% black holes; the smaller black hole fraction did not result in significantly different behavior. They also tested alternative initial configurations, starting some models with the black holes in primordial binaries with other black holes, or in primordial binaries with lower-mass stars.

The results of these simulations may be summarized as follows. Of a total of 204 black holes, 62 ($\sim$ 30%) were ejected from the model clusters in the form of black-hole binaries. A total of 124 ($\sim$ 61%) black holes were ejected single, and one escaping black hole had a low-mass star as a companion. The remaining 17 ($\sim$ 8%) black holes were retained by their parent clusters. The binding energies E_b of the ejected black hole binaries ranged from about $1000\,kT$ to $10000\,kT$ in a distribution more or less flat in $\log E_b$, consistent with the assumptions made by Hut et al. (1992). The eccentricities e followed a roughly thermal distribution [$p(e) \sim 2e$], with high eccentricities slightly overrepresented; the mean eccentricity was $\langle e \rangle = 0.69 \pm 0.10$. The 17 binaries with the lowest binding energies ($\log_{10} E_b < 3.5$) had on average higher eccentricities ($\langle e \rangle = 0.78 \pm 0.05$) than the more tightly bound binaries ($\langle e \rangle = 0.62 \pm 0.11$). About half of the black holes were ejected while the parent cluster still retained more than 90% of its birth mass, and $\gtrsim$ 90% of the black holes were ejected before the cluster had lost 30% of its initial mass. These findings are in good agreement with previous estimates that black hole binaries are ejected within a few Gyr, well before core collapse occurs (Kulkarni et al. 1993; Sigurdsson & Hernquist 1993). Recently Merritt et al. (2004) confirmed these findings with their own simulations using `NBODY6++` (Hemsendorf et al. 2002; and can be downloaded from ftp://ftp.ari.uni-heidelberg.de/pub/staff/spurzem/nb6mpi/) a parallel version of `NBODY6` (Aarseth 1999).

Portegies Zwart & McMillan (2000) have performed additional calculations incorporating a realistic (Scalo 1986) mass function, the effects of stellar evolution, and the gravitational influence of the galaxy. These model clusters generally dissolved rather quickly (within a few hundred Myr) in the galactic tidal field. We found that clusters which dissolved within $\sim$ 40 Myr (before the last supernova) had no time to eject their black holes; however, those that survived for longer than this time were generally able to eject at least one close black hole binary before dissolution.

Based on these considerations, we conservatively estimate the number of ejected black hole binaries to be about $10^{-4}N$ per star cluster, more or less independent of the cluster lifetime.

8.3.4.1 Characteristics of the Black Hole Binary Population

The energy of an ejected binary and its orbital separation are coupled to the dynamical characteristics of the star cluster. For a cluster in virial equilibrium, we have

$$kT = \frac{2E_{\text{kin}}}{3N} = \frac{-E_{\text{pot}}}{3N} = \frac{GM^2}{6Nr_{\text{vir}}}, \tag{8.44}$$

where M is the total cluster mass and r_{vir} is the virial radius. A black hole binary with semi-major axis a has

$$E_b = \frac{Gm_{\text{bh}}^2}{2a}, \tag{8.45}$$

and therefore

$$\frac{E_b}{kT} = 3N \left(\frac{m_{\text{bh}}}{M}\right)^2 \frac{r_{\text{vir}}}{a}. \tag{8.46}$$

In computing the properties of the black hole binaries resulting from cluster evolution, it is convenient to distinguish three broad categories of dense stellar systems: (1) YoDeCs, (2) globular clusters, and (3) galactic nuclei. Table 8.1 lists characteristic parameters for each category. The masses and virial radii of globular clusters are assumed to be distributed as independent Gaussian with means and dispersions as presented in the table; this assumption is supported by correlation studies (Djorgovski & Meylan 1994). Table 8.1 also gives estimates of the parameters of globular clusters at birth (bottom row), based on a recent parameter-space survey of cluster initial conditions (Takahashi & Portegies Zwart 2000; Baumgardt & Makino 2003); globular clusters that have survived for a Hubble time have lost $\gtrsim 60\%$ of their initial mass and have expanded by about a factor of three. We draw no distinction between core-collapsed globular clusters (about 20% of the current population) and non-collapsed globulars—the present dynamical state of a cluster has little bearing on how black hole binaries were formed and ejected during the first few Gyr of the cluster's life.

The above-described process causes globular clusters to be depleted of black holes. At most one binary containing two black holes can be present in any globular cluster. In §8.4.4 we will further discuss the consequences of the ejected black holes and black hole binaries, as the latter may become important sources for gravitational wave detectors.[16]

---

[16] Today's globular clusters may undergo a similar ejection phase, which can be seen in the high proportion of recycled pulsars and the interestings possibility of a high merger rate of white dwarfs (Shara & Hurley 2002).

8.4   Discussion and Further Speculations

Now we have discussed the basic principles of star cluster dynamics with black holes. In this section we further discuss some of the consequences and observables by which this theory can be tested.

8.4.1   Turning an Intermediate Mass Black Hole in an X-Ray Source

An IMBH in a stellar cluster with velocity dispersion σ dominates the potential well within its radius of influence $R_i = Gm_{bh}/\sigma^2$; inside R_i the orbits are approximately Keplerian, and the stars are distributed according to a power law (see §8.2.1.1 and also (Bahcall & Wolf 1976). N-body calculations show that $\alpha = 1.5$ (Baumgardt et al. 2004), and we assume this value in the following.

Stars can reach an orbit with peribothron of order of the tidal radius by energy diffusion or by angular momentum diffusion. However, stellar collisions become the dominant dynamical process within the collision radius $r_{coll} \sim (m_{bh}/m)r$ (Frank & Rees 1976), disrupting stars within that region, and making energy diffusion within r_{coll} implausible. For this reason we concentrate on tidal capture of stars on very eccentric orbits. When the star arrives at peribothron a certain amount of energy ΔE_t is invested in raising tides, causing the orbit to circularize. It is hard to know how the star dissipates the tides after the repetitive encounters with the IMBH. Two extreme models of "squeezars" (stars that are "squeezed" by the tidal field) are studied by Alexander & Morris (2003), namely "cold squeezars", which puff up, or "hot squeezars", which are heated only in their outer layers and radiate their excess energy effectively.

Hopman et al. (2003) calculated that the probability that an intermediate mass black hole captures a stellar companion via this process (assuming the hot squeezar model) has a reasonable probability, making it likely that we could in fact observe one or two of such objects. Once captured the companion star is likely to fill its Roche–lobe while still on the main-sequence.

From the moment of first Roche-lobe contact the evolution of the binary is further determined by mass transfer from a lower mass (secondary) star to the much more massive (primary) IMBH. This process is driven by the thermal expansion of the donor and the loss of angular momentum from the binary system. Mass transfer then implies that the donor fills its Roche lobe ($R_{don} = R_{Rl}$) and continues to do so; i.e., $\dot{R}_{don} = \dot{R}_{Rl}$. We also assume that, as long as $L_{disc} < L_{Edd}$ all the mass lost from the donor via the Roche lobe is accreted by the black hole, $\dot{M} = -\dot{m}$. If the mass transfer rate exceeds the Eddington limit, the remaining mass is lost from the binary system with the angular momentum of the accreting star.

We estimate the x-ray luminosity during mass transfer using the simple model of Körding, Falcke, & Markoff (2002). They argue that in the hard state ($\dot{m} > \dot{m}_{\mathrm{crit}}$) $L_x \simeq 0.1\dot{m}c^2$, where at lower accretion rates the source becomes transient (see also Kalogera et al. 2004) with an average luminosity of $L_x \simeq 0.1c^2\dot{m}^2/\dot{m}_{\mathrm{crit}}$. For $\dot{m}_{\mathrm{crit}}$ we use the relation derived by Dubus et al. (1999).

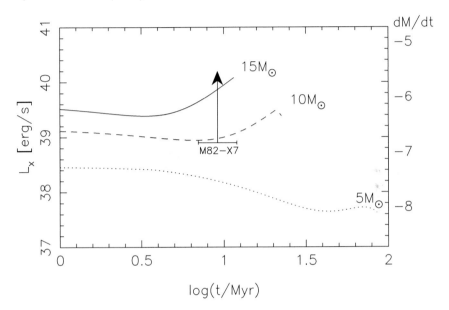

Figure 8.22. Estimated x-ray luminosity (log erg/s) as a function of time for a 1000 $M_\odot$ accreting black hole. The solid, dashed and dotted curves are for a 15 $M_\odot$, 10 $M_\odot$ and a 5 $M_\odot$ donor which started Roche-lobe overflow at the zero-age main-sequence. (Figure from Hopman, Portegies Zwart & Alexander 2004.)

8.4.2   Speculation on the Formation of Supermassive Black Holes

A million-solar-mass star cluster formed at a distance of $\lesssim 30\,\mathrm{pc}$ from the galactic center can spiral into the galactic center by dynamical friction before being disrupted by the tidal field of the galaxy (see Gerhard 2001). Only the densest star clusters survive to reach the center. These clusters are prone to runaway growth and produce massive compact objects at their centers. Upon arrival at the galactic center, the star cluster dissolves, depositing its central black hole there. Black holes from inspiraling star clusters may subsequently merge to form a supermassive black hole. Ebisuzaki et al. (2001) have proposed that such a scenario might explain the presence of the central black hole in the Milky Way galaxy.

If we simply assume that bulges and central supermassive black holes are formed from disrupted star clusters, this model predicts a relation between black hole and bulge masses in galaxies similar to the expression (Equation (8.40)) connecting the mass of an IMBH to that of its parent cluster. However, the ratio of stellar mass to black-hole mass might be expected to be smaller for galactic bulges than for star clusters, because not all star clusters produce a black hole and not all star clusters survive until the maximum black hole mass is reached. We would expect, however, that the general relation between the black hole mass and that of the bulge remains valid.

Figure 8.17 shows the relation between the black hole mass and the bulge mass for several Seyfert galaxies and quasars. The expression derived in § 8.3.1.3 and the results of these N-body calculations (§ 8.3.2) are also indicated. The solid and dashed lines (Equation (8.40)) fit the N-body calculations and enclose the area of the measured black hole mass–bulge masses. On the way, the solid curve passes though the two black hole mass estimates, for M82 star cluster MGG-11. We note that the observed relation between bulge and black hole masses has a spread of two orders of magnitude. If this bold extrapolation really does reflect the formation process of bulges and central black holes, this spread could be interpreted as a variation in the efficiency of the runaway merger process. In that case, only about one in a hundred star clusters reaches the galactic center, where its black hole is deposited.

The dashed curve is an extrapolation beyond $10^7 \, M_\odot$ the range where we think that Equation (8.40) is applicable. There appears to be a very interesting relation for galactic nuclei between the velocity dispersion and the mass of the central black hole (Ferrarese & Merritt 2000; Gebhardt et al. 2000, 2001; Merritt & Ferrarese 2001). I do not wish to claim that the process described in Equation 8.40 and the formation of supermassive black holes in galactic nuclei are equivalent (see however Ebisuzaki et al. 2001), but just point out how well the same theory explains the relation between nuclear black hole mass and the velocity dispersion in the nucleus. At the moment of writing there are more than a dozen alternative explanations for this interesting relation (see for example Haehnelt et al. 2000; Dokushaev et al. 2002; Boroson et al. 2002; Green et al. 2004), and this number is growing almost daily.[17]

8.4.3   Is the Globular Cluster M15 a Special Case?

The presence of an intermediate mass black hole in the core collapsed globular cluster M15 has been vividly debated over the last several decades (Illingworth & King, 1977a, 1977b) and is still discussed to the present

---

[17] We saw similar interesting wild-grown theories in the gamma-ray burst community a few years back.

day (van der Marel et al. 2002). If M15 contains an intermediate mass black hole, as has recently been suggested by some observers (Gerrisen et al, 2002; later retracted and subsequently corrected in Gerrisen 2003) and argued against by theorists (Baumgardt et al. 2003), it is unlikely to have formed via the scenario discussed in § 8.3.1. The cluster's initial relaxation time probably exceeded our upper limit of $\sim 100\,\mathrm{Myr}$. The current half-mass relaxation time of M15 is about $2.5\,\mathrm{Gyr}$ (Harris 1996), which is far more than our $100\,\mathrm{Myr}$ limit for forming a massive central object from a collision runaway (see § 8.3.1 for details).

An intermediate-mass black hole in the globular cluster M15 may have been formed by a different scenario. An alternative is provided by Miller & Hamilton (2001), who describe the formation of massive ($\sim 10^3\,\mathrm{M}_\odot$) black holes in star clusters with relatively long relaxation times. In their model the black hole grows very slowly over a Hubble time via occasional collisions with other black holes, in contrast to the model described here, in which the runaway grows much more rapidly, reaching a characteristic mass of about 0.1% of the total birth mass of the cluster within a few megayears.

One possible way around M15's long relaxation time may involve the cluster's rotation. Gebhardt (2000, 2001, private communication) has measured radial velocities of individual stars in the crowded central field, down to two arcsec of the cluster center. He finds that, both in the central part of the cluster ($R < 0.1R_{\mathrm{hm}}$) and outside the half-mass radius, the average rotation velocity is substantial ($v_{\mathrm{rot}} \gtrsim 0.5\langle v \rangle$). Rotation is quickly lost in a cluster (Baumgardt et al. 2003), so to explain a current rotation, M15's initial rotation rate must probably have been even larger than observed today (see Einsel & Spurzem 1999). Hachisu (1979; 1982) found, using gaseous cluster models, that an initially rotating cluster tends to evolve into a "gravo-gyro catastrophe" which drives the cluster into core collapse far more rapidly than would occur in a non-rotating system. If the gravo-gyro-driven core collapse occurred within $25\,\mathrm{Myr}$, a collision runaway might have initiated the growth of an intermediate mass black hole in the core of M15.

8.4.4 The Gravitational Wave Signature of Dense Star Clusters

The peak amplitude of the gravitational wave form produced by black hole inspiral is (Peters 1964)

$$h = 8.0 \times 10^{-20} \left(\frac{M_{\mathrm{chirp}}}{\mathrm{M}_\odot} \right)^{5/6} \left(\frac{20\,\mathrm{ms}}{P_{\mathrm{orb}}} \right)^{1/6} \left(\frac{1\,\mathrm{Mpc}}{d} \right). \tag{8.47}$$

Here the "chirp" mass is $M_{\mathrm{chirp}} = (m_1 m_2)^{3/5}/(m_1 + m_2)^{1/5}$ for a binary with component masses m_1 and m_2. The frequency of the gravitational wave is $2/P_{\mathrm{orb}}$, where P_{orb} is the orbital period. The first LIGO interferometer is expected to achieve a sensitivity h of 10^{-20} to 10^{-21} at its most sensitive frequency around 100 Hz (Abramovici et al. 1992), corresponding to an orbital period of about 20 ms. The details of the waveform and the recovery of the signal from the noisy data complicate matters somewhat. With a sensitivity of $h = 5 \times 10^{-21}$, black hole binaries will be detectable out to a distance of about 100 Mpc, and intermediate mass black holes to several kpc distances. Within a volume of $\frac{4}{3}\pi d^3$ we then expect a detection rate for the first generation of interferometers of about $0.3\, h_0^3$ per year. For $h_0 \sim 0.72$ (Freedman et al. 2001), this results in about one detection event per decade. LIGO-II is tentatively expected to see out to an effective distance about ten times farther than LIGO-I and be operational between 2005 to 2007 (K. Thorne, private communication). This would result in a 1000 times higher detection rate, or several events per month.

The current best estimate of the maximum distance within which LIGO-I can detect an inspiral event is

$$R_{\text{eff}} = 18\,\text{Mpc}\ \left(\frac{M_{\text{chirp}}}{\text{M}_\odot}\right)^{5/6} \tag{8.48}$$

(K. Thorne, private communication). For neutron star inspiral, $m_1 = m_2 = 1.4\,\text{M}_\odot$, so $M_{\text{chirp}} = 1.22\,\text{M}_\odot$ and $R_{\text{eff}} = 21$ Mpc. For black-hole binaries with $m_1 = m_2 = m_{\text{bh}} = 10\,\text{M}_\odot$, we find $M_{\text{chirp}} = 8.71\,\text{M}_\odot$ and $R_{\text{eff}} = 109$ Mpc.

8.4.4.1 The Gravitational Wave Signature for Intermediate-Mass Black Holes

In § 8.4.1 we discussed a model for producing x-rays from an intermediate-mass black hole, in this section we continue that discussion but in relation to gravitational wave detectors.

Figure 8.23 presents the evolution of the gravitational wave signal for several systems that start mass transfer at zero age. Assuming a standard distance of 10 Kpc, the binaries that start mass transfer at birth emit gravitational waves at a strain of about $\log h \simeq -20.2$ (almost independent of the donor mass). The frequency of the gravitational waves is barely detectable for the LISA space based antennae (see Figure 8.23). During mass transfer the binary moves out of the detectable frequency band, as the orbital period increases. Once the donor has turned into a white dwarf, as happens for the $5\,\text{M}_\odot$ and $10\,\text{M}_\odot$ donors, the emission of gravitational waves brings the two stars back into the relatively high frequency regime and it becomes detectable again for the LISA antennae. This process, however, takes far longer than a Hubble time.

The binary in which a $2\,\text{M}_\odot$ main-sequence star starts to transfer to a $1000\,\text{M}_\odot$ black hole, however, remains visible as a bright source of gravitational waves for its entire lifetime (see Figure 8.23). The reason for this striking result is the curious evolution of the donor as it transfers mass

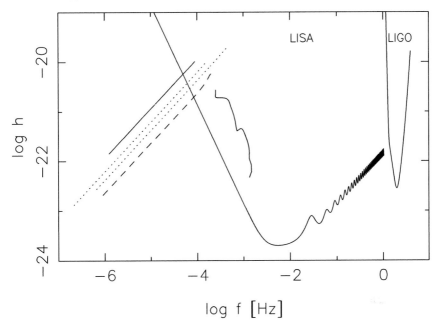

Figure 8.23. A selected sample of binaries in the gravitational wave strain and frequency domain with the LISA and LIGO noise curves over-plotted. We assumed a distance to the source of 10 kpc. The lower solid curve to the right is for a $2\,M_\odot$ donor, the dashes for a $5\,M_\odot$ donor, the dotted curves are for a $10\,M_\odot$ donor with a $100\,M_\odot$ black hole (lower dotted curve) and $1000\,M_\odot$ black hole (upper dotted curve). The upper solid curve is for a $15\,M_\odot$ donor. Except for the lower dotted curve all are calculations for a $1000\,M_\odot$ donor. Notice the enormous difference for the $2\,M_\odot$ donor, which is caused by the lack of growth in size of the donor.

at a slow rate. The $2\,M_\odot$ main-sequence donor becomes fully mixed after the hydrogen fraction exceeds about 66%. As a result, the star remains rather small in size and therefore the orbital period during mass transfer remains roughly constant. This binary remains visible in the LIGO frequency regime for its entire main-sequence lifetime; only the gravitational wave strain drops as the donor is slowly consumed by the intermediate mass black hole. Mass transfer in this evolution stage is rather slow, causing the x-ray source to be transient. Such a transient could result in an interesting synchronous detection of x-rays and gravitational waves.

The star cluster contains many binaries, some of which may have orbital periods small enough, and component masses large enough to be visible by advanced gravitational wave detectors. In Figure 8.25 we show the population of binaries in gravitational-wave space (strain h versus fre-

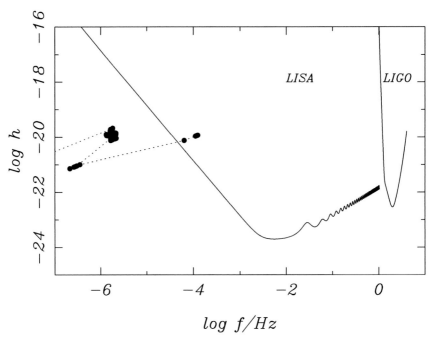

Figure 8.24. Evolutionary track in gravitational wave space (frequency and strain) for an intermediate-mass black hole formed in a young and dense star cluster at a distance of 1 kpc. The initial conditions were to mimic the M82 star cluster MGG-11 (Portegies Zwart et al. 2004). The bullets indicate the moments of output one million years apart. They are connected with a dotted line. The intermediate-mass black hole acquires a stellar mass black hole in a close binary near the end of the simulation at about 90 Myr which puts it in the LISA band.

quency f) for the compact binaries for one of our MGG-11 simulations (see § 8.3.3) at an age of 100 Myr. This simulation model had initially 20% of its 128 k stars in a binary system. The majority of systems is not even on this graph, as they comprise main-sequence stars, giants or other "large" stellar objects. Only binaries with at least one compact objects are presented, and of these only the white dwarf binaries make it in the LISA band. Many of these systems, however, will be unobservable in realistic star clusters, as the star cluster is much farther than the here-adopted 1 kpc or the system ends up in the confusion-limited noise of the white dwarf population from the Milky Way galaxy (see e.g., Hogan & Bender 2001; Seto 2002; Nelemans et al. 2004; but see Benaquista et al 2001).

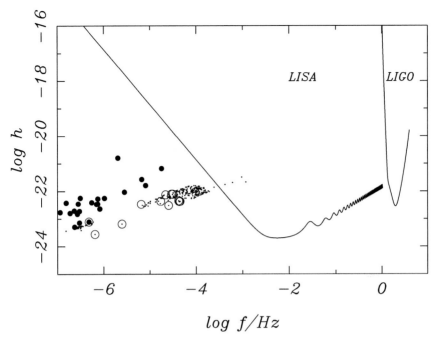

Figure 8.25. Population of compact stars in binaries in gravitational wave space (frequency and strain) in a simulated star cluster at a distance of 1 kpc. The initial conditions were to mimic the M82 star cluster MGG-11 (Portegies Zwart et al. 2004). The cluster was 100 Myr when this image was taken. The bullets indicate binaries with one or two black holes, open circles indicate neutron stars and small dots are for white dwarfs. Symbols that overplot exactly indicate close binaries with two different compact objects. Note that it is quite common for a circle to overlap with a small dot: in these cases they are (ns, wd) binaries. Only a few white dwarf binaries may be visible in the LISA band; black hole and neutron star binaries are generally too wide to be detectable. The intermediate-mass black hole is not shown here, but in Figure. 8.24.

8.4.4.2 The Merger Rate of Stellar Mass Black Holes

In § 8.3.4 we discussed the evolution of relatively low-density clusters through phase B and C. These clusters tend to eject their black holes, in part in the form of binaries consisting of two black holes. The orbital parameters of these binaries are such that merger may occur in a relatively short time-scale; well within a Hubble time. We now will discuss the merger rate per unit volume $\mathcal{R}$ which we predict from various types of star clusters.

A conservative estimate of the merger rate of black-hole binaries formed in globular clusters is obtained by assuming that globular clusters in other galaxies have characteristics similar to those found in our own. Using the galaxy density in the local universe (see Table 8.2) the result is

$$\mathcal{R}_{GC} = 5.4 \times 10^{-8} h^3 \text{ yr}^{-1} \text{ Mpc}^{-3}. \tag{8.49}$$

Irregular galaxies, starburst galaxies, early type spirals and blue ellip-tical galaxies all contribute to the formation of young dense clusters. In the absence of firm measurements of the numbers of YoDeCs in other galaxies, we simply use the same values of S_N as for globular clusters (see Table 8.2). The space density of such clusters is then

$$\phi_{\text{YoDeC}} = 3.5 \, h^3 \text{ Mpc}^{-3}, \tag{8.50}$$

and the black hole merger rate is

$$\mathcal{R}_{\text{YoDeC}} = 2.1 \times 10^{-8} h^3 \text{ yr}^{-1} \text{ Mpc}^{-3}. \tag{8.51}$$

For purposes of comparison, we also estimate the contribution to the total merger rate from galactic nuclei, neglecting obvious complicating fac-tors such as the presence of central supermassive black holes (Sage 1994), which may inhibit the formation, hardening and ejection of black hole bi-naries. The space density of galactic nuclei is only[18] about

$$\phi_{GN} = 0.012 \, h^3 \text{ Mpc}^{-3}, \tag{8.52}$$

and the corresponding contribution to the black hole merger rate is

$$\mathcal{R}_{GN} = 2.5 \times 10^{-9} h^3 \text{ yr}^{-1} \text{ Mpc}^{-3}, \tag{8.53}$$

which is negligible compared to the other rates.

Based on the assumptions outlined above, our estimated total merger rate per unit volume of black hole binaries is

$$\mathcal{R} = 7.5 \times 10^{-8} h^3 \text{ yr}^{-1} \text{ Mpc}^{-3}. \tag{8.54}$$

However, this may be a considerable underestimation of the true rate. First, as already mentioned, our assumed number ($\sim 10^{-4} N$) of ejected black hole binaries is quite conservative. Second, the correlation between orbital eccentricity and binding energy and the excess of high-eccentricity binaries mentioned earlier both favor more rapid inspiral, causing a larger fraction of the black hole binaries to merge. Third, the observed population of globular clusters naturally represents only those clusters that have survived until the present day. The study by Takahashi & Portegies Zwart (2000) and Baumgardt & Makino (2003) indicates that $\sim 50\%$ of globular clusters dissolve in the tidal field of the parent galaxy within a few billion years of formation. We have therefore underestimated the total number of globular

---

[18] Each galaxy may contain hundreds of globulars but has only one nucleus.

clusters, and hence the black-hole merger rate, by about a factor of two. Fourth, a very substantial underestimation stems from the assumption that the masses and radii of present-day globular clusters are representative of the initial population. When estimated initial parameters (Table 8.1, bottom row) are used, the total merger rate increases by a further factor of six. Taking all these effects into account, we obtain a net black hole merger rate of

$$\mathcal{R} \sim 6 \times 10^{-7} h^3 \ \text{yr}^{-1} \ \text{Mpc}^{-3}. \tag{8.55}$$

We note that this rate is significantly larger than the current best estimates of the neutron-star merger rate (e.g. Phinney 1991; Tutukov & Yungelson 1994; Burgay et al. 2003; and many others). Since black hole mergers are also "visible" to much greater distances, we expect that black hole events will dominate the LIGO detection rate.

8.4.4.3 Gravitational Inspiral

An approximate formula for the merger time of two stars due to the emission of gravitational waves is given by Peters (1964):

$$t_{\text{mrg}} \approx 150 \, \text{Myr} \left(\frac{\text{M}_\odot}{m_{\text{bh}}}\right)^3 \left(\frac{a}{\text{R}_\odot}\right)^4 (1 - e^2)^{7/2} . \tag{8.56}$$

The sixth column of Table 8.1 lists the fraction of black hole binaries that merge within a Hubble time due to gravitational radiation, assuming that the binary binding energies are distributed flat in $\log E_b$ between $1000 \, kT$ and $10,000 \, kT$, that the eccentricities are thermal, independent of E_b, and that the universe is $\sim 13 \, \text{Gyr}$ old ($H_0 = 72 \pm 8 \ \text{km s}^{-1} \, \text{Mpc}^{-1}$; Freedman et al. 2001). The final column of the table lists the contribution to the total black hole merger rate from each cluster category.

For black hole binaries with $m_1 = m_2 \simeq 10 \, \text{M}_\odot$ we expect a LIGO-I detection rate of about $3 \, h^3$ per year. For $h \sim 0.72$ (Freedman et al. 2001), this results in about 1 detection event annually. LIGO-II should become operational by 2007, and is expected to have R_{eff} about ten times greater than LIGO-I, resulting in a detection rate 1000 times higher, or 2–3 events per day.

Black hole binaries ejected from galactic nuclei, the most massive globular clusters (masses $\gtrsim 5 \times 10^6 \, \text{M}_\odot$), and globular clusters that experience core collapse soon after formation tend to be very tightly bound, and merge within a few million years of ejection. These mergers therefore trace the formation of dense stellar systems with a delay of a few Gyr (the typical time required to form and eject binaries), making these systems unlikely candidates for LIGO detections, as the majority merged long ago. This effect reduces the current merger rate somewhat, but more sensitive future gravitational wave detectors may be able to see some of these early universe events. In fact, we estimate that the most massive globular clusters contribute about 50% of the total black hole merger rate. However, while their black hole binaries merge promptly upon ejection, the longer relaxation times of these clusters mean that binaries tend to be ejected much later than in lower-mass systems. Consequently, we have retained these binaries in our final merger rate estimate (Equation (8.55)).

Finally, we have assumed that the mass of a stellar black hole is $10\,M_\odot$. Increasing this mass to $18\,M_\odot$ decreases the calculated merger rate by about 50% — higher-mass black holes tend to have wider orbits. However, the larger chirp mass increases the signal to noise, and the distance to which such a merger can be observed increases by about 60%. The detection rate on Earth therefore increases by about a factor of three. For $6\,M_\odot$ black holes, the detection rate decreases by a similar factor. For black hole binaries with component masses $\gtrsim 12\,M_\odot$ the first generation of detectors will be more sensitive to the merger itself than to the spiral-in phase that precedes it (Flanagan & Hughes 1998a, 1998b). Since the strongest signal is expected from black hole binaries with high-mass components, it is critically important to improve our understanding of the merger waveform. Even for lower-mass black holes (with $m_{bh} \gtrsim 10\,M_\odot$), the inspiral signal comes from an epoch when the black holes are so close together that the post-Newtonian expansions used to calculate the wave forms are unreliable. The wave forms of this "intermediate binary black hole regime" (Brady et al. 1998) are only now beginning to be explored.

8.5   Concluding Remarks

We approach the end of the chapter on the ecology of black holes in dense star clusters, as part of the Como Advance School in Astrophysics. Here we will summarize a few of the key concepts discussed in this chapter.

Star clusters go through three (or four) evolutionary phases called A, B and C, each of which is dominated by either stellar mass loss or relaxation (see § 8.2). In this discussion we assumed that external influences are not particularly competitive in effect, but if they are, we call them phase D. Phase D generally results in an early termination of the cluster.

Black holes tend to behave differently in each of these phases. In phase A, it is possible to build up a star of $\sim 1000\,M_\odot$ that ultimately collapses to a black hole of intermediate mass (see § 8.3.1).

This black hole may subsequently capture a stellar companion by tidal effects and turn into a bright x-ray source (see § 8.4.1).

In clusters where a runaway collision is prevented by stellar evolution, the most massive stars collapse to stellar mass black holes in usual supernovae (see § 8.2.2).

The stellar mass black holes sink to the cluster center by dynamical friction, pair off in binaries and eject each other from the stellar system.

This scenario elegantly explains the absence of black hole x-ray transients in globular clusters (see § 8.3.4).

The ejected black hole binaries spiral-in due to gravitational wave radiation and ultimately coalesce. Upon coalescence they produce a strong burst of gravitational waves that can be detected to a distance of $\sim 100\,\mathrm{Mpc}$ (see § 8.4.4.2).

Acknowledgments

First of all I am grateful to the organizers (Monica Colpi, Francesco Haardt, Vittorio Gorini, Ugo Moschella and Aldo Treves) for the wonderful week in Como. It is a great pleasure to name and thank my collaborators who helped develop the work on which this chapter is based; they are: Tal Alexander, Ortwin Gerhard, Alessia Gualandris, Mark Hemsendorf, Clovis Hopman, Nate McCrady, David Merritt, Slawomir Piatek, Piero Spinnato, Rainer Spurzem. Special thanks go to Jun Makino and the University of Tokyo for allowing me to use his great GRAPE-6 hardware, andto Piet Hut and Steve McMillan for their valuable help and support. In addition I would like to acknowledge the hospitality of American Museum of Natural History, das Astronomisches Rechen Institute, Drexel University, the Institute for Advanced Study, the University of Basel and of course to Tokyo University for the use of their GRAPE-6 facilities. This work was supported by NASA ATP grants NAG5-6964 and NAG5-9264, by the Royal Netherlands Academy of Sciences (KNAW), the Dutch Organization of Science (NWO), and the Netherlands Research School for Astronomy (NOVA).

References

Aarseth S A 2003 *Gravitational N-body simulations* Cambdridge University press 2003

Aarseth S J 1963 *MNRAS* **126** 223

Aarseth S J 1999 *PASP* **111** 1333

Aarseth S J and Hoyle F 1964 *Astrophysica Norvegica* **9** 313

Aarseth S J and Lecar M 1975 *ARA&A* **13** 1

Abramovici A, Althouse W E, Drever R W P, Gursel Y, Kawamura S, Raab F J, Shoemaker D, Sievers L, Spero R E and Thorne K S 1992 *Science* **256** 325

Alexander T and Morris M 2003 *ApJL* **590** L25

Antonov V A 1962 *Solution of the problem of stability of stellar system Emden's density law and the spherical distribution of velocities* Leningrad: Leningrad University 1962

Apple A W 1985 *SIAM J Sci Stat Comp* **6** 85

Bahcall J N and Wolf R A 1976 *ApJ* **209** 214

Barnes J and Hut P 1986 *Nat* **324** 446

Baumgardt H and Makino J 2003 *MNRAS* **340** 227

Baumgardt H, Hut P and Heggie D C 2002 *MNRAS* **336** 1069

Baumgardt H, Hut P, Makino J, McMillan S and Portegies Zwart S 2003a *ApJL* **582** L21

Baumgardt H, Makino J and Portegies Zwart S 2003 in *Scientific Highlights of the IAU XXVth General Assembly* (Eds D Richstone P Hut)

Baumgardt H, Makino J, Hut P, McMillan S and Portegies Zwart S 2003b *ApJL* **589** L25

Benacquista M J, Portegies Zwart S F and Rasio F 2001 *Class Quantum Grav* **18** 4025

Binney J and Tremaine S 1987 *Galactic dynamics* Princeton NJ: Princeton University Press 1987 747 p

Blaauw A 1961 *BAN* **15** 265

Boroson T A 2002 *Bulletin of the American Astronomical Society* **34** 1265

Bouvier J, Rigaut F and Nadeau D 1997 *A&A* **323** 139

Brady P R, Creighton T Cutler C and Schutz B F 1998 *Phys Rev D* **57** 2101

Burgay M, D'Amico N, Possenti A, Manchester R N, Lyne A G, Joshi B C, McLaughlin M A, Kramer M, Sarkissian J M, Camilo F, Kalogera V, Kim C and Lorimer D R 2003 *Nat* **426** 531

Chernoff D F and Weinberg M D 1990 *ApJ* **351** 121

Cohn H 1980 *ApJ* **242** 765

Coleman Miller M and Hamilton D P 2001 *MNRAS* submitted

Davies M Benz W and Hills J 1991 *ApJ* **381** 449

Dehnen W 1993 *MNRAS* **265** 250

Deiters S and Spurzem R 2000 in ASP Conf Ser 211: Massive Stellar Clusters p 204

Deiters S and Spurzem R 2001 *Astronomical and Astrophysical Transactions* **20** 47

Djorgovski S and Meylan G 1994 *AJ* **108** 1292

Dokuchaev V I and Eroshenko Y N 2001 *Astronomy Letters* **27** 759

Drukier G A 1996 *MNRAS* **280** 498

Dubus G, Lasota J, Hameury J and Charles P 1999 *MNRAS* **303** 139

Dwek E, Arendt R G, Hauser M G, Kelsall T, Lisse C M, Moseley S H, Silverberg R F, Sodroski T J and Weiland J L 1995 *ApJ* **445** 716

Ebisuzaki T, Makino J, Tsuru T G, Funato Y, Portegies Zwart S, Hut P, McMillan S, Matsushita S, Matsumoto H and Kawabe R 2001 *ApJL* **562** L19

Eckart A and Genzel R 1997 *MNRAS* **284** 576

Eggleton P P, Fitchett M J and Tout C A 1989 *ApJ* **347** 998

Einsel C and Spurzem R 1999 *MNRAS* **302** 81

Ergma E and van den Heuvel E P J 1998 *A&A* **331** L29

Fellhauer M, Kroupa P, Baumgardt H, Bien R, Boily C M, Spurzem R and Wassmer N 2000 *New Astronomy* **5** 305

Ferrarese L and Merritt D 2000 *ApJL* **539** L9

Figer D F, McLean I S and Morris M 1999 *ApJ* **514** 202

Flanagan É É and Hughes S A 1998a *Phys Rev D* **57** 4535

Flanagan É É and Hughes S A 1998b *Phys Rev D* **57** 4566

Frank J and Rees M J 1976 *MNRAS* **176** 633

Freedman W L, Madore B F, Gibson B K, Ferrarese L, Kelson D D, Sakai S, Mould J R, Kennicutt R C, Ford H C, Graham J A, Huchra J P, Hughes S M G, Illingworth G D, Macri L M and Stetson P B 2001 *ApJ* **553** 47

Fregeau J M, Cheung P, Zwart S F P and Rasio F A 2004 ArXiv Astrophysics e-prints

Fregeau J M, Gürkan M A, Joshi K J and Rasio F A 2003 *ApJ* **593** 772

Fregeau J M, Joshi K J, Portegies Zwart S F and Rasio F A 2002 *ApJ* **570** 171

Fryer C L and Kalogera V 2001 *ApJ* **554** 548

Fukushige T and Heggie D C 1995 *MNRAS* **276** 206

Fukushige T, Ito T, Makino J, Ebisuzaki T, Sugimoto D and Umemura M 1991 *Publ. Astr. Soc. Japan* **43** 841

Galleti S, Federici L, Bellazzini M, Fusi Pecci F and Macrina S 2004 *A&A* **416** 917

Gebhardt K, Bender R, Bower G, Dressler A, Faber S M, Filippenko A V, Green R, Grillmair C, Ho L C, Kormendy J, Lauer T R, Magorrian J, Pinkney J, Richstone D and Tremaine S 2000 *ApJL* **539** L13

Gebhardt K, Bender R, Bower G, Dressler A, Faber S M, Filippenko A V, Green R, Grillmair C, Ho L C, Kormendy J, Lauer T R, Magorrian J, Pinkney J, Richstone D and Tremaine S 2001 *ApJL* **555** L75

Gerhard O 2001 *ApJL* **546** L39

Gerssen J, van der Marel R P, Gebhardt K, Guhathakurta P, Peterson R C and Pryor C 2002 *AJ* **124** 3270

Gerssen J, van der Marel R P, Gebhardt K, Guhathakurta P, Peterson R C and Pryor C 2003 *AJ* **125** 376

Ghez A M, Klein B L, Morris M and Becklin E E 1998 *ApJ* **509** 678

Goodman J 1987 *ApJ* **313** 576

Goodman J and Hut P 1989 *Nature* **339** 40

Green R F, Nelson C H and Boroson T 2004 in Coevolution of Black Holes and Galaxies from the Carnegie Observatories Centennial Symposia Carnegie Observatories Astrophysics Series Edited by L C Ho 2004 Pasadena: Carnegie Observatories http://wwwociwedu/ociw/symposia/series/symposium1/proceedingshtml

Gualandris A, Zwart S P and Eggleton P P 2004 ArXiv Astrophysics e-prints

Guhathakurta P, Yanny B, Schneider D P and Bahcall J N 1996 *AJ* **111** 267

Gürkan M A, Freitag M and Rasio F A 2004 *ApJ* **604** 632

Hachisu I 1979 *Publ. Astr. Soc. Japan* **31** 523

Hachisu I 1982 *Publ. Astr. Soc. Japan* **34** 313

Haehnelt M G and Kauffmann G 2000 *MNRAS* **318** L35

Harris W E 1996a *AJ* **112** 1487

Harris W E 1996b VizieR Online Data Catalog **7195** 0 http://vizier.u-strasbg.fr/viz-bin/VizieR

Heger A, Fryer C L, Woosley S E, Langer N and Hartmann D H 2003 *ApJ* **591** 288

Heggie D and Hut P 2003 *The Gravitational Million-Body Problem: A Multidisciplinary Approach to Star Cluster Dynamics*: Cambridge University Press 2003 372 pp

Heggie D C 1975 *MNRAS* **173** 729

Heggie D C 1992 *Nat* **359** 772

Heggie D C and Mathieu R 1986 *MNRAS* in P Hut S McMillan (eds) Lecture Not Phys 267 Springer-Verlag Berlin

Heggie D C and Ramamani N 1995 *MNRAS* **272** 317

Heggie D C, Giersz M, Spurzem R and Takahashi K 1998 Highlights in *Astronomy* **11** 591

Hemsendorf M, Sigurdsson S and Spurzem R 2002 *ApJ* **581** 1256

Heyl J, Colless M, Ellis R S and Broadhurst T 1997 *MNRAS* **285** 613

Hogan C J and Bender P L 2001 *Phys Rev D* **64** 062002

Hopman C, Portegies Zwart S F and Alexander T 2004 *ApJL* **604** L101

Hurley J R, Pols O R and Tout C A 2000 *MNRAS* submitted

Hut P and Inagaki S 1985 *ApJ* **298** 502

Hut P, McMillan S and Romani R W 1992 *ApJ* **389** 527

Illingworth G and King I R 1977a *ApJL* **218** L109

Illingworth G D and King I R 1977b *BAAS* **9** 343

Ito T, Ebisuzaki T, Makino J and Sugimoto D 1991 *Publ. Astr. Soc. Japan* **43** 547

Joshi K J, Nave C P and Rasio F A 2001 *ApJ* **550** 691

Kalogera V, Henninger M, Ivanova N and King A R 2004 *ApJL* **603** L41

Kawai A, Fukushige T, Makino J and Taiji M 2000 *Publ. Astr. Soc. Japan* **52** 659

Kawai A, Makino J and Ebisuzaki T 2004 *ApJS* **151** 13

King I R 1966 *AJ* **71** 64

Körding E, Falcke H and Markoff S 2002 *A&A* **382** L13

Kroupa P 2001 *MNRAS* **322** 231

Kroupa P and Weidner C 2003 *ApJ* **598** 1076

Kroupa P, Tout C A and Gilmore G 1990 *MNRAS* **244** 76

Kulkarni S R, Hut P and McMillan S 1993 *Nat* **364** 421

Maeder A 1992 *A&A* **264** 105

Makino J 1991 *ApJ* **369** 200

Makino J 1996 *ApJ* **471** 796

Makino J and Aarseth S J 1992 *Publ. Astr. Soc. Japan* **44** 141

Makino J and Hut P 1990 *ApJ* **365** 208

Makino J and Taiji M 1998 *Scientific simulations with special-purpose computers : The GRAPE systems*: John Wiley & Sons c1998

Makino J, Fukushige T, Koga M and Namura K 2003 *Publ. Astr. Soc. Japan* **55** 1163

Makino J, Taiji M, Ebisuzaki T and Sugimoto D 1997 *ApJ* **480** 432

Massey P and Hunter D 1998 *ApJ* **493** 180

McLaughlin D E 1999 *ApJL* **512** L9

McMillan S L W 1986a *ApJ* **306** 552

McMillan S L W 1986b *ApJ* **307** 126

McMillan S L W and Portegies Zwart S F 2003 *ApJ* **596** 314

Merritt D and Ferrarese L 2001 *ApJ* **547** 140

Merritt D, Piatek S, Zwart S P and Hemsendorf M 2004 ArXiv Astrophysics e-prints

Meynet G, Maeder A, Schaller G, Schaerer D and Charbonnel C 1994 *A&AS* **103** 97

Mezger P G, Duschl W J and Zylka R 1996 *A&A Rev* **7** 289

Miller G E and Scalo J M 1977 *BAAS* **9** 566

Miller G E and Scalo J M 1979 *ApJS* **41** 513

Nelemans G, Tauris T M and van den Heuvel E P J 2001 in *Black Holes in Binaries and Galactic Nuclei* p 312

Nelemans G, Yungelson L R and Portegies Zwart S F 2004 ArXiv Astrophysics e-prints (astro-ph/0312193)

Okumura S K, Makino J, Ebisuzaki T, Fukushige T, Ito T, Sugimoto D, Hashimoto E, Tomida K and Miyakawa N 1993 *Publ. Astr. Soc. Japan* **45** 329

Olson K M and Dorband J E 1994 *ApJS* **94** 117

Peters P C 1964 *Phys Rev* **136** 1224

Phinney E S 1991 *ApJL* **380** L17

Pinfield D J, Jameson R F and Hodgkin S T 1998 *MNRAS* **299** 955

Portegies Zwart S F and McMillan S L W 2000 *ApJL* **528** L17

Portegies Zwart S F and Verbunt F 1996 *A&A* **309** 179

Portegies Zwart S F, Baumgardt H, Hut P, Makino J and McMillan S 2004, *Nature* **428** 724

Portegies Zwart S F, Hut P, Makino J and McMillan S L W 1998 *A&A* **337** 363

Portegies Zwart S F, Makino J, McMillan S L W and Hut P 1999 *A&A* **348** 117

Portegies Zwart S F, Makino J, McMillan S L W and Hut P 2001 *ApJL* **546** L101

Portegies Zwart S F, Verbunt F and Ergma E 1997 *A&A* **321** 207

Raboud D and Mermilliod J C 1998 *A&A* **329** 101

Rubenstein E P 1997 *PASP* **109** 933

Rubenstein E P and Bailyn C D 1997 *ApJ* **474** 701

Rubenstein E P and Bailyn C D 1999 *ApJL* **513** L33

Sage L 1994 *Nat* **369** 345

Salpeter E E 1955 *ApJ* **121** 161

Sanders R H and Lowinger T 1972 *AJ* **77** 292

Scalo J M 1986 *Fund of Cosm Phys* **11** 1

Schaller G, Schaerer D, Meynet G and Maeder A 1993 VizieR On-line Data Catalog: J/A+AS/96/269 Originally published in: 1992A&AS96269S **96** 269

Schödel R, Ott T, Genzel R, Hofmann R, Lehnert M, Eckart A, Mouawad N, Alexander T, Reid M J, Lenzen R, Hartung M, Lacombe F, Rouan D, Gendron E, Rousset G, Lagrange A-M, Brandner W, Ageorges N, Lidman C, Moorwood A F M, Spyromilio J, Hubin N and Menten K M 2002 *Nat* **419** 694

Seto N 2002 *MNRAS* **333** 469

Shara M M and Hurley J R 2002 *ApJ* **571** 830

Sigurdsson S and Hernquist L 1993 *Nat* **364** 423

Spinnato P F, Fellhauer M and Portegies Zwart S F 2003 *MNRAS* **344** 22

Spitzer L 1971 in Pontificiae Academiae Scientiarum Scripta Varia Proceedings of a Study Week on Nuclei of Galaxies held in Rome April 13-18 1970 Amsterdam: North Holland and New York: American Elsevier 1971 edited by DJK O'Connell p443 p 443

Spitzer L 1987 *Dynamical evolution of globular clusters* Princeton NJ: Princeton University Press 1987 191 p

Spitzer L J and Hart M H 1971a *ApJ* **164** 399

Spitzer L J and Hart M H 1971b *ApJ* **166** 483

Springel V, Yoshida N and White S D M 2001 *New Astronomy* **6** 79

Sugimoto D, Chikada Y, Makino J, Ito T, Ebisuzaki T and Umemura M 1990 *Nat* **345** 33

Taiji M, Makino J, Fukushige T, Ebisuzaki T and Sugimoto D 1996 in IAU Symp 174: Dynamical Evolution of Star Clusters: Confrontation of Theory and Observations p 141

Takahashi K and Portegies Zwart S F 2000 *ApJ* **535** 759

Timmes F X, Woosley S E and Weaver T A 1996 *ApJ* **457** 834

Tutukov A V and Yungelson L R 1994 *MNRAS* **268** 871

van Albada T S 1968 *BAN* **19** 479

van den Bergh S 1984 *PASP* **96** 329

van den Bergh S 1995 *AJ* **110** 2700

van der Marel R P, Gerssen J, Guhathakurta P, Peterson R C and Gebhardt K 2002 *AJ* **124** 3255

von Hoerner S 1960 *Zeitschrift Astrophysics* **50** 184

von Hoerner S 1963 *Zeitschrift Astrophysics* **57** 47

Wandel A 1999 *ApJL* **519** L39

Wandel A 2001 *ApJ* submitted

Watters W A, Joshi K J and Rasio F A 2000 *ApJ* **539** 331

White N E and van Paradijs J 1996 *ApJL* **473** L25

Index

Chapter 5

Chapter 8